THE
COMPANY
OFFICER

THE COMPANY OFFICER

Clinton H. Smoke

Delmar Publishers

an International Thomson Publishing company I(T)P®

Albany • Bonn • Boston • Cincinnati • Detroit • London • Madrid
Melbourne • Mexico City • New York • Pacific Grove • Paris • San Francisco
Singapore • Tokyo • Toronto • Washington

NOTICE TO THE READER

Cover/insert photos courtesy of Evan Lauber

Emergency Management Online College Courses, Disaster Planning: http://beat1.spjc.cc.fl.us/em/em.html

Delmar Staff
Publisher: Alar Elken
Acquisitions Editor: Mark Huth
Developmental Editor: Jeanne Mesick
Production Coordinator: Toni Bolognino

Art and Design Coordinator: Michele Canfield
Editorial Assistant: Dawn Daugherty
Marketing Manager: Mona Caron

COPYRIGHT © 1999
By Delmar Publishers
an International Thomson Publishing Company, Inc.

The ITP logo is a trademark under license.

Printed in the United States of America

For more information, contact:

Delmar Publishers
3 Columbia Circle, Box 15015
Albany, New York 12212-5015

International Thomson Publishing Europe
Berkshire House
168-173 High Holborn
London, WC1V7AA
United Kingdom

Nelson ITP, Australia
102 Dodds Street
South Melbourne,
Victoria, 3205 Australia

Nelson Canada
1120 Birchmont Road
Scarborough, Ontario
M1K 5G4, Canada

International Thomson Publishing France
Tour Maine-Montparnasse
33 Avenue du Maine
75755 Paris Cedex 15, France

International Thomson Editores
Seneca 53
Colonia Polanco
11560 Mexico D. F. Mexico

International Thomson Publishing GmbH
Königswinterer Straße 418
53227 Bonn
Germany

International Thomson Publishing Asia
60 Albert Street
#15-01 Albert Complex
Singapore 189969

International Thomson Publishing Japan
Hirakawa-cho Kyowa Building, 3F
2-2-1 Hirakawa-cho, Chiyoda-ku,
Tokyo 102, Japan

ITE Spain/ Paraninfo
Calle Magallanes, 25
28015-Madrid, Espana

_ 2 3 4 5 6 7 8 9 10 XXX 04 03 02 01

Library of Congress Cataloging-in-Publication Data

Smoke, Clinton.
    The company officer / by Clinton Smoke.
        p.   cm.
    Includes bibliographical references and index.
    ISBN 0-8273-8472-6 (softcover)
    1. Fire extinction—Vocational guidance.   2. Fire fighters—
Certification—United States.   I. Title.
TH9119.S65 1998
363.37′023′73—dc21   98-4200
                                                CIP

Contents

Appendixes

Preface

ABOUT THIS BOOK

The purpose of this book is to assist you in becoming a fire officer. If you are seeking certification under the auspices of NFPA 1021, the *Standard for Fire Officer Professional Qualifications,* you will find comfort in knowing that this book was specifically written with that *Standard* in mind. The present *Standard* is written in a job performance requirement (JPR) format. This format identifies prerequisite knowledge levels, in addition to the specific requirements, for each of the four officer levels. Mastering the content of this book will help demonstrate your competency in the prerequisites as well as the specific requirements for certification for Fire Officer I.

Becoming certified as a fire officer is important, but becoming proficient as a company officer in the fire service is even more important. The *Standard* covers minimum requirements, and this book addresses these requirements. Additionally, in light of what we read in the press these days, we felt that several topics, namely ethics, diversity, and harassment should also be included. All of this information is vital in both volunteer and career departments; it is offered here to help you effectively perform your duties as a company officer. Whether you are reading this book to improve your skills, or taking a structured program leading to certification, we hope that you will find this book informative, interesting, and useful.

In addition to satisfying the requirements for fire officer, we have, when possible, tried to make the material in this text useful and interesting for a broader audience or emergency responders. The issues for administrators, managers, and supervisors of rescue squads and emergency medical service organizations that are independent of a fire department are essentially the same as those in the fire service. Where appropriate, this book is intended to help them as well as those of the fire service.

The administration of fire officer certification programs varies widely. Some states have no structured program, whereas other states offer a comprehensive program that goes well beyond the scope of this text. In Virginia, the Department of Fire Programs, a state agency, provides training and certification for fire service personnel. For firefighter, fire officer, and others, the state follows the NFPA *Standards.* For Fire Officer I, certification is accomplished by completing a 64-hour course and taking several tests. Local fire departments set their own policies regarding the need for certification among their members, but throughout northern Virginia, most of the departments require Fire Officer I certification for all personnel moving into supervisory positions. (For most, this means promotion to lieutenant.) I have the privilege of teaching some of these rising stars in conjunction with my duties at Northern Virginia Community College. This book is the result of a perceived need for a single up-to-date comprehensive text for teaching a certification course for fire officers.

HOW TO USE THIS BOOK

This book is intended to be used as a text to help the reader gain competency in the requirements of NFPA 1021. That *Standard* lists the requirements for Fire Officer I under six broad categories: administration, community and government relations, human resource management, inspection and investigation, emergency service delivery, and safety. This book is organized into chapters covering those major topics. The early chapters deal with communications, organizations, management, and leadership. These tools are then applied to the process of preparing for and managing emergency incidents.

In the section that follows, a sentence or two summarizes each chapter. Please take a moment to look over this material and get an overview of the entire text. Seeing the big picture will help you understand our design.

Chapter 1 starts with a discussion of the vital role of the company officer. It also examines the challenges that face company officers and outlines reasons for their being well prepared when they move into this important position.

Chapter 2 is about communications. We think that this topic is so important that we put it right at the front of the book. It is a tool you will be using throughout the remainder of the course.

Chapter 3 addresses organizations and the company officer's role in the organization. Understanding your organization helps define your place and purpose in that organization. This chapter sets the stage for the discussions of management and leadership that follow.

Chapter 4 briefly reviews the field of management science and examines the company officer's role as a manager.

Chapter 5 provides some guidelines for practical applications of modern management principles at the company level.

Chapter 6 briefly examines leadership principles and the company officer's role as a leader.

Chapter 7 offers some effective tools for making leadership work at the company level.

Chapter 8 addresses the important role of the company officer in administering the department's occupational safety and health programs at the company level.

Chapter 9 acknowledges that the prevention of injuries, fatalities, and property loss due to fire is a part of the job too, and that company officers should be doing more to support their department's efforts to prevent these losses from occurring.

Chapter 10 reviews how fire destroys, and how we can be smarter in dealing with the threat of fire.

Chapter 11 provides a brief overview of fire-cause determination, so that company officers can accurately determine the cause of fires, especially when arson is indicated.

Chapter 12 covers the company officer's responsibilities for keeping themselves and their company trained, fit, and ready to deal with expected emergencies. "Waiting for the big one" is no longer an acceptable activity. Fire service personnel are expected to use every waking moment to prepare for events and to take action to reduce the consequences of these occurrences.

Finally, Chapter 13 deals with management of the emergency scene. When fires or other emergency events do occur, the company officer is likely the first on the scene. His assessment of the situation and his initial actions will have a major impact on all of the events that follow. When done well, the rest of the event may go well. When done poorly, it places the department, its members, the victims, and their property at risk.

Each chapter includes a list of objectives; satisfy yourself that you have learned each objective. Throughout the text you will find key terms and notes of particular importance in the sidebars. The shaded boxes contain interesting information that directly relates to the content in the chapters. There is a series of questions at the end of each chapter. You should be able to answer each question, without looking at the text. Finally, each chapter includes a list of additional readings. Each of the suggested readings offers additional information on the topic covered in that particular chapter. The goal is to give you as much information as possible to help you in your duties as a company officer.

Your mastery of the material can be demonstrated in several ways. A traditional approach is to take a test. An alternate approach is to ask you to personally complete all of the material in the companion *Student Workbook* to a satisfactory level. The *Workbook* covers all of the elements of NFPA 1021 for Fire Officer I. To validate your efforts, the authority providing your certification may ask for further evidence of your personal competence by asking you to take an oral or written test. If you have seriously studied the material, and answered all of the questions, you should be completely ready for that challenge. Good luck in your career!

ABOUT THE AUTHOR

Clinton Smoke is a professor and head of the Fire Science Program at Northern Virginia Community College in Annandale, Virginia. His fire service background includes more than 20 years of volunteer and paid experience in several fire departments. He also served over 20 years active military duty in the Coast Guard, during which time he commanded four cutters. Other assignments included duties as the senior instructor at the Coast Guard's Officer Candidate School and as head of the service's Vessel Safety Branch. Professor Smoke has a B.A. in Fire Science Administration, an M.S. in Communications Management, and an MBA.

The following list is a comparison of the requirements for Fire Officer I as listed in NFPA 1021 and the contents of this book.

NFPA 1021 Section	Chapter(s) in Text
General Requirements	
2-1	
2-1.1	3, 4, 5, 8, 9, 10
2-1.2	2, 13
Human Resource Management	
2-2	
2-2.1	6, 7
2-2.2	8, 13
2-2.2.1	2, 13
2-2.2.2	2, 6, 7
2-2.3	6, 7, 8, 12
2-2.3.1	2, 6, 7
2-2.3.2	2, 6, 7
2-2.4	2, 6, 7, 12
2-2.4.1	2
2-2.4.2	2, 7, 12
2-2.5	6, 7
2-2.5.1	2, 6, 7, 8
2-2.5.2	7
2-2.6	6, 7
2-2.6.1	2, 6, 7
2-2.7	4, 5
2-2.7.1	6, 7
2-2.7.2	4, 5
Community and Government Relations	
2-3	
2-3.1	2, 13
2-3.2	2, 4, 5
2-3.2.1	2, 6, 7
2-3.2.2	2
2-3.3	2, 3, 4
2-3.3.1	2
2-3.3.2	6, 7

Administration	
2-4	
2-4.1	4, 5
2-4.2	4, 5
2-4.2.1	2
2-4.2.2	6, 7
2-4.3	4, 5
2-4.3.1	4, 5
2-4.3.2	2
Inspection and Investigation	
2-5	
2-5.1	11
2-5.2	11
2-5.2.1	10
2-5.2.2	2, 10, 11
2-5.3	11
2-5.3.1	11
2-5.3.2	11
Emergency Service Delivery	
2-6	
2-6.1	12, 13
2-6.2	12
2-6.2.1	10, 12
2-6.2.2	2
2-6.3	13
2-6.3.1	13
2-6.3.2	2, 10, 12, 13
2-6.4	13
2-6.4.1	10, 13
2-6.4.2	2, 7, 8, 13
Safety	
2.7	8
2-7.1	8
2-7.2	8
2-7.2.1	8
2-7.2.2	8
2-7.3	8
2-7.3.1	8
2-7.3.2	2, 8

Acknowledgments

Many people made this work possible. I am pleased to share any credit that may be due with them; on the other hand, any errors or omissions are mine alone. Much of this material is the result of suggestions from students. I am grateful to them. This book is dedicated to them and to all of the great company officers in the fire service.

To Alan Brunacini of Phoenix, Arizona, my thanks for your many great ideas and for allowing me to use some of them here. To Ms. Vina Drennan, my thanks for allowing me to share your message. To the National Fire Protection Association, my thanks for sharing your information and for your permission to use NFPA 1021 herein. Likewise, my appreciation to the U.S. Fire Administration and to TriData Corporation of Arlington, Virginia, for your assistance.

To Dave Diamantes, Robert Gagnon, and Robert Klinoff, all of whom have books by Delmar, my thanks for helping me along the way. To Roy Nester and Jay Iacone of the Fairfax County Fire and Rescue Department, and to T. R. Koonce of J. Sargent Reynolds Community College, my thanks for all of your great suggestions. To Scott Boatwright and Michael Regan, of the Fairfax County Fire and Rescue Department, my thanks for great photographs. To the men and women of the Arlington County Fire Department and the Fairfax County Fire and Rescue Department, my thanks for allowing me to take your pictures and for your patience while I did.

To my wife, Judith, my thanks for all your support and for all of your constructive suggestions over the years. To my secretary, Linda Land, and to the Library Staff at Northern Virginia Community College, my thanks for your help in this undertaking. Finally, my thanks to Mark Huth, Jeanne Mesick, Dawn Daugherty, and the great fire science team at Delmar for their assistance in producing the book you are reading.

We would also like to thank all of those who contributed their expertise in reviewing the manuscript or time in acquiring the photographs. Our special thanks go to:

Michael Arnhart
High Ridge Fire Department
High Ridge, MO

James Barnes
Baltimore County Fire Department
Towson, MD

Jimmy Bryant
Shreveport Fire Training Academy
Shreveport, LA

Dennis Childress
Orange County Fire Authority
Rancho Santiago Community College
Orange, CA

Mike Connors
Naperville Fire Department
Naperville, IL

Robert Hancock
Hillsborough County Fire & Rescue
Tampa, FL

Tom Harmer
Titusville Fire & Emergency Services
Titusville, FL

Rudy Horist
Elgin Community College
Elgin Fire Department
Elgin, IL

Robert Laford
Mt. Wachusett Community College
Petersham, MA

Murrey E. Loflin
Virginia Beach Fire Department
Virginia Beach, VA

Jack K. McElfish, Chief
Richmond Fire Department
Richmond, VA

Mike Morrison
Cosumnes River Community College
Elk Grove Community Services District Fire
    Department
Elk Grove, CA

Timothy Pitts
Guilford Technical Community College
Greensboro Fire Department
Greensboro, NC

Chris Reynolds
Hillsborough County Fire & Rescue
Hillsborough County Community College
Tampa, FL

Captain Kevin P. Terry and members of the
    Fuller Road Fire Department
Albany, NY

Scott Vanderbrook
Oveido Fire Rescue Department
Oveido, FL

Cindy Willett
Baltimore County Fire Department
Towson, MD

Tom Wutz
Deputy Chief Fire Services Bureau
Office of Fire Prevention and Control
Albany, NY

Foreword

I am pleased that my friend Clinton Smoke asked me to write this foreword for his new text-book. Clinton has been teaching and writing about fire department operations for a long time. He brings that practical experience to his new book along with his excellent communications skills. He has been associated with many important and progressive developments during his career. I have enjoyed our friendship and mutual involvement in many of these projects.

How we select, prepare, and support company officers is a challenge that is critical to the day-to-day delivery of fire department services. Throughout my entire career I have heard fire department managers (me included) articulate that challenge by saying "if I could 'have' any group within my department, make that group our company officers." To "have" a group means to somehow capture the consistent effectiveness of that group. Another indication of how important the company officer is to our overall effectiveness is hearing an experienced fire chief say that the quickest and most accurate way to evaluate the effectiveness of a fire com-pany is simply to look at the profile of the company officer in charge of that unit.

A major reason the company officer is so important is that he or she is the only boss who has continuous access to our workforce. Company officers directly supervise the workers who deliver services to our customers. The company officer is the up-close and personal work-ing boss who is always with and part of the crew of Engine 1 throughout the entire shift. There is no other middle person between the workers and the work. He rides in the right front seat, he sits at the head of the dinner table, he controls the TV remote (if he has enough seniority), he holds the IV bag and the flashlight, and he backs up the nozzle. When the workers do good, the boss smiles (big deal); when there is a problem, the company officer helps fix it. He can make a bad day better by whispering the answer to a worker who gets disconnected from the answer. He is the on-line safety officer for the crew and has the most direct control over how (and if) the crew goes home at the end of the shift with all their body parts and faculties intact.

Another major reason that the company officer is so critical to us is that he manages the business of our business, right at the point where the customer's problem occurs. He is in Mrs. Smith's kitchen at 3:00 A.M. when she smells smoke and at that point, he is the entire depart-ment to her. He is able to create a positive feeling and memory in Mrs. Smith based on how quickly his company responds, how well they solve her problem, and most of all, how nice they are to her. Simply, between 3:00 A.M. and about 3:20 A.M., we get one chance to win or lose, based on how well Engine 1 performs at Mrs. Smith's house. The memory that Mrs. Smith has of the night her kitchen caught fire is the product of how Engine 1's crew made her feel; how Engine 1's crew feels is the product of how their company officer treats them. This is how the company officer gets to connect the inside (crew) with the outside (customer) and this creates our most important relationship.

I am involved both professionally and personally with company officers, so my comments and feelings are not those of a casual observer. I was a company officer (at Engine 1) in an earlier life. The Phoenix Fire Department now employees 255 fire captains (our company officer title). Nearly one third are on duty as I write this foreword, and today they will respond to 450 plus requests for service. How they perform will regulate how 327 on-duty firefighters and thousands of customers will live, or die.

On a personal basis, my three children are all Phoenix firefighters. My son Nick was a company officer for 10 years before he was recently promoted to battalion chief. My son John is a captain on a busy downtown ladder company. My firefighter daughter Candi is going through the company officer promotion preparation process. Throughout the years, I have followed the standard going-on-duty send off with them by saying, "be careful today," and the regular post-shift question, "what'd you guys do yesterday?" This book is directed at how to effectively manage the activity that happens between in between these two rituals—the duty shift. Pay attention as you go through this material. It is the most important part of what we do.

Alan V. Brunacini
Fire Chief
Phoenix Fire Department

Chapter

1

The Company Officer's Role—Challenges and Opportunities

Objectives

Upon completion of this chapter, you should be able to describe:

- The duties of a company officer.
- The national standard for competency for fire officers. NFPA 1021
- Career development opportunities for company officers.
- Strategies for success as a company officer.

INTRODUCTION

It is a dark and stormy night. The big fire truck moves slowly along the narrow mountain road, red lights flashing, spotlights drilling holes into the darkness along either side. The firefighters are looking for a car that was reported to have run off the road in this remote area.

Everyone in the truck is a little uneasy. It is raining hard. Trees and wires are down, and there is always a chance of flooding when it rains this hard. While the inside of the truck's cab is relatively dry, comfortable, and safe, there is an uncomfortable feeling of tension. Riding in the cab are two firefighters, a driver, and you, the company officer. In difficult conditions like these, you are expected to be calm and to reassure the others. But you have some anxieties, too. Not only are you sharing the same concerns about the weather as your young firefighters, but you are responsible for managing them and the event.

Suddenly one of your firefighters yells, "Stop!" Your driver puts on the brakes and brings the big vehicle to a safe stop. As you look out into the darkness, the spotlight illuminates a wrecked vehicle against a tree. The car door is open and it looks like someone is inside. Now, not only are you responsible for those who came with you, but you are expected to assess and manage this entire situation, assist those who might be trapped in the car or lying injured on the ground, and keep everyone, including your own personnel, from further injury.

You think about the situation for a moment. You pick up the radio microphone; you are not yet sure of what to say. As you open the door, the cab lights come on, revealing the anxiety of your colleagues as they look to you for direction. A lot of responsibility is in your hands right now.

For the company officer, it is just another day at work.

THE ROLE OF THE COMPANY OFFICER

■ **Note**

A company officer is a first-line supervisor responsible for the performance and safety of assigned personnel in an emergency service organization.

What is the role of the company officer? A company officer is a first-line supervisor responsible for the performance and safety of assigned personnel in an emergency service organization. The company officer should be capable of leading his or her personnel during both emergency and nonemergency activities.

The company officer's job is as varied as all of the activities of the organization. Company officers (as shown in Figure 1-1) are often the senior representatives of the fire department at the scene of countless emergencies ranging from automobile accidents to fires, from treating sick and injured persons to mitigating hazardous materials spills. In nearly every case, the company officer is one of the first persons on the scene and is also likely to be the last to leave.

But there is more to being a company officer than managing emergency situations. When the company is in the station, the company officer essentially runs a small part of the fire department, rescue squad, or emergency medical services (EMS) organization. At the company level, the company officer manages the resources: people, equipment, and time.

Figure 1-1 *The company officer. One of the most exciting jobs in the fire service, the company officer manages resources during both emergency and nonemergency activities.*

In most departments, companies spend less than 10% of their time dealing with emergencies. That means they spend 90% of their time in other activities, including providing fire safety public education programs for citizens, preplanning for fires and other emergencies, training, and maintaining good physical fitness. Regardless of the activity, the company officer is expected to plan, manage, and lead the company.

These duties may not sound as important as those in which the company is on the scene of an emergency, but they are. The company officer is also expected to be a full-time leader and instructor for the personnel assigned to the company, so human relations skills are very important for company officers. Look at the following list: How many of these activities deal with human relations?

The Many Roles of the Company Officer		
Coach	Innovator	Public relations representative
Communicator	Instructor	Referee
Counselors	Leader	Role model
Decision maker	Listener	Safety officer
Evaluator	Manager	Student
Firefighter	Motivator	Supervisor
Friend	Planner	Writer

As shown in the list, the company officer is expected to be able to perform many roles. Company officers must learn how to perform these roles while retaining their previous skills and knowledge. Company officers do not normally have the luxury of standing around at the scene of emergencies with a radio in one hand, directing the firefighters' activities. They frequently find that they must help fight fires or help diagnose and treat patients, along with the rest of the troops. In addition, they are expected to manage and supervise their subordinates. Many argue that the company officer has one of the most demanding jobs in the organization.

Leading others is the company officer's principal job. The capabilities, efficiency, and morale of the company are direct reflections of the company officer's leadership ability (see Figure 1-2). The personnel are the company

■ **Note**
Leading others is the company officer's principal job.

Figure 1-2 *Working with others is part of the company officer's job.*

officer's responsibility. If the company officer does not take care of them, no one else will.

In the fire service we use the term *company* to describe the work teams. Clearly there is a need for teamwork at the company level. With the reduced staffing that many fire departments face, the company officer is not only the leader of the team but one of the players too. Company officers are expected to lend a hand when needed, whether it be advancing hose, forcing doors, or similar activities that do not compromise their roles as leaders. There is nothing wrong with helping out. But be careful: When an officer takes a position as a firefighter on the attack hose line, one has to ask, "Who is in charge? Who is in charge of the firefighters' accountability and safety?"

Although company officers are often required to help with resolving the problem at hand, the officer's primary job is to lead. Many new officers find the most difficult part of becoming an officer is accepting the fact that they are no longer firefighters. The best firefighters are often promoted to become officers; we would not want it any other way. But these same individuals sometimes find that the transition from *doer* to *leader* is difficult. They want to be involved with the work. In most cases they are very good at performing firefighting tasks and find a great sense of enjoyment in doing them. They find it difficult to give up the physical tasks and to think about supervision and management. For many, the physical tasks are easier and more enjoyable than the mental effort and the paperwork that accompanies the officer's job.

The opposite situation occurs from time to time when an officer will not have anything to do with the firefighters or their work. These company officers seem to forget where they came from and what life was like as firefighters. There are various reasons for this situation, but sometimes firefighters see these officers as placing their own careers ahead of those of the firefighters. Firefighters have the knowledge, skills, and abilities to serve the citizens of their communities. Officers should remember that firefighters are the backbone of the service-delivery system. The officer has to take care of them, nurture them, and help them develop their full potential. That is the company officer's job.

THE COMPANY OFFICER'S ROLE IN THE DEPARTMENT

In later chapters, we discuss organization and management principles. For now we use a simple graphic device (as shown in Figure 1-3) to represent a typical fire service organization. Notice that the company officer plays an important role in the organizational structure. For firefighters, the company officer is not only their supervisor, but their next link in the chain of command. That link connects them with the rest of the department. The company officer may also be considered their link to the public, for it is the company officer who likely speaks for the company and the department both at the scene of an emergency or during nonemergency situations.

■ **Note**
Company officers are expected to lend a hand when needed, whether it be advancing hose, forcing doors, or similar activities that do not compromise their roles as leaders.

■ **Note**
Firefighters are the backbone of the service-delivery system. The officer must take care of them, nurture them, and help them develop their full potential.

Figure 1-3 *The company officer occupies a vital link in the fire department's structure.*

To the fire chief and other senior officers, the company officer represents the company. If the chain of command is being followed, the company officer is both the voice of the chief to the firefighters and the voice of the firefighters to the chief.

Two other factors affect the company officer's position in the organization. First, we often take relatively inexperienced personnel (at least in a leadership role) and place them in one of the most challenging positions in the fire service. Second, we often place these new company officers at a worksite that is remote from their supervisor for all of their normal nonemergency activities and for many minor emergencies too. Although this practice is not unique to the fire service, we should consider the benefits to the individual officer and to the organization of being closer to the supervisor.

Where a company is colocated with fire department headquarters, the company officer has greater access to his or her supervisor and should always keep the lines of communication open (as shown in Figure 1-4). There are certainly pros and cons of being a company officer in such an environment, but many companies and many company officers in such environments tend to be above average and these officers are often selected for advancement ahead of their peers. Could it be that these officers are better because they have access to their supervisor?

Finally let us look at the role of the company itself. In fire departments or other emergency response organizations, 80% to 90% of the employees work at the company level, making the organization's mission statement come alive for the citizens. That mission implies that the department is there to provide a service to the community. Service delivery is the sole reason these organizations exist. To put it another way, most of the organization's service delivery, or to use a more current term, most of the organization's customer service, is delivered by personnel at the company level.

Figure 1-4 *Keep the boss informed of your plans and activities.*

PROMOTION TO COMPANY OFFICER

In the past, departments used a variety of methods to select their company officers. Some had a selection process, using tests and other assessment tools to determine the individuals best qualified for promotion. Other departments based the promotion process on seniority, implying that the person with the longest service was the best qualified for advancement. (A few departments carried this process all the way to the fire chief's position.) Some departments were not so subtle: relatives and friends were promoted ahead of others.

Regardless of the promotion system in place, many of these officers were not fully prepared for the new duties they faced. We should not criticize these individuals; rather, we should admire those who were successful despite these conditions. We should be understanding of all of those who have struggled; in many cases, they were placed into positions without the experience or training needed for the job. Imagine the consequences of letting other professionals, such as civil engineers or physicians, practice under similar conditions.

Regardless of whether we are serving in a paid or volunteer capacity, we would like to be considered professionals. The public is entitled to find qualified, competent, and professional personnel in emergency service organizations. Fire departments should be sure that their personnel are professionals. This can be accomplished by certification.

certification
a document that attests that one has demonstrated the knowledge and skills necessary to function in a particular craft or trade

For purposes of this text, **certification** means that an individual has been tested by an accredited examining body on clearly identified material and found to meet a minimum standard. Certification provides a yardstick by which to measure competency in every type of department from the largest to the smallest, and from all-paid to all-volunteer. There are other reasons to certify:

- Protection from liability
- Recognition of demonstrated proficiency
- Recognition of professionalism
- Budget and salary justification

The traditional way to prepare for entering a profession is to attend college or a technical school. Community college-based fire science programs have been around for more than 30 years. They were created to help individuals enter and move up the fire service organizational ladder. A 1969 publication entitled *Guideline for Fire Service Education Programs in Community and Junior Colleges* identified basic courses that should be included in all fire science programs. These included communications skills, human relations, government, fire prevention, firefighting strategy and tactics, and building construction, among others. Those topics are still included in fire science programs today, and they are also part of the national competency standards addressed in the next section. Interestingly, that report listed nearly 150 associate degree programs in existence. Today, there are more than twice that many.

We have already suggested that it would be difficult to provide a concise yet complete definition of the company officer's job, but we should try. One way to do this might be to ask a committee of respected fire service professionals to define the company officer's job and to define the competency requirements that address the knowledge, skills, and abilities needed to perform that job. Then we can develop a training program to help prepare individuals to move effectively into the supervisory ranks.

A fire department or other emergency-response organization is a complex business involving science, good business practices, and human relations. For the most part, fire protection is a local problem that must be handled by the citizens of the community as they see fit. However, many national organizations and federal agencies have a role in fire protection as well. The same can be said for EMS, hazardous material response and urban rescue operations.

Officers should be aware of the existence and role of these organizations, for they have had a significant impact on changing the character of the fire service over the last quarter century. Ten of these organizations gathered together for a meeting in 1970. The leaders of these organizations became known as the Joint Council of National Fire Service Organizations.[1] Their meeting was significant because it was the first time they had organized to present a united effort to address the nation's fire problems. One of the council's first tasks was to develop nationally accepted standards for firefighters, fire officers, and others. The importance of this role is clearly seen in the very first paragraph of the council's goals statement.

■ **Note**

Officers should be aware of the existence and role of these national and federal organizations, for they have had a significant impact on changing the character of the fire service over the last quarter century.

Goals of the Joint Council of National Fire Service Organizations

- To develop nationally recognized standards for competency and achievement of skills development, technical proficiency, and academic knowledge appropriate to every level of the fire service career ladder

- To make the public aware of the significant contributions made by the fire service of this nation in protecting life by providing increased financial and moral support to aid the fire service in carrying out its mission

- To make public officials at every level of government more aware of their responsibilities in providing increased financial and moral support to aid the fire service in carrying out its mission

- To reassess public fire protection in light of contemporary demands, ensuring appropriate fire protection for all communities at a reasonable cost

- To establish realistic standards of educational achievement and to provide to every member of the fire service equal educational opportunities commensurate with professional requirements

- To identify and establish nationwide information systems that will enable improved analysis of the fire problem with particular emphasis on the life and safety factors for the public and its firefighters

- To encourage and undertake the research and development necessary for the prompt and successful implementation of these goals

THE NATIONAL COMPETENCY STANDARD FOR FIRE OFFICERS

A significant step in the certification process occurred when the Joint Council established the National Professional Qualifications System in 1972. That system provided a means of representing the career opportunities in the fire service and specified a particular body of knowledge required for each level and for each specialty area in the fire service career ladder. The standards apply to volunteer and paid personnel alike. Volunteers, just as paid personnel, need to know what to do and how to do it at the scene of any emergency. The lives of their colleagues as well as the lives and property of the citizens of their own community are often at stake.

These standards started and remain a product of a peer group process and are not the result of any regulatory action. The present standards provide nationally recognized competency criteria for firefighters, fire service instructors, fire officers, inspectors, investigators, public fire educators, and others.

codes
a systematic arrangement of a body of rules

standard
a rule for measuring or a model to be followed

The present professional certification standards are part of the National Fire Protection Association's (NFPA) **codes** and standards-making process. There are almost three hundred NFPA codes and **standards** dealing with every aspect of fire protection. These documents have a profound impact on improving safety where we live and work. They are widely adopted for legislation and regulation at the federal, state, and local levels. Many federal agencies, including the Occupational Safety and Health Administration (OSHA), National Institutes of Health (NIH), and the Department of Energy (DOE), reference NFPA's codes and standards in their own regulations. Many insurance companies use NFPA documents for guidelines in assessing risks and setting premiums. NFPA's codes and standards are the result of more than two hundred committees consisting of more than five thousand individuals who serve voluntarily. Built into the codes- and standards-making process is an opportunity for the public to comment on documents at every stage of their development. Only after the public comment phase is complete is the document brought before the membership of NFPA for formal adoption. Once this phase is completed the standards are published and available for voluntary adoption by local jurisdictions.

Ten of the NFPA documents deal with the qualifications of those who serve in the fire service. As public organizations, fire service organizations are open to public scrutiny and are held accountable for their actions. There is considerable value in being able to demonstrate that the personnel of these agencies are certified as meeting the competency standards of an entity that has itself been evaluated by an independent, thorough, objective, and public process and approved or accredited as meeting the requirements of the process.[2]

NFPA published the first national standard for fire service officers in 1976. It defined the knowledge, skills, and abilities needed by the company officer (see Figure 1-5). Over the years the document has been revised several times, each

■ **Note**

The present version of NFPA 1021, Standard for Fire Officer Professional Qualifications, outlines the requirements for fire officers at four levels of competency.

time improving the list of requirements for those seeking certification as fire officers.

The present version of NFPA 1021, *Standard for Fire Officer Professional Qualifications*, outlines the requirements for fire officers at four levels of competency. The first level, Fire Officer I, focuses on the needs of the first-line supervisor, clearly including company officers. For Fire Officer II, the requirements focus on the management aspects of the company officer's job and help prepare individuals to move into staff assignments. The requirements for Officer III deal mostly with administration and management, preparing officers to move into positions of increasing responsibility at the midmanagement level within their organizations. At the top of the officer certification ladder (see Figure 1-6), the requirements for Officer IV satisfy the needs of senior staff and chief officers. NFPA 1021 is shown as Appendix A of this book.

Figure 1-5 *Good writing skills are important for the company officer.*

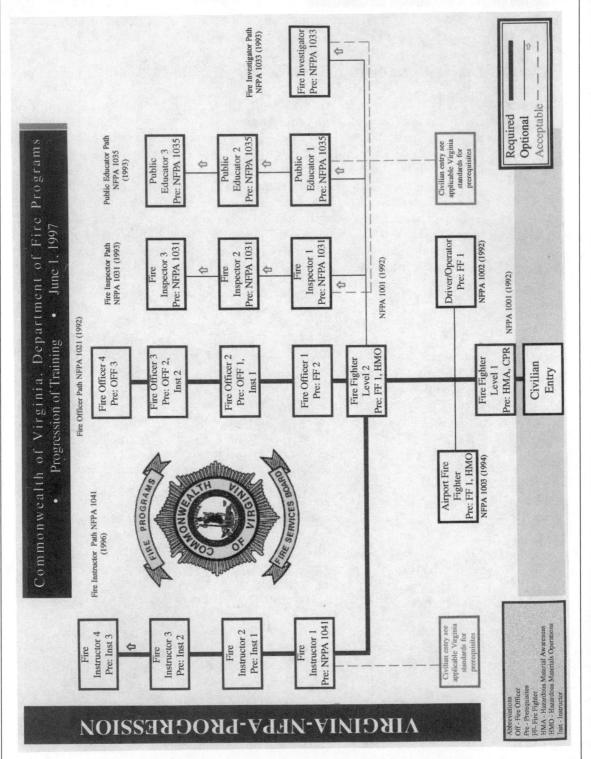

Figure 1-6 *Certification ladder. Courtesy of the Virginia Department of Fire Programs.*

CAREER DEVELOPMENT: AN OPPORTUNITY FOR COMPANY OFFICERS

The fire and rescue service is unique. No other occupation serves the public in so many ways. The ability to react to the public's needs and resolve any problems encountered places a high level of responsibility on every employee, including the newest firefighter. For these reasons, demanding entry-level requirements and rigorous recruit training are vitally important. In order to maintain these skills and learn new ones, a continuing program of training and education should be provided to all members. This training helps employees maintain proficiency at their present levels, meet certification requirements, learn new procedures, and keep up with emerging technology.

Training to maintain skills should be provided as part of an employee's job. Preparing for advancement should involve some individual initiative on the part of the employee. In other words, if you want to get promoted, you have to put a little sweat equity into the process. Although advancement brings greater prestige, it also brings added responsibilities. One should have an opportunity for promotion, but at the same time, one should take some personal initiative.

The company officer's job is vast and varied. It covers many diverse topics. Company officers are expected to walk into the new job, know all the answers, and be ready to go to work. In reality, life is not quite like that. We learn something new every day and yet we really never know all the answers. However, we can prepare to make the transition from firefighter to fire officer with study and effort.

There have been many ways of promoting officers over the last century, but we are seeing more and more departments adopt an organized career development program (see Figure 1-7) for their employees. Such programs usually require courses from various disciplines from a local community college. The program content should be carefully considered so that once it is in place, it remains constant. Thus employees can know exactly what is required for advancement. To help employees prepare for advancement, and more importantly to prepare to capably serve in the new position after advancement, the new skills, knowledge, and abilities should be mastered before the employee becomes eligible for promotion to the next rank.

Clearly, and appropriately, this policy places a burden on those who seek promotion. The department can ease this burden by encouraging participation through tuition assistance and arranging with a local community college for courses to be scheduled at convenient times and locations. Where employees work 24-hour days, this means asking the college to offer courses twice a week on consecutive days or nights to accommodate the employees' work schedules. Thus, employees can be assured of both the convenience and the opportunity to take courses without having to take time off from duty.

Some personnel may view these training and education requirements as barriers to their advancement; others will see them as opportunities. Taking the courses helps them develop new skills, knowledge, and abilities so that they can perform duties in additional areas, perform current duties in a more effective and

■ **Note**

To help employees prepare for advancement, and more importantly to prepare to capably serve in the new position after advancement, the new skills, knowledge, and abilities should be mastered before the employee becomes eligible for promotion to the next rank.

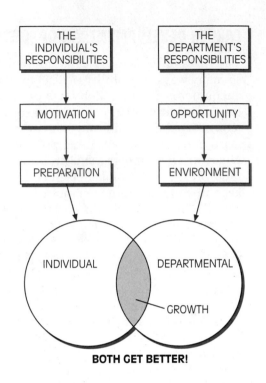

Figure 1-7 *Career development is a shared responsibility. Both you and your community have an obligation and both will benefit from your growth. Courtesy of the Phoenix, Arizona, Fire Department.*

safe manner, and learn new procedures as they are introduced. Much of the training and education required of a career development program can also help in one's personal life, both now and after retirement. The information obtained while learning computer skills, improving communications skills, or becoming more effective in interpersonal relations does not have to be left in the station—it can be used off duty as well as on, throughout the rest of your life (see Figure 1-8).

In the final analysis, career development provides employees with an established program of advancement opportunity. For the individual, advancement means increased responsibilities. As advancement occurs, the department and ultimately the citizens of the community benefit from having knowledgeable, productive, and effective employees. A well planned career development program is a good deal for everyone: It benefits the employee, the department, and the community.

However, these benefits imply some responsibilities. The employer should encourage participation, provide an environment that provides opportunities for preparing for advancement, and establish a valid system to select candidates for advancement. The employee must be motivated to prepare for and accept the additional responsibilities and duties that come with promotion.

Figure 1-8
Communicating effectively with others is an important task for company officers.

■ **Note**
Be aware that a program leading to fire officer certification is only the first step; additional training/education is strongly recommended.

High standards should be set for promotion to the first supervisory ranks. These standards reflect the significant demands placed on supervisory personnel while performing their duties. The primary purpose of the requirements is to ensure that personnel are fully prepared to face the complex challenges of the new positions they are assuming. Most of the requirements focus on administration, management, and supervisory issues.

LIEUTENANTS[3]. Lieutenants should be certified as Fire Officer I in accordance with NFPA 1021. Be aware that a program leading to fire officer certification is only the first step; additional training/education is strongly recommended. The following should be considered: Suppression officers should consider taking courses in building construction and fire suppression methods. Yes, both of these topics are introduced in firefighter training programs, but you need more information on both of these topics because they are at the core of your job as a firefighter. Both of these topics were addressed in the original list of courses for professional development in the fire service and both are addressed in semester-long courses in nearly every fire science program. In addition all officers should take a college-level English class. A shortcoming of too many officers is the lack of good communication skills.

CAPTAINS. Captains should be certified as Fire Officer II. In addition to the requirements of lieutenant, a second semester of English is recommended, either English Composition II or a technical writing course. Also consider taking a college mathematics course. Many adults forget basic math skills if they are not used. Math is useful for planning and managing budgets, in working with hydraulics, in determining dosages, and so forth. Two additional classes are recommended. The first is a class in administration and management. There are many options here. For example, a fire science program may offer a course in fire service administration while the college's EMS program may offer a course entitled EMS Administration. Either would be appropriate. Where neither is available, a course in principles of management would be a good alternative. The other course is in supervision or human relations supervisory skills. Again, most fire science and EMS degree programs have such a course, but where these may not be available, look for a course in supervisory practices, usually offered by the business or management department.

Many of these courses are part of a college-level degree program. Completion of these courses and the other requirements listed here will place you well along the path toward a college degree. Prior college efforts and relevant courses taken at the National Fire Academy should always be accepted.

To make the system work best, these courses should be completed before you take the test for advancement. To validate the process, the test or assessment process should carefully consider the content of all of these courses in their selection criteria. Because of the up-front effort on the part of the candidates, fewer candidates may appear for the test, but it is more likely that all will be qualified. On the test day, all that is required is to select the best qualified and rank them. Once promoted, these individuals should be ready to assume the duties of their new position.

STRATEGIES FOR SUCCESS AS A COMPANY OFFICER

Be a professional. Being a professional means that you are dedicated and committed to the job, learning all that you can and giving as much as you can. Work to make yourself, your company, and your department the very best possible, whether you are paid or a volunteer.

Set personal goals. Every time you go to the station, make some contribution to the improvement of the place and your organization.

Continuously work on your own training and education and encourage the same in others. Training is a part of the job, so obtain all of the training and education that you can. Set personal training and education goals, and use your time and energy to attain them.

Be loyal to your colleagues and your department. Speak well of your job, your department, and your associates. Although the Constitution guarantees freedom of speech, good judgment suggests that this freedom must be tempered with reason and sometimes with constraint.

Be a role model. You will likely be able to recall the first officer or first instructor you had as a firefighter, EMT, or rescue squad member. These individuals leave a lasting impression with us—hopefully positive. For those who are in your charge, the same holds true; be a role model for them. Help them to develop a professional attitude about the job, and help them to be all that they can be. Being a good role model does not mean that you must be perfect. It means you try hard and work to improve so that you can do better tomorrow. Being a role model means being a professional. Professionalism encompasses attitude, behavior, communication style, demeanor, and ethical beliefs. It is as simple as A, B, C, D, and E:

Attitude is at the core of your performance. A good positive attitude suggests that you are working to be a role model.

Behavior is how you act. You are watched by others, both while you are on duty and when you are off duty. Your actions reflect upon yourself, your department, and your profession (see Figure 1-9).

Figure 1-9 *The successful company officer is a role model for others.*

Communication is how you get your ideas across to others. Communication can be oral, written or nonverbal. We are in a people business; we work with people and we serve people. Today's emergency service organizations spend a great deal of time working to improve the human relations side of us, so that we work together better and serve our citizens better.

Demeanor embraces all three of the previously listed items. Focus your energies on the mission of your organization and use positive attitude, good personal behavior, and effective communications to accomplish the goals of your organization.

Ethics deals with conforming to the highest professional standards of your organization. It is more than just words and it is more than just acting out words. It is doing the right thing, every time, every day.

Summary

Being an officer in the fire service is one of the greatest jobs on earth. At times you are your own boss. You also supervise others. Often you are doing many of the things you came into the fire service to do, without having to do some of the dirty work. There is no higher calling. Being a company officer is not easy, but it has its rewards. One of those rewards is that we get to work with others, often as leaders, to do the very things that motivated us to enter this profession in the first place—saving lives and property.

There are significant rewards associated with the position of company officer, but there are also significant responsibilities. Our purpose here is to help you prepare to meet these responsibilities.

Review Questions

1. List five of the roles of the company officer.

2. Where does the company officer fit into the fire department organization?

3. List some of the rewards of being a company officer.

4. List some of the challenges that face company officers.

5. What is the name of the document that identifies competency standards for fire officers?

6. Why should fire officers be certified?

7. List three of the five strategies for success discussed in this chapter.

8. What is the company officer's primary job?

9. What is the company officer's role in leading the company?

10. What is the company officer's role in the overall organization?

Additional Reading

Bennett, Jack A., "Effectiveness at the Company Officer Level," *American Fire Journal,* August 1996.

Broadwell, Martin M., "The Case for Pre-Supervisory Training," *Training,* October 1996.

Clark, Burton A., "Higher Education and Fire Service Professionalism," *Fire Chief,* September 1993.

Favreau, Donald F., *Guidelines for Fire Service Educa-tion Programs in Community and Junior Colleges* (Washington, DC: American Association of Junior Colleges, 1969).

NFPA 1000, *Fire Service Professional Qualifications Accreditation and Certification Systems* (Quincy, MA: National Fire Protection Association, latest Edition).

Notes

1. Among the organizations represented were the International Association of Fire Chiefs, the International Association of Firefighters, the International Fire Service Training Association, the International Society of Fire Service Instructors, the National Fire Protection Association, and the National Volunteer Fire Council. The council, having accomplished its initial goals, disbanded in 1992.

2. NFPA 1000, Fire Service Professional Qualifications Accreditation and Certification Systems, paragraph A-1-3.1

3. We use the terms *lieutenant* and *captain* to denote the first two supervisory ranks. These terms are traditionally used, but other terms are used as well. If your organization uses a different term, use your term. The list should still be valid.

Chapter 2

The Company Officer's Role in Effective Communications

Objectives

Upon completion of this chapter, you should be able to describe:

- The communications process.
- The need for effective personal communications.
- Ways to improve your oral communications.
- Ways to improve your listening skills.
- Ways to improve your writing skills.

INTRODUCTION

Effective communications are a vital part of your professional life. Consider these generally accepted truths:

- Those who can communicate effectively get more out of life.
- Those who cannot communicate effectively will not reach their full potential.

Given these statements, you should realize that good communications skills are essential in both your work and in your personal life. Many fire officers will gladly sign up for a weekend of training on a technical aspect of firefighting. Many would also benefit from making a similar commitment to improving their communications skills. Effective communications skills will also add value to your personal life.

This chapter appears at the front of the text for several good reasons. First, it is one of the most important topics in the book. Many fire chiefs report that the one weakness prevalent among their officers is their inability to communicate effectively. A lack of good communications skills will hurt even the best of individuals. You may have the best ideas, but to implement them you must communicate them to others. You have to communicate to make recommendations to your superiors and to motivate your subordinates. You have to communicate to a variety of audiences, including the public you serve. When you represent yourself, your communications shortcomings will only hurt yourself. However, when you represent your department, community, or profession, your inabilities to communicate your ideas and knowledge has a much greater impact.

The second reason for placing this material in the front of the text is that communications is a tool that is needed to complete the requirements for certification. According to NFPA 1021, *Standard for Fire Officer Professional Qualifications,* good oral and written communications skills are a prerequisite for most of the certification requirements for Fire Officer I.[1]

■ **Note**
Good communications skills are essential in both your work and in your personal life.

■ **Note**
A lack of good communications skills will hurt even the best of individuals.

THE IMPORTANCE OF COMMUNICATIONS IN OUR WORK AND LIFE

Being able to communicate effectively is important for both you and your fire department. Think about how much of your time is spent communicating; it is an important part of your life.

Formal versus Informal Communications

To begin our discussion of personal communications, we start by distinguishing between formal and informal communications. Because we are focusing on the needs of the job as an officer in the fire service, we are focusing on the formal aspects of both oral and written communications.

Formal communications are conducted according to established standards. They tend to follow the customs, rules, and practices of the industry or workplace. The fire service has its established practices. The medical community has its practices. The legal profession has quite a different set of practices. Formal communications transmit official information. Formal orders and directives, standard operating procedures, and official correspondence are examples of formal communications. Such formal communications usually have legal standing within the organization.

Informal communications are simpler and more spontaneous. We write with less formality when sending memos, short notes, and e-mail to others. Likewise we tend to be less formal during social events such as meals and other situations where we might be together with colleagues.

Written communications can be either formal or informal. We have already cited some examples of formal communications (e.g., formal orders). But many organizations use informal written communications as well. We usually communicate up and down the organization using formal communications. When we need to communicate across the organization, we usually use a more informal communications style.

For example, you might ask a peer in another section for information or an opinion on a new idea. Because this communication is based more on friendship than on any formal relationship, informal communications are more comfortable.

Informal communications work well in established organizations where there are stable relationships among the work units and among the individual employees. When work units are emerging, when there is conflict, or when there is a lack of trust, we tend to see greater use of formal communications.

Forms of Personal Communications

■ **Note**

The spoken word is the most commonly used method of personal communications.

The spoken word is the most commonly used method of personal communications. It is fast, practical, and usually effective. Oral communications can be one way or two way. A recorded message on an answering machine is an example of one-way communication. A conversation on the phone is an example of two-way communications. Oral communication can be accomplished with one **receiver** or many, such as the audience in a conference or classroom. It can be done in daylight or total darkness. However, verbal communications are limited—they may be inhibited by barriers such as language difficulties or background noise, and may not be documented.

The second form of personal communications is written communications. Written communications provide a record that can be used for future reference, showing that the message was sent, and in some cases there is evidence to show that the message was received.

receiver

in communications, the receiver is the intended recipient of the message

ELEMENTS OF THE COMMUNICATIONS PROCESS

Regardless of whether you are speaking or writing, the communications process is generally thought to include five elements, called the *communication model*.

Communication Model

Sender. The process starts with the sender. The sender is the person who has the information and who wishes to send it to another. The process of selecting the method of communications, forming thoughts, and expressing these thoughts in words is an important part of this process. All of this activity is accomplished by the sender.

Message. The sender selects the message. The message is the information, which is usually communicated using words, although certain symbols, such as the signs used in mathematics, are also considered part of the message. The proper selection of the message is vital to the communications process.

Medium. The sender also selects the method of transmission, sometimes referred to as the medium. The medium is the means by which the message is transmitted. Mediums include oral and written communications, signs, and graphic representations. A stop sign is a medium. When standing in the center of an intersection, a police officer's upraised hand is a medium. The sound of a siren might also be a medium. All send the same message. Other mediums of communication include books, the spoken word, images on the screen in class.

Receiver. The receiver is the recipient of the message. While the receiver is thought to play a passive role, we shall soon see that the receiver plays an active role in the communications process.

Feedback. The final step in the communications process is called **feedback**. Feedback tells the sender that the **message** was received and understood. Feedback comes in many ways (see Figure 2-1a and 2-1b). Compliance with a request is an obvious example of feedback. If you ask someone to turn on the light, and they do, that is feedback. The act of turning on the light indicates that they received the message, that they understood the message, and that they complied with the request.

In this situation, the feedback was obvious. Sometimes getting feedback is much more difficult. The lack of feedback leaves the sender wondering if the message was received, if it was understood, and if anyone is going to respond. We discuss feedback later in this chapter.

feedback
reaction to a process which may alter or reinforce that process

message
in communications, the message is the information being sent to another

Figure 2-1a *Effective communications skills is an important tool for officers.*

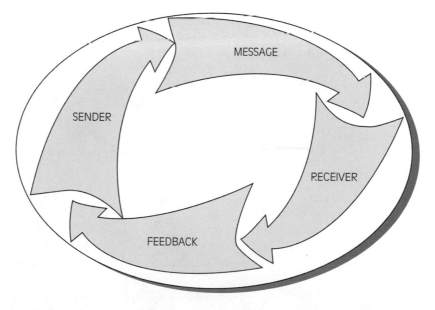

Figure 2-1b *The communications model represents a continuous process.*

The third form of communications is communicating without words. A smile or a frown convey more than words ever can. A hand extended to a friend or even to a stranger makes a clear statement. When appropriate, a pat on the back can send a clear message of support and confidence.

There are other forms of communications of course. For example, symbols or signs (see Figure 2-2) can be used to convey a thought. Some are personal, some are not.

Figure 2-2 *Signs are a form of communication. Some are more personal than others.*

ORAL COMMUNICATIONS

Of all those listed, oral communication is easiest and most used. It is easy because it is effective. It is effective because we get feedback from watching the listener. As a result, most of us prefer face-to-face contact when speaking with another. Face-to-face contact allows the feedback process to work at its best. Good eye contact allows us to look directly at the other individual and watch that person's facial expressions and gestures. A nod of understanding encourages the speaker to continue. On the other hand, a frown should cause the speaker to pause and cover the material in another way. Even the most effective speaker should realize that there will be times when her message is not understood.

Although most of us like to communicate face-to-face, that is not always possible. When you use a phone (see Figure 2-3) or radio, you still communicate with

Figure 2-3 *Communicating by phone or radio can be more difficult than communicating face-to-face.*

sender
a part of the communications process, the sender transmits a thought or message to the receiver

barrier
an obstacle; in communications, a barrier prevents the message from being understood by the receiver

words, but you cannot see the other person. You cannot see the facial expressions and gestures, therefore you are deprived of some of the forms of feedback that are essential to the communications process. When using such forms of communications it is important for you (the **sender**) to make sure that the receiver has received and understands the correct message.

In any case, the speaker should stop from time to time and ask the listener a question. This practice allows the sender and the receiver to trade places for a moment. This makes the communications process easier for both parties. This works, regardless of whether you are speaking to one or one hundred.

When you make a habit of pausing occasionally and asking a question, several things happen. As we have already indicated, this practice allows you (the sender) and the receiver to momentarily trade places. The change is usually welcomed on both sides. Also, you will get questions, and with those questions, you get additional feedback that helps to confirm understanding.

Another advantage of this process is that the receiver will anticipate that a pause is coming, and will hold his questions until the appropriate opportunity is presented. So, this practice allows the sender and the receiver to have a short break from their respective activity, it allows for feedback, and it allows the listener a chance to ask questions, if needed.

The listener should have to concentrate only on listening. At times that can be difficult. Sometimes this means removing the barriers that get in the way of good communications. It also means that the receiver should avoid anticipating what the sender will say or interrupting the speaker with questions.

Barriers to Effective Communications

We have discussed some of the characteristics of effective communications. Now let us consider some of the barriers that make communications difficult. Any obstacle in the communications process is called a **barrier**. Consider a barrier to be like a filter. One or more filters, or barriers, reduces the information flow between the sender and the receiver. There are several types of barriers (see Figure 2-4).

Physical barriers. Physical barriers are environmental factors that prevent or reduce the sending and receiving of communications. Walls, distance, and background noise are examples of physical barriers. These are usually obvious barriers.

Personal barriers. Personal barriers are less obvious. They arise from the judgments, emotions, and social values we place on people. These factors cause a psychological barrier, which can be just as real as a physical barrier. Personal barriers act as filters in nearly all of our interactions with others. We see and hear what we want to see and hear, and we remain selectively "tuned out" to that which we do not wish to see or hear.

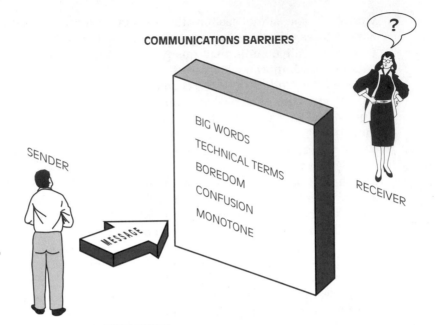

COMMUNICATIONS BARRIERS

SENDER

MESSAGE

BIG WORDS
TECHNICAL TERMS
BOREDOM
CONFUSION
MONOTONE

RECEIVER

?

Figure 2-4 *There are many barriers to effective communications.*

Semantic barriers. Semantic barriers arise from language problems. Our language is filled with words that have multiple meanings and words that have vague meanings. Consider the word *round*. There are several dozen meanings to this word in a standard dictionary. In fact there are more than five hundred words in common use that have more than twenty meanings. Consider the word *dummy*. A dummy is a rescue training tool. A dummy has a different meaning to those in the publishing industry. The word dummy is also used in the game of bridge. None of these terms are offensive. But just call someone a dummy and see what happens. No wonder we sometimes have trouble communicating!

You should be aware of the impact of barriers and change the conditions that impose the barriers when the communications process is adversely affected. Some things can be easily controlled. You can control the choice of words, the use of technical terms or trade jargon, and the speed of delivery. You can sometimes control the time and place of the communication. You can enhance understanding and retention by repeating vital information and showing the same information in some graphic form. All of these techniques will have considerable impact on the listener.

So it is with a new firefighter who may be working for you. New firefighters may be bright and highly motivated, but when you are teaching them something new and you talk over their heads or speak too fast for their comprehension, you will not be effectively communicating. Place yourself in the role of that new firefighter. Provide the information in small bites and ask questions to be sure that the communications process is still working. Other things to consider include the age, educational levels, possible language differences, the background and experience, and the working relationship. Overcoming communications barriers can be achieved by being adaptive to the audience, having a specific purpose, and by staying focused, brief, and clear.

When you are speaking to groups of people and when you are speaking to individuals outside of the fire department, the process is even more challenging. For example, consider the delivery of the department's public life safety education message. One day you might be asked to speak to a group of third graders. That evening you might be asked to speak to a group of senior citizens. The next day, you may go into a community where English is the second language. Although the purpose of your visit is similar on each occasion, the way you communicate your safety message should be quite different.

Part of the Communications Process Is Listening

We have indicated the importance of good, effective communications and that the sender has quite a responsibility to select the medium and the message, to check for feedback, and to be aware of the potential barriers. Communicating also involves listening. In fact, listening may be the most important part for the company officer. Understanding others requires an active role on the part of the listener. This role is called **active listening**. As a listener, you can show the sender that you are actively listening by focusing all of your attention on her (the speaker) and by showing genuine interest in her message. This usually means looking at the speaker, doing nothing but listening, and allowing your facial expressions to indicate interest and understanding (see Figure 2-5). Alert facial expression and good posture indicate a good listener.

When the message is unclear or not understood, the listener should ask questions when appropriate. Questions and comments usually indicate interest and encourage the speaker to expand on an area that was not clearly presented.

While questions are signs of good listening, interrupting the speaker with too many questions or offering solutions to the speaker's problems before he has finished is not good practice. These activities show that the listener is impatient to bring the conversation to a close.

Certainly there are times when the process must be hurried and even interrupted. That is part of life in the fire service. When interruptions occur, both parties should make an effort to resume the conversation as soon as circumstances permit. The real mark of a good leader is a comment that one sometime hears from one of the firefighters: "I like the captain. He listens to me when I have a problem."

Consider the following situation: You walk into your supervisor's office with a question. If your boss stops work, looks at you, and gives you 100% of his attention, it is likely that he is actively listening. On the other hand if one or more of these conditions is not met, it is likely that he is not actively listening. Which situation would you rather encounter?

■ **Note**
Understanding others requires an active role on the part of the listener.

active listening
the deliberate and apparent process by which one focuses attention on the communications of another

■ **Note**
Alert facial expression and good posture indicate a good listener.

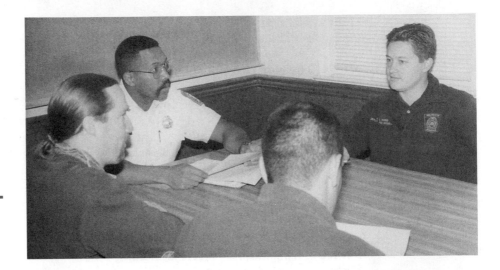

Figure 2-5 *Good listening skills are part of communicating.*

When you are actively listening, three important things happen:

- First, you actually hear what is being said.
- Second, you are more likely to remember what was said.
- Third, you show respect for the sender.

Listening to others is an important part of the your job and probably takes up a good part of your day as company officer. Good listening skills are important. Research has shown that we take in a very small part of what we hear and that we remember only a small part of what we take in. Listening is an active process that requires considerable effort.

At several points in this text we depart from preparing you to become an officer in the fire service and offer some friendly advice. This is one of those occasions. Consider the information above about active listening in your personal life. How well do you communicate in an informal setting? How well do you listen to your friends, your spouse, or your children? They also deserve 100% of your listening talent. Try some active listening at home or in a social setting.

Communications Is a Part of Leadership

As a supervisor, the way you communicate with your people has a lot to do with your leadership style. Suppose a subordinate stops at your office with a question. If you take time to focus on that person, show respect for and interest in her as a person, and actively listen to her idea or question, you are doing a lot to enhance both the communications process and the personal relationship. Not only are you listening but you are showing a genuine interest in the visitor and her thoughts. Whether you do this or not, the time and energy spent in the conversation is about the same. But look at the difference in the outcomes. With active listening you hear better, understand better, and you say, "I care about you."

Regardless of whether you are communicating through the spoken word or through your writing, there is a need to consider the human relations aspect in every communications activity. Human relations are discussed in greater detail in Chapters 6 and 7, but we should note here that consideration for the other person is important to the communications process.

When we communicate, we have an opportunity to show respect for the person or persons with whom we are communicating. That respect, or the lack of it, will be clearly apparent in the speaker's tone as well as the words that are used. When the sender uses words and phrases that show respect and consideration for the receiver's interests and point of view, it is far more likely that the message will be accepted, even when it brings bad news.

Although you should show a concern for the human relations aspect of communications, there must come time in the conversation when you have to get to the point of the meeting. Getting to the point does not mean that there is any lack of respect, nor does it mean that the sender has to be blunt or abrupt. This problem is especially true when dealing with performance issues in the workplace. We are often afraid of hurting the other person's feelings.

It is important to remember the human relations aspect of communications, but you have to communicate a message. That requires effective communications and the ability to focus on specifics. After a brief warmup, get to the heart of the message and tell it like it is. Avoid generalities and exaggerations. "I've told you a million times, don't exaggerate!" is a perfect example. In the workplace, a supervisor's comment, "You are late again! It seems that you are late every day." is likely to start a confrontation rather than a productive session in which solutions can be achieved.

As a supervisor, you (sender) should be able to talk about performance issues without talking about the personality or personal traits of the subordinate (receiver). If the subordinate's work is not up to standard, the supervisor should talk about the standard and the fact that the subordinate is not meeting the standard. The supervisor might show how this impacts on others and what this does to the organization's overall performance. These are facts. What does this do to the conversation? In this case, the employee's performance becomes the issue, not the employee's personality, attitude, or ability.

■ **Note**

When we communicate, we have an opportunity to show respect for the person or persons with whom we are communicating.

■ Note
Remember to make
your communications a
two-way process.

The sender should be able to express his or her own feelings and reactions, when appropriate, especially when dealing with performance problems. For example, the supervisor might say, "I am disappointed with this work."

Finally, remember to make your communications a two-way process (see Figure 2–6). As the supervisor, you can usually ask a questions that will bring more information to light or help the other person understand the issue. Likewise when writing, always close with an offer to continue the discussion or provide additional information.

Making Practical Use of Effective Communications

Consider the following example: "Bill, I have noticed that on our last three calls you were the last one getting to the truck. On our most recent call the driver and I had to wait 30 seconds for you to arrive and get properly seated. Those delays may seem minor, but as you well know, there are times when every second counts. Although no one else has noticed it, this concerns me. Is there a problem? Can I do anything that might help you get there a little faster?"

Notice that the officer has not directly accused Bill of anything wrong, nor has the officer threatened Bill in any way. The officer has only stated his observations and expressed a personal concern about the situation. The officer asked Bill two open-ended questions. Hopefully, these will give Bill an opportunity to recognize that there is a problem, to realize that the boss is both concerned and

Figure 2-6 *When discussing an employee's performance, remember to let the employee tell his side as well.*

willing to help. In this case the officer has shown his concern for his firefighter and given Bill a chance to identify and solve his own problems.

Contrast that action with the following: "Bill, you are too slow on all of our alarms! I guess you just don't think that others are very important. That is, if you can think. . . ."

In summary, show respect, get to the point, focus on specifics, express your own personal feelings, and make the discussion a two-way process.

At this point you may likely say, "Well, that is okay in the fire station, but when we get out on the fireground things change." Surely the tempo picks up, and the communications tend to be very directive in nature. "Do this! Do that!" Some of that is justified by the urgency of the situation. But even under these conditions, your leadership style is heard in the tone of your voice and in what you say. If you remain polite and reasonable in your communications, you will get just as much, if not more, from the troops.

One of the company officer's many jobs is to facilitate the communications process. Firefighters tend to be interested in matters that are going on elsewhere within their department and they should also be interested in what is going on elsewhere in the fire service. You should facilitate this process. By virtue of your rank and experience you probably have more access to what is happening at headquarters and elsewhere. You should be reading the directives from headquarters and the trade magazines. In both cases, appropriate information should be passed on to the troops as soon as practical.

If a firefighter has a question, you should take the initiative to help her find the answer (see Figure 2-7). That answer may be to simply (and politely) tell her where to look for help. Some questions will require a call to headquarters. In many cases, it certainly would be far more appropriate for you to make the call.

■ **Note**

Show respect, get to the point, focus on specifics, express your own personal feelings, and make the discussion a two-way process.

Figure 2-7 *When you do not know the answer, admit it. Then make an effort to find the answer.*

Although most firefighters show surprisingly little hesitation to call headquarters, you are more likely to know whom to call and what questions to ask. Some departments allow the firefighters to call headquarters; other departments prohibit it. Find out what is appropriate in your organization and communicate this information to the troops.

Other Ways to Improve Your Speech Communications

Most adults speak well. However, many adults would benefit from some further training on this most commonly used form of communications. One way to improve your communications skills is to watch yourself on video tape. Make a speech and record it. Then sit down by yourself and critically watch what you did. Be forewarned: It can be a painful experience.

Effective oral communications requires some effort. Your speech should be clear and distinct. The pace should be such that the listener has time to hear and understand what is being said. Common problems in oral communications include not looking at the person to whom you are speaking and distracting activities. Distracting activities can range from using a word or phrase to the point that it becomes annoying or some physical activity such as jingling keys or coins.

As a part of your officer training, you should be looking for ways to improve your oral communications skills. Taking a training course leading to becoming a fire instructor is a good start. Other options include courses by Dale Carnegie and Toastmasters International. Both programs offer opportunities for individuals to come together to become friends and gain experience in communicating verbally with others in a friendly environment. College courses are also available.

IMPROVING WRITING SKILLS

■ **Note**
Effective fire officers must be able to write.

Effective fire officers must be able to write. They must be able to express their ideas to their supervisors. Most of the significant recommendations and requests they submit are written. It is also important that officers be able to express their ideas to their subordinates. This is how most policies, procedures, and directives are published. Finally, officers have to complete a variety of reports (see Figure 2-8). All of this activity requires writing skill.

Principles of Effective Writing

Use the following principles whenever you employ writing in your communications.

Consider the reader. First and foremost, consider who the target reader is. Use plain language. Use terms and abbreviations that readers will clearly understand. If in doubt, spell out the term at its first use followed by the abbreviation or acronym. Technical terms, jargon, and abbreviations are like hurdles on the track: they slow the reader rather than helping.

Figure 2-8 *Writing reports is another important job for the company officer.*

Emphasis. Memos, letters, and directives should usually be limited to one topic. This practice allows the writer and the reader to focus on just one issue and discuss it as needed until the issue is understood and resolved. This rule also helps satisfy the next item.

Brevity. Brevity is desired. Consider the material you get in the mail every day. The letters that are brief and to the point usually get read right away. Brevity is often misunderstood. Brevity does not mean too short; it just means not too long. It means that it is long enough to do the job.

Simplicity. Use everyday words when possible. Use the same words you would use while speaking. When writing, we frequently resort to bigger words, thinking that it will impress the reader. Our efforts should be to impress the reader with the information we are sharing, not writing style. In many cases, the use of larger words actually slows the communications process. Here is an example: "Long, rambling, and elaborate dissertations will usually languish in the receiver's mailbox until they die of old age."

Here is another: "It is the policy of the fire chief that all activities of this fire department shall be carried out in a manner that will ensure the protection and enhancement of the environment through control and abatement of environmental pollution." It might have been easier to say, "Don't pollute."

USING EASIER WORDS AND PHRASES

Instead of	Try
commence	start
facilitate	help
enhanced method	better
finalize	end
improved costs	less expensive
in order to	to
in the event of	if
it is requested	please
it is the recommendation of this office	we recommend
it is necessary that you	you should
optimum	best
prioritize	arrange
prorogate	issue
utilize	use

On the other hand, use specific terms when appropriate. A report that contains ". . . it was a large fire." leaves too much to the imagination of the reader. The message here is use language that is appropriate for the occasion and the reader.

■ **Note**
Use language that is appropriate for the occasion and the reader.

Objectivity. Most of your writing requires an impersonal viewpoint. In most situations where facts are being reported, there is little place for personal bias. Some reporting formats invite a personal opinion, and at that point the writer should clearly establish that these statement are her own opinion. For example, a standard reporting format for an accident investigation includes a provision for the writer's personal opinion. After determining and outlining the facts, the writer has more knowledge about the accident than anyone else on earth. The writer's opinion is valuable at this point and should be accepted in light of his expertise. However, under normal conditions, we are simply part of the communications process: Our own bias should not be apparent in our writings.

■ **Note**
Our own bias should not be apparent in our writings.

Mechanical accuracy. This term describes the various rules of good written communications. Most of us have been taught the rules, but some of us need to review them from time to time. We will cover the most common problems here. At the end of the chapter is a list of additional references that may be useful.

Underlining or italics are generally used to identify the titles of books, plays, works of art, magazines, and newspapers. An article within one of these publications is usually set off in quotation marks. For example, we might have a sentence that reads: Those interested in firefighter safety issues should read the report entitled "U. S. Firefighter Injuries," published annually in the *NFPA Journal*.

There are generally accepted rules for using numbers in text. For the numbers one through ninety-nine, you spell out the word. Starting with 100, you use the digits.

Many writers make mistakes using capital letters; most of the mistakes result from using capitals when they are not needed, such as when the word is not associated with any specific person. Capitals are usually used when referring to a specific individual, place, or thing, while lowercase letters generally refer to the generic term. For example the training officer, the lieutenant, and the fire chief, but Captain Wright and Chief Coffman. Consider the following examples:

a captain	Captain Joe Friday
a chief	Chief Jim Flynn
a lake	Lake Michigan
a street	Main Street
a textbook	*The Company Officer*

Capitals are used at the start of a

sentence

proper noun

line of verse

name of a human race

name of a government body

noun used to refer to the Deity

direct quote when within a sentence

name of a planet, constellation, or star

holiday, a day of the week, or a month of the year.

Another communications problem is the use of passive voice. Consider the following conversation:

Doctor: When did you first notice your use of verbs in the passive voice?

Patient: The utilization was first noticed by me shortly after the fire service was entered. The fire service was also entered by my brother. The same condition has been noticed by him.

Doctor: Did you know that most of the verbs we read and speak with are active voice?

Patient: Well, it is believed by me that most verbs are made passive by fire service writers. In the letters and reports that have been prepared by this speaker, passive voice has been utilized extensively.

Are problems caused? Yes, problems are caused. It would be better to say, writing in the passive voice causes problems. Using passive voice usually results in wordy, roundabout, and confusing writing. Avoid these problems by putting the subject of the sentence (the doer) ahead of the verb. For example: The station was inspected by the fire chief. Who did the action? The fire chief. This sentence would be better if it read: The chief inspected the station. There are times when the passive voice may be the only choice. For example, if the person who performed the act is unknown, the passive voice may be appropriate. For example: The fire was started by unknown causes. Also the passive may be more effective when the receiver of the action is more important than the doer. For example: Any person who violates this rule will be fired. Your writing will be more interesting and more effective when you use the active voice.

We have limited this discussion to the most common problems. There are others, of course. Fire departments would do well to publish a guide with some good examples of correspondence and report writing. The guide should also include information regarding the details of style that may be unique to the organization. Regardless of whether such information is available from your agency, it would be smart to obtain one of the several generally accepted writer's guides. These can be found in bookstores everywhere. Several are listed at the end of this chapter.

In addition to these standard references, most dictionaries contain several pages of good information on grammar. In addition, a good modern dictionary is invaluable for checking the spelling and the exact meaning of words. Do not be ashamed to look at these references when you have a question. The fact that so many references are available suggests that many writers need to look things up from time to time.

■ **Note**

Fire departments would do well to publish a guide with some good examples of correspondence and report writing.

Putting It All Together

The following are some suggestions for effective writing.

Organize your thoughts. Organization may be the most important part of writing. Unfortunately for many writers it is also the most difficult part of writing. Like building a house, organizing does not have to be done all at once. Start with what you know, and add the other facts as they come to mind. Outlines help some writers. Word processing helps too. The ability to easily arrange and rearrange sentences and paragraphs has been one of the significant improvements brought by the word processor. Rewrite! Paper is cheap. Time and good impressions are priceless.

■ **Note**

When you write, think about the one sentence you would keep if you were allowed to keep only one.

Start with facts, explain as necessary, then stop! When you write, think about the one sentence you would keep if you were allowed to keep only one. That sentence should be in the first paragraph and may even be the first sentence of the first paragraph. With this approach you put requests before justifications, answers before explanations, conclusions before discussions, and summaries before details.

You might be saying, "Wait a minute! That's not what we were taught in school." In today's faced-paced business world, managers want answers and information in the shortest possible form. If your answer is reasonable, they do not need a lot of detail. Those who can provide information in this style are appreciated.

Consider the busy reader. If the answer can be provided in one or two sentences, put those at the top of the page. If that information satisfies the reader's needs, they may stop as soon as have what they need. If they need more explanation or justification or want to read more about the way you obtained your information, they can read on. While it is important to provide the "big bang" up front, it is also important to provide the other information. If facts are presented, they should be documented. If conclusions are drawn, they should be supported with logic.

Use short paragraphs. Long paragraphs often swamp ideas and overwhelm the reader. Cover one topic before starting another, and let the topic take several paragraphs if necessary. The white space between the paragraphs helps to separate your thoughts, makes the reading easier, and encourages your reader to continue.

Listings are a useful technique to provide information in a quick and easy-to-read format. This is often done with bullets. This technique is used throughout this book.

Use a natural writing style. Write as if you were speaking. Imagine that the reader is sitting next to you and that you are speaking aloud the same words you are writing. Use personal pronouns, everyday words and short sentences. Try to avoid "I" except when speaking about yourself or expressing an opinion that is clearly your own. When representing your organization, the use of we, our, and us is more appropriate.

Be direct. Try to avoid roundabout sentences. Again, think about the situation in which you are speaking and the reader is sitting next to you. Consider the following two sentences as examples:

It is necessary that your reply be received by the first of next week.
We want your reply by the first of the week.

Which statement are you more likely to say in conversation? Which statement is easier to understand?

Review Your Work

When it comes to writing, several habits will help. First, avoid waiting until the last minute to do your work. The pressure of the deadline will force you to make

Figure 2-9
Computers make writing a lot easier. But you still need to understand the rules to be an effective writer.

compromises that will not be in your best interests. By doing the work in advance, you also have the opportunity to review and edit your work before submitting it.

With computers in widespread use (see Figure 2-9), the ability to use word processing software is available to nearly everyone. Use this technology to enhance your writing. Write your letter or memo. Review it on the screen and run the spell check. Then print a draft copy. Read it and mark up the copy with corrections or areas where you think it needs improvement. Wait about 24 hours (at least until the next day) and repeat the process. In nearly every case you will make additional corrections. When you reread the document you will want to add a word or two or rearrange several sentences to enhance the reader's understanding. Make the corrections and print a second draft. If time permits, repeat the process again.

Although your second reading is always important, for real benefit try to get someone else to read the document. That person does not have to be an editor or a grammar teacher, but someone who has reasonably good communications skills and who has the patience to read very carefully. Good proofreading is not like reading the sports pages of the paper; it requires a slow deliberate reading. Some writers even read aloud to themselves when proofreading. In so doing they will hear mistakes more readily than if they were just reading in the usual manner. In any case, you (or someone else) should read the document carefully and make corrections where necessary. These steps will significantly enhance your writing. Good writers perform both all of the time.

■ Note

Take care with the final
form and appearance of
your product.

Take care with the final form and appearance of your product. In many cases the paper you submit is like a sales effort; it represents you and your organization. You want it to make a good impression. Be sure that the final product is clean and well groomed. Make sure that the printer is working property. When you submit good work, you can be proud of your accomplishments.

GOOD EXAMPLES OF POOR WRITING

From the files of our Human Resources Section:

"I recommend this candidate with no qualifications whatsoever."

"In my opinion you will be most fortunate to get this candidate to work for you."

"I cannot recommend this candidate too highly."

"I am pleased to say that this candidate is a former colleague of mine."

From the files of our Accident Investigation Section:

"I was legally parked when I backed into another vehicle."

"The man had no idea of which way to run, so I ran over him."

"Coming home at night I drove into the wrong driveway and collided with a tree I don't have."

"I collided with stationary truck that was coming the other way."

"The guy was all over the road. I had to swerve several times before I hit him."

"I had been driving forty years when I feel asleep at the wheel and had an accident."

And finally, from the writing of a college student:

"Abraham Lincoln wrote the Gettysburg Address while traveling from Washington to Gettysburg on the back of an envelope. He also freed the slaves by signing the Emasculation Proclamation. On the night of April 14, 1865, Lincoln went to the theater and got shot in his seat by one of the actors in a moving picture show. The believed assinator was John Wilkes Booth, a supposedly insane actor. This ruined Booth's career.[2]

Other Ways to Improve Your Writing Skills

In Chapter 1 we suggested that you might benefit from taking a writing course. Let us mention that again: a good course, workshop, or other formal learning activity, in which you are asked to write, will always be beneficial. Many high schools, libraries, community centers, and colleges offer short courses as well as the traditional semester-long classes in writing skills for adults. Most colleges offer a variety of communications courses and many provide a testing service to help you to identify your own personal needs. Take advantage of these opportunities.

In addition, bookstores and libraries have a wealth of good information on this topic. Several useful books of particular value for fire service writers are listed at the end of this chapter. Take advantage of as much of this information as you can and work to improve your communications skills. You will be more effective at work and in other activities where communications are important. You may even want to try writing an article for one of the trade journals (they are all looking for fresh authors) or something even more ambitious.

Summary

Good communications skills have a positive impact on every aspect of your work. (See Figure 2-10.) Being able to communicate effectively enhances your leadership ability, helps you gain respect from your supervisors and peers, and makes you more effective in talking to the public, the media, and others. In all cases, your ability to communicate creates an impression about you, your organization, and your profession. Work to be an effective communicator.

Figure 2-10
Technology has allowed us to communicate more quickly and easily. But effective communications still require that you be able to express yourself, both in oral and in written communications.

Review Questions

1. What is communications?
2. What is the role of the company officer in communications?
3. What are the five parts of the communications model?
4. What are the common barriers to communications?
5. How can these barriers be minimized?
6. What is meant by the term *active listening*?
7. How does your communications style indicate your leadership style?
8. Name several of the principles of effective writing.
9. What steps can you take to improve your writing?
10. What is the company officer's role in the communications process?

Additional Reading

Barnett, Marva T., *Writing for Technicians,* Third Edition (Albany, NY: Delmar Publishers, 1987).

The Chicago Manual of Style, Fourteenth Edition (Chicago: The University of Chicago Press, 1996).

Comstock, Thomas W., *Communicating in Business and Industry,* Second Edition (Albany, NY: Delmar Publishers, 1990).

Dils, Jan, *Writing Fire and Non-fire Report Narratives* (Published by the author, P. O. Box 50544, Pasadena, CA 91115) 1990.

Fowler, H. Ramsey, and Jane E. Aaron, *Little, Brown Handbook,* Seventh Edition (New York: Addison Wesley Education Publisher, Inc., 1998).

Hess, Karen M., and Henry M. Wrobleski, *For the Record: Report Writing for Fire Services* (Shelter Cove, CA: Innovative Systems, 1989). Distributed by Burgess International Group, Inc., 7110 Ohms Lane, Edina, MN 55435.

Noonan, Peggy, *Simply Speaking* (New York: Harper Collins, 1997).

Turabian, Kate L., *A Manual for Writers of Term Papers, Theses and Dissertations,* Fifth Edition (Chicago: The University of Chicago Press, 1996).

Notes

1. There are several ways to assess proficiency in communications. One alternative would be to ask you to successfully complete a college-level English class. This idea has a lot of merit, and many departments use this approach.

2. Lederer, Richard, *Anguished English*. (Wyrick and Company, Charleston, SC) 1987.

Chapter

3

The Company Officer's Role in the Organization

Objectives

Upon completion of this chapter, you should be able to describe:

- Organizational structure.
- Authority and responsibility.
- Lines of authority.
- Duties and responsibilities.
- Line and staff organizations.
- The role of the company officer in the organization.

INTRODUCTION

There are more than 32,000 fire departments in this country. It has been suggested that no two of them are alike. Fire departments range in size from small departments with a dozen members and one vehicle to departments with thousands of members and hundreds of vehicles, but they are all organizations. In spite of the vast variation, many characteristics are common to all fire departments. Understanding the characteristics of effective organizations will enhance your role in your organization.

Although large cities are protected by large and often well-known fire departments, most of America is protected by small fire departments, many of them comprised entirely of volunteer members (see Figure 3-1). Most of the departments are staffed by volunteer members, have one or two pieces of equipment, and serve communities of less than 10,000 people. Even in this environment, good leadership and management skills are important.

GROUPS AND ORGANIZATIONS

Formal Organizations

Groups exist whenever two or more people share a common goal. Organizations are groups of people. Typically, these people share a common goal, have formal rules and have designated leaders. Obviously this is a rather simplistic description of an organization, but organizations are really nothing more than just large groups of people. But organizations can easily become quite complicated. Just think about all the one-on-one relationships that might be possible in an organization of five or ten people. As organizations grow, the number of relationships multiplies quickly. Even with twenty members, the possible combinations of relationships start to mount, and as the number increases to 100 or 200, the number of possible relationships goes up very quickly.[1]

We would like to think that each of these relationships is positive and that the outcome of any interaction between two or more members of the organization has positive results, but we know from experience that this is not always the case. Good communications plays an important role in organizations. Understanding the relationships within the organization is also important to the success of the individuals, their relationships, and the success of the entire organization.

Understanding organizations is more than just understanding our fellow employees. We look at the human aspect of organizational relationships in Chapters 6 and 7, but here let us look at organizations from a theoretical point of view and examine some of the principles of organizations. In due time we will see how people fit into and behave within organizational structures.

When studying any organization we should start by looking at the organizational chart. Organizational charts can be a lot more than a bunch of boxes connected with lines. If the chart is prepared properly, the actual structure of the

Figure 3-1 *A typical volunteer fire department.*

■ **Note**

The organizational chart indicates the formal lines of authority and responsibility.

formal organization is reflected in the chart. The organizational chart indicates the formal lines of authority and responsibility, lines of communications, and so forth. We could take everyone in the organization to a large field, arrange them neatly according to the way they fit into the organization, and take their picture, but this action is usually not practical. An organizational chart provides a similar graphical representation on paper.

An organizational chart is only a graphic representation of what the organization should look like. In larger organizations there are always changes, as new activities are added and people are transferred from one assignment to another. And there are always a few vacancies due to leave, promotions, and retirements. In spite of these fluctuations, it is always a good idea to have an organizational chart and to have it posted where employees can see it. Keeping it up to date and putting the employees' names in the blocks brings the chart to life and helps everyone associate the organization with the real people who occupy its ranks. From the chart we should be able to see who works together, as well as see the lines of communications and authority within the organization.

Understanding organizational structure and one's place in the organization is important.

Informal Organizations

So far we have talked only about the formal organization. In any workplace there is also an informal organization. Informal organizations are not bad. Informal organizations exist because certain people like certain other people. They chat over a cup of coffee or have lunch together. Sometimes employees even take time to get together after work for softball or bowling. Trying to define this type of organization and represent it with a chart would probably be impossible. Although we cannot represent it on paper, informal organizations really do exist concurrently with most formal organizations.

AUTHORITY FOR THE LOCAL FIRE DEPARTMENT

■ **Note**

With 32,000 fire departments in the country, the types of departments vary considerably. However, they generally fall into one of several major types.

With 32,000 fire departments in the country, the types of departments vary considerably. However, they generally fall into one of several major types.

Public Fire Department

public fire department
a part of local government

A **public fire department** is a part of local government, with the head of the department (fire chief) responsible to the chief administrative officer or elected official of the locality. An alternative arrangement is often called a fire bureau. In this arrangement, fire, police, and other emergency response organizations are placed together into one department, often called the *public safety department*.

County Fire Department

County fire departments are becoming popular in many areas, especially where the rapid influx of residents is turning rural country into suburban communities. In such cases fire protection is often provided by independent volunteer companies. The county government may provide centralized support services for these departments, including communications, training, and purchasing of equipment. In many areas, the county government also takes care of the various activities generally referred to as fire prevention.

staffing
providing human resources

There may be a small paid staff to coordinate these activities. As the county grows, it usually becomes necessary to start **staffing** the volunteer stations with paid personnel during the day when most of the volunteers are at work. This staffing is especially important where emergency medical services (EMS) are provided. With further community growth, additional personnel are hired until the stations are staffed by paid personnel around the clock. Depending upon local circumstances and politics, the entire program may be under the control of one person, called the *fire chief*. The relationship of the fire chief to the independent fire companies can range from harmonious to hostile.

Fire District

Another popular organizational approach is to establish a fire district. In this case, the citizens of several communities may combine to provide the funding needed to establish a fire district to provide fire protection. The funding may come directly from the citizens or be funneled through the local governments in the way of taxes. There is usually a commission or other elected group that oversees the department's activities. This type of organization is popular in many western states where people are moving into rural areas.

Most of the fire departments in America are volunteer companies. Many are funded through local tax money or by subscriptions, whereas others are financed in whole or in part through their own money-raising efforts. Many departments raise their operating funds through dinners, bingo nights, and meeting room rentals.

However the department is organized and funded, it must be authorized by the local government. In many states this is done first at the state level through a provision in the state code. The following is typical:

> Any number of persons may form themselves into a company for the purpose of extinguishing fires. The company may develop rules and regulations consistent with state laws and local ordinances for purposes of governing itself. The principal officer of the department shall be called the fire chief.
>
> Every member of the company shall attend any alarm according to the rules and regulations of the department and endeavor to extinguish the fire.
>
> When the department is in the process of answering an alarm or operating at the scene of an emergency incident, the chief or other officer in charge of such fire department shall have the authority to maintain order, direct the actions of firefighters, keep bystanders at a safe distance, facilitate the speedy movement of firefighters and their equipment, and cause an investigation to be made into the circumstance surrounding the cause and origin of all fires and explosions. This authority includes the right to enter property at any time of the day or night to facilitate fire extinguishment and to remain on said property for a reasonable period to conduct an investigation.

As a member of the fire service, it is important to know the laws and the basis for the authority of your fire department. When a fire department is not legally organized or when a fire department does not operate in full accord with state laws, local ordinances, and its own regulations, it operates outside of the scope of these principles and the government it represents. As such, it exposes itself, and its members, to considerable legal risk.

ORGANIZATIONAL STRUCTURE

Like other organizations, a fire department consists of people working together in a coordinated effort to achieve a common goal (see Figure 3-2). To function effectively, a fire department must have an organizational plan that shows the relationship between the various divisions and activities. Work should be divided

■ **Note**
However the department is organized and funded, it must be authorized by the local government.

■ **Note**
Like other organizations, a fire department consists of people working together in a coordinated effort to achieve a common goal.

Figure 3-2 *Although the fire service is a formal organization, informal relationships exist as well. Here a deputy chief (the shift commander) enjoys lunch with company members.*

among the divisions and the individuals within these units according to a plan. The plan should be based on functional activities such as prevention, emergency operations, and support activities.

Coordination of these activities is important. In small departments, coordination can usually be easily accomplished because of the relatively few number of people involved. However, as we have already noted, as the size of the department increases, the number of relationships increases and the effort required to coordinate their efforts grows accordingly.

There should be a job description for every position in the organization. The job description should identify the skills required, the principal work expected, and the working relationships required of the position. In addition to the job description, organizations usually have rules and regulations that define the policies of the organization, the conduct that is expected, and standard procedures for handling frequently occurring situations. Many people find all of this organizational structure rather distasteful. The alternative may be less pleasant and certainly less effective. Many organizations have lost much time and money because they lacked the essential tools for good organizational management.

■ **Note**
There should be a job description for every position in the organization.

THE PLAIN CITY FIRE DEPARTMENT (PCFD)

What does it take to organize a fire department?

Suppose that your community needs fire protection. Someone says, "Let organize a fire department." The local government has been planning for this event and has some money set aside to help get started.

- How do you start a fire department?
- How do you establish organized fire protection?

Let us look at what is represented in an organization and what represents organized fire protection. Based on your previous experiences, you have been selected to help start the Plain City Fire Department. Knowing that it takes several firefighters to staff one company, you get a couple of friends to join you in this venture. And because you want to be able to provide both fire and EMS services, you realize that you will actually need to divide the company according to function. Because a major highway is being built through the city, it will soon be necessary to have separate stations on each side of the highway. Even though the department is small, you can see some of the issues of organizations starting to arise.

- Who is in charge of the department?
- Who is in charge of each station?
- Who takes care of things like training, vehicle maintenance, and buying equipment?
- Do you try to get specialized personnel or do you cross train all members to provide both EMS and suppression services?
- Do the EMS and fire functions share resources?

The Plain City Fire Department could be a paid or a volunteer organization. At this point it does not really make any difference. Firefighting personnel are comprised of paid, volunteer, and paid-on-call firefighters. When one or more of these groups is combined into one department, we have what is referred to as a *combination department*. Many factors determine the type of arrangement that might work best in a given community. Community size and financial resources must always be considered. It is also important to consider the services that are expected, the risks that can be expected, the frequency of incidents, and the availability of the volunteer firefighters. All of these factors must be considered by a community to determine what works best.

Increasing demands for service and limited funding have prompted chiefs to attempt new approaches to service delivery. For example, fire chiefs have

Figure 3-3
Organizational structure of the Plain City Fire Department.

tried various combinations of mixing firefighters and EMS personnel, using volunteers to staff stations when paid personnel are on vacation, and using paid personnel to staff stations, when volunteers are away from their community at work, to provide effective and efficient emergency services. These initiatives have occasionally met with resistance. Tradition, turf battles, and political issues arise when new ideas are being considered and when change is introduced. Strong leadership and open communications are vital to the success of any organization.

To effectively manage the Plain City Fire Department it is necessary to create some sort of organizational hierarchy. Most organizations have a pyramid structure, with one person in charge, and an increasing number of subordinates at each level as you move downward in the organization. There are good reasons for this traditional form of structure. Here we see our familiar pyramid representing the organizational structure of the Plain City Fire Department (see Figure 3-3).

The Fire Chief

Let us examine the pyramid from the top. In the fire service, the top of the organization is usually referred to as the *fire chief*. Some organizations are using other terms, such as "director," to designate the top official in the fire department, but we use the traditional term fire chief. The fire chief is the head of the agency. The fire chief can be selected by any one of several methods, but once installed, the fire chief is expected to perform the duties as the leader of the fire department.

Fire chiefs are usually tasked in their job description with providing all the services required, both for the department itself and for the public they serve. Few fire chiefs can perform all of these functions all of the time so they delegate some of the work. When done properly, the fire chief has an organization to undertake

delegate
to grant to another a part of one's authority or power

these activities in a structured manner. To make things work well, the chief **delegates** some of the responsibility to the subordinates in the organization. The fire chief is usually supported by staff that includes other officers and an administrative support staff to make the entire operation function effectively. The size and function of the senior staff will vary based on the size of the department and on something we will call "organizational attitude."

THE ROLE OF THE CHIEF

Fire departments tend to be very conscious of rank within their own organization. Clearly the fire chief *is* the boss, and all others are subordinate. But when they are out of their own town, all fire chiefs seem to be considered equal. You can see this at a fire chiefs' conference. The chief of a single-station fire department might wear a uniform that is just like the uniform worn by the chief in a city of 250,000. An assistant chief (or other rank) may be a far more conspicuous player in a state or national organization than the chief of the same department. When these personnel work together in committees and other activities they have an equal voice and are respected based on their contribution, not their organizational rank or number of fire trucks back home. The system tends to recognize their contribution rather than their rank.

Wouldn't it be nice if it was like that everywhere?

In these days of increased concern over the cost of government, many fire departments are taking a hard look at the size of their senior staff, especially in paid fire departments. In many older departments, fire chiefs have become dependent upon large staffs to help them run the department. Yet the cost of maintaining that staff is significant. One senior officer may be paid as much as several firefighters. When forced to reduce budget, chiefs are forced to make hard decisions between eliminating one senior staff position or eliminating several firefighters. Departments with more than 10% of their employees in staff assignments should look at the way they are doing business.

The Assistant Chief

The senior staff represents middle management. This is where the vice presidents are located in the corporate world. In the fire service we have assistant, deputy, and battalion chiefs. These officers fill many of the middle-management positions within the department. They serve as division officers and heads of major components of the headquarters staff. In many departments, captains and lieutenants are also assigned to the headquarters staff as section heads.

WHAT DOES THE ASSISTANT CHIEF DO?

There are several roles for the assistant chief. Being an assistant chief is sort of like being the vice president of the United States: It is a nice job and everyone is polite to you, but most folks would rather speak to the boss.

There are at least four types of relationships the chief can have with the assistant, based on the fire chief's style.

Fire Chief Type 1. In this arrangement, the assistant chief is clearly the second in command. Most things go to the chief but it all goes through the assistant chief first. (At least it had better go through the assistant first!) In this arrangement, the assistant chief has a tight rein on the department's activities. Because of the closeness of the relationship between the two leaders, few members see any difference when the chief is away.

Fire Chief Type 2. In this arrangement, the assistant chief runs the department. The assistant coordinates the efforts of the staff and makes nearly all the decisions regarding the day-to-day operation of the organization. The fire chief shows up for ceremonies and represents the department outside of the organization, for example with other departments, agencies, and of course, before the elected officials of the community.

Fire Chief Type 3. In this arrangement, the fire chief probably promoted a long-time friend to the position of assistant chief. The assistant chief has been in the department longer than anyone else. The assistant is on a lot of committees, deals with disciplinary problems, and shows up at some public functions, filling in for the fire chief when the chief is out of town.

Fire Chief Type 4. In this arrangement, the assistant really is not in the chain of command, in spite of what the organization chart may indicate. The assistant chief gets the not-so-fun tasks of listening to discipline problems and attending committee meetings.

In reality there is probably a little of all four types in most fire departments. There is much to be said for the type 1 or 2 styles. Depending upon the size of the department, the chief would do well to use an assistant who can run the department. This practice puts to use several of the rules we have discussed, especially those regarding the chain of command and the division of labor. The chief is the one to visit with the elected officials, meet with counterparts from other departments and other parts of local government, and represent the department at formal functions.

The extent to which the chief chooses to remain in control of the details of the organization says a lot about that individual's personality. With type 1, they maintain most of that control, whereas with type 2, they trust the staff and delegate most of the control. In either case, the assistant is clearly in the chain of command.

■ **Note**
Regardless of where you are in the organization, the treatment of your second in command is important.

We realize that this is not a text for fire chiefs, although we hope that you are interested in being a fire chief some day. However, there is an important lesson here for everybody in the organization. Regardless of where you are in the organization, the treatment of your second in command is important. Keeping that person in the information flow, delegating the fun-to-do as well as the not-so-fun jobs, and preparing that person for the position you hold is a significant part of what organizational relationships are all about and should involve every aspect of training, management, and leadership, for yourself and for your unit. Delegating some of this activity not only prepares another to take your place, but likely increases his or her understanding and support of the work at hand. It also provides you with a very well-qualified assistant.

Battalion Chiefs

Battalion chiefs usually represent the middle management layer of a fire department. In larger organizations, they may have responsibility for part of the department and supervise three to six companies. In smaller communities, they may be responsible for all on-duty emergency response personnel. In either case, they will probably become incident commanders at an emergency response of any consequence.

Company Officers

Moving down the pyramid to the next level, we have the company officers. Most company officers are supervisors of firefighters and as a result are referred to as first-line or front-line supervisors. As we have already noted, company officers wear many hats. Company officers are expected to be proficient in their jobs as firefighters. They are also expected to have the managerial ability to understand the organization and the supervisory skills needed to effectively lead their subordinates.

Firefighters

Throughout the fire department organization are firefighters. Most firefighters are assigned to companies and work at fire stations. However, firefighters can also be found throughout the headquarters staff. In some staff positions the firefighter's technical knowledge is essential, however healthy firefighters should be assigned to the operational positions within the organization as much as possible.

THE ORGANIZATIONAL CHART

The pyramid diagram was adequate for the discussion in Chapter 1, but you can quickly see the limitations of this representation, even for our small fire

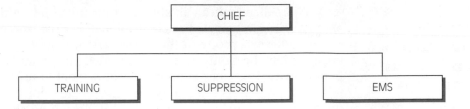

Figure 3-4 *A more traditional organizational chart representing the Plain City Fire Department.*

line authority
a characteristic of organizational structures denoting the relationship between supervisors and subordinates

accountability
being responsible for one's personal activities; in the organizational context, accountability includes being responsible for the actions of one's subordinates

responsibility
being accountable for actions and activities; having a moral and perhaps legal obligation to carry out certain activities

department. It might be more appropriate to represent our department as shown in Figure 3-4. This chart more accurately represents the organizational relationships of the department. For this small department it represents the line authority. An officer with **line authority** manages one or more of the functions that are essential to the department's mission. When we see an organizational chart, we usually think of the authority that one has over the activities that fall under them. In addition to authority, they also have **accountability** and **responsibility.**

We have taken care of the essential activities of the department. However, as the new department takes shape, the chief recognizes the need for prevention and training. So a prevention officer and a training officer are established (see Figure 3-5). We could debate the placement of these activities, but consider that the prevention and training positions are staff assignments. We discuss the details of line and staff authority in the next section.

Let us go back to the Plain City Fire Department. Suppose that there is no real organization, that everyone is the same rank and has the same responsibilities, and that the department is called to a structural fire. What do you think would happen?

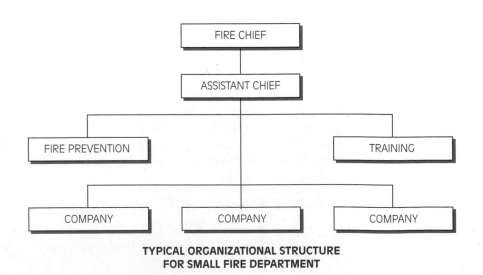

Figure 3-5 *Chart of the Plain City Fire Department showing fire prevention and training as staff functions.*

TYPICAL ORGANIZATIONAL STRUCTURE FOR SMALL FIRE DEPARTMENT

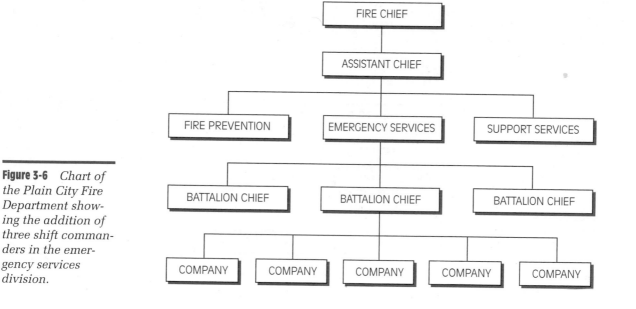

Figure 3-6 *Chart of the Plain City Fire Department showing the addition of three shift commanders in the emergency services division.*

As a department grows, we see a continuation of the evolution that has already started. Figure 3-6 is an example of an organizational chart for a medium-size fire department.

As our hypothetical community grows, so does the need for fire protection. The department shown in Figure 3-6 can no longer keep up with the demand for services. The fire chief and the senior staff are overwhelmed with just keeping up with the growth, and they find that they cannot really efficiently supervise those personnel assigned in the field. In addition, delays in the arrival of senior officers responding to large-scale emergencies are affecting the way the department deals with these emergencies.

As the department grows, the number of relationships also continues to grow. In addition, as the organization grows, there are more options for further growth (see Figure 3-7). We see that the department has added a senior position, a person who is called the *shift commander*. This position takes some of the burden of actually running the department 24 hours a day off the senior staff. The shift commander has command authority over all but the largest events. Even more importantly, the shift commander has authority to deal with most of the administrative issues that arise within the department that were formerly approved by officers at headquarters.

Another alternative is to divide the community into three geographical divisions and place a senior officer in charge of each division (see Figure 3-8). The

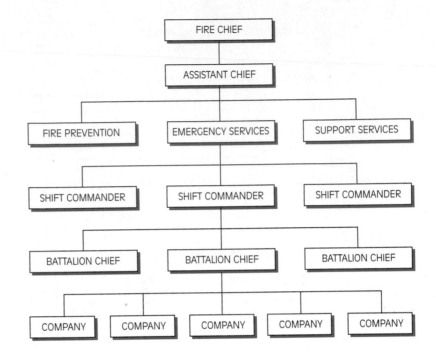

Figure 3-7 *As more companies are added in the Plain City Fire Department, another layer of managers is needed.*

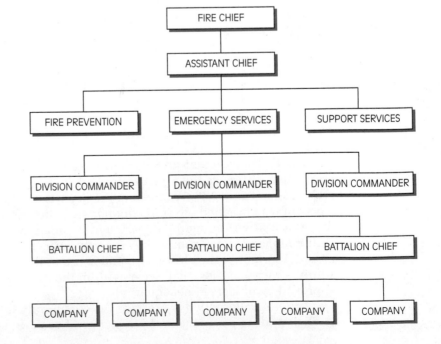

Figure 3-8 *As an alternative to shift commanders, the fire department's resources in Plain City might be geographically divided among three division commanders. The division commanders would work a normal work week. Each division commander would have three battalion chiefs, possibly one for each shift.*

line functions
refer to those activities that provide emergency services

staff functions
refer to those activities that support those providing emergency services

division commander has complete responsibility for all fire department functions within a certain part of the city.

Both of these approaches are widely used. Each has its advantages and each has its disadvantages. Our purpose here is not to reorganize your department but to introduce ways that departments are organized and some of the reasons that these organizations are essential. A third alternative would be to establish a separate EMS support organization.

Staff versus Line Organization

In the field of industrial management, **line functions** refer to the activities that accomplish the manufacturing process. Many fire service textbooks draw a parallel here, suggesting that the fire service has generally considered the line function to be synonymous with providing emergency services. This service-delivery activity is called by various names including suppression, operations, and emergency services. This definition of the line organization suggests that everything else is a supporting role, performed by staff personnel.[2]

Staff functions are those activities that *support* the firm's basic purpose. Staff functions do not normally get directly involved with delivering emergency services. However, the support staff is essential in the overall operation of the department, and without it the fire department would quickly grind to a halt. Traditional staff functions (see Figure 3-9) include fire prevention and support

Figure 3-9 *Staff functions are important to the life of any organization.*

services. Each of these major functions may be represented by a division that performs a variety of activities, based on the size of the department.

In a large fire department, the prevention division may have separate sections dealing with inspections, investigations, life safety education, plans review, and systems testing. Likewise, in a large fire department, support services may include communications, payroll, personnel, research, recruiting, resource management, safety and health, and training.

On the other hand, in small communities, many of the support activities common to all government agencies are gathered together under a central agency. Purchasing and personnel activities are good examples. As the fire department grows, it eventually becomes beneficial to perform these activities within the fire department's organization.

■ **Note**

Titles vary from department to department.

Before we conclude this discussion, it is important to note that titles vary from department to department. In some departments the second level officers are deputies and in other departments they are referred to as assistant chiefs. Where there is only one level of middle management, they could be called by either title. In smaller departments the second-in-command officer may be a battalion chief. Titles are important but sometimes confusing. Checking the organizational chart usually clears up these issues.

THE COMPANY OFFICER

■ **Note**

The company officer is usually the first supervisory level in a fire service organization.

The company officer is usually the first supervisory level in a fire service organization. In many ways, company officers have an opportunity to exercise more influence than any other rank when one considers the role of the company for all of the internal activities of the department as well as the activities associated with the department's delivery of emergency services. The company officer provides the tone of the company in terms of morale, preparedness, and service delivery.

The Fire Chief's Perspective of the Company Officer

Some chief officers see their company officers as a part of the organization's management team. Many chiefs allow and even encourage their company officers to get involved in the decisions that have strategic impact on the department. Such chiefs value the input from their company officers and support their ideas. This implies trust in one another. Effective team building is encouraged in this organizational environment. This environment also places a value on education. Education enhances your ability to think through complex issues and to effectively express your ideas to others.

The Firefighter's Perspective of the Company Officer

Regardless of the relationship of the company officer to the chief and the other members of the department's senior staff, firefighters usually see their company officer as part of the department's management team. We have already indicated that the company officer has a greater impact on company members than any other officer in the department. After all, it is the company officer who enforces department policies, grants the firefighters' requests, and evaluates their performance. Indeed the company officers hold a lot of the strings that control the firefighters' work and attitude.

Firefighters informally evaluate their company officers based on how well they (the officers) perform their duties. The firefighters evaluate how well the company officers can perform their jobs, how well they apply policy, and how fair they are in their actions among the members of the company. They can assess whether the officer can teach and coach the members. They can evaluate if the company officer is up to date on new ideas and techniques and composed under pressure. They can evaluate communications skills. The list goes on and on. Unfortunately, firefighters seldom get to express their opinions on these matters.

Most company officers would like to be respected by both their bosses and their subordinates. Some company officers are focused on pleasing the boss and looking good. Other company officers focus their energies on the welfare of their subordinates, sometimes at the expense of their relationship with their own supervisors. Being successful in both areas can be quite a challenge.

> **■ Note**
>
> How the company officer fits into the overall organizational structure impacts on many others.

How the company officer fits into the overall organizational structure impacts on many others. It certainly impacts on the daily lives of the firefighters who work for that officer. Eventually it will impact on the company's readiness to deal with anything other than the simplest emergencies. Good companies become standouts in the organization. Companies that perform poorly tend to pull from the talents of other companies and drag down the average of all the others. Poor performing companies eventually affect the entire department and possibly even the community they serve.

PRINCIPLES OF GOOD ORGANIZATIONS

There are several principles of organizations. Successful organizations seem to follow most of these "rules" quite well. Sometimes organizations are successful, even though they do not follow all of these rules. However, most organizations run into problems when these principles are forgotten.

> **■ Note**
>
> Some positions perform duties directly in support of the mission whereas others perform duties in support of those who are out on the street.

Officers in the fire service should be concerned about the organization in which they function and the job they perform within that organization. Most positions are essential to the department's mission: few fire departments have positions that exist just to fill a block on an organizational chart. Some of those positions perform duties directly in support of the mission whereas others perform duties in support of those who are out on the street. Both are essential!

Let us look at the organization we call our fire department. Even though it is likely that you did not invent it, and even though it is almost as likely you will not change it right away, you should understand the organization and how your particular position fits into the overall scheme of things.

Scaler Organization Structure

scaler principle
organizational concept that refers to the interrupted series of steps or layers within an organization

The **scaler principle** relates to the continuous chain of command in the organization. Like playing a scale on a musical instrument where every note is sounded, the scaler principle suggests that every level in the organization is considered in the flow of communications, authority, and responsibility. Most fire departments are established along the traditional lines of a scaler organization (see Figure 3-10).

Unity of Command

unity of command
the organizational principle whereby there is only one boss

This concept is frequently confused with chain of command. **Unity of command** refers to the concept of having one boss. In examining the organizational structures in this chapter, you will note that each subordinate reports to only one boss. This concept is known as unity of command. Unity of command is an essential organizational concept.

Figure 3-10 *The badges shown represent all of the steps from firefighter to chief, as does the principle of scaler organizations. Photo courtesy of the Fairfax County Fire and Rescue Department.*

Division of Labor

Organizations break the accountability process into small units that can be managed. A fire company, such as an engine or truck company, is usually assigned a specific task, and is usually assigned to cover a specific part of the community. Every company has a job to do at the scene of a fire or other emergency. Every company has a place where it is expected to do that job. These are essential components of the fire department's organization and are essential to fire department management. When companies are divided into geographic areas and assigned to specific tasks, they illustrate the first principle of organization, division of labor.

As the department becomes larger, the administrative activities must be organized into manageable sections. This activity also illustrates the division of labor. Division of labor allows an organization to divide large jobs into specific smaller tasks. Most fire departments provide emergency medical service. Where this occurs, a fairly senior officer probably coordinates the department's overall EMS policy. Likewise at the successive levels in the organization, there are officers who *manage* the EMS program and *supervise* some or all of the EMS responders' activities. For departments of any size, this management/supervisory activity of field personnel would be impossible for one person to manage. However, by logically dividing the activity among several people, the task is quite manageable.

Assignment of Specific Responsibilities

■ Note
Primary responsibility is vested in one person.

Another good principle of organizations is that no specific responsibility should be assigned to more than one person. Now, this does not mean that we no longer work in teams nor does it mean that we no longer have backup personnel. We are saying that the primary responsibility is vested in one person. What happens when that person is out for the day? Clearly the assignment is moved to another employee, but our standard procedures normally take care of this. Typically every riding position on a truck or engine company has a specific duty upon arrival at a fire or other emergency event. It is important to have the position covered and it is important that the person in that position is aware of his or her responsibilities before fulfilling that assignment. Likewise, those in staff assignments should have an alternate trained and prepared to fill their duties in when they are absent.

Control at the Proper Level

Supervisors should not get lost in a maze of details. Violations of this rule are a common complaint. The fire chief should not be dealing with the details that are rightly the job of the company officer. Likewise, the company officer should avoid overmanaging at the station level. This practice is often referred to as micromanaging.

Span of Control

span of control
organizational principle that addresses the number of personnel a supervisor can effectively manage

The principle of span of control is simple: Do not have too many subordinates. The number of people you can effectively supervise is usually referred to as the **span of control.** The number of people you can effectively supervise varies of course, based on many factors, but generally for our purposes that number is between four and seven.

Delegation

delegation
the act of assigning duties to subordinates

Along with the other principles, delegation plays into the workings of an organization. As the organization expands we assign certain duties to subordinates. This activity is called **delegation.** It is important that the delegation be mutually understood and that authority be provided along with the responsibilities.

Delegation allows dividing or sharing the work and allows subordinates to take on part of the supervisor's work. However, the supervisor is still ultimately responsible for the assignment. We discuss the art of delegating in Chapter 5. For our purposes here, it is an important part of the overall concept of organizational structure.

Good supervisors take the heat from their supervisors without passing it down to the subordinates. That person, usually the leader, acts as a buffer in the communications process, knowing that the troops will get the job done without using pressure from the chief's position to make it happen. Consider the following sentences. Which sentence will do more to strengthen the officer's position?

"The chief said that we need to do more preplanning."

"I think that we should be doing more preplanning."

A final thought about delegation. You should feel free to delegate to your subordinates. Delegation is an important empowerment tool and generally makes the subordinate feel better about his or her job.

■ Note

Delegation is an important empowerment tool and generally makes the subordinate feel better about his or her job.

THE ORGANIZATION IS TRANSPARENT TO THE AVERAGE CUSTOMER

Most citizens are not concerned about your organization's structure or rules. Nor does the average citizen know or care whether the department has an assistant chief in support services or whether it uses a fully integrated incident management system. They do know and care about the service you offer, and they expect the service to be fast and dependable. Understanding your organization's structure may not be important to the citizens, but it should be important to its members.

Flat and Tall Organizations

Linked to the span of control principle is the concept of flat and tall organizations. When the span of control is smaller, that is, when supervisors have fewer people to supervise, we tend to see organizations with a narrower or "taller" structure. This works well, but with the concern over the cost of doing business these days, organizations, especially government organizations, are taking a hard look at the span of control of their managers and the size of the middle management structure. Many organizations are eliminating positions at the middle management level. This results in an organization that appears flatter. Reduced costs are one benefit of flatter organizations; another benefit is that communications is usually improved because there are fewer layers in the organization, fewer levels of management between the fire chief and the firefighters. In the early 1980s, Thomas Peters and Robert Waterman published *In Search of Excellence,* a book that became a best seller. It is still in bookstores and still worth reading. The authors looked at some of America's best-run companies and found eight practices that seemed to be characteristic of these companies. Some of the items on their list may fit better into our upcoming chapters on management and leadership, but several are directly tied to the subject of organizational style.

■ **Note**
Keep the company organization simple.

One of the eight items they listed was called "simple form, lean staff." The idea is to keep the company organization simple. We have already talked about the popular practice of reducing the layers in the organization and reducing middle management positions. And the value of a small staff has already been discussed.

Along with the smaller staff is the idea of keeping things as simple as possible. This means that you do not have a staff specialist for every conceivable idea and problem that might come along. You have a few good people and you give them authority over a broad range of management issues.

Autonomy and Entrepreneurship

A couple of big words describe a very simple but often overlooked principle. Again, the idea comes from *In Search of Excellence.* The authors use the words *autonomy* and *entrepreneurship* to describe a management style. This concept follows the characteristics discussed in the preceding paragraph. With a well-developed organization that is lean and simple, the next step is to develop an understanding of who is responsible for the various functions. If all decisions have to flow up to the fire chief, why have a staff? The manager (fire chief) should assign specific responsibilities, and give the person who gets the assignment a feeling that the activity is theirs to manage, a feeling of entrepreneurship, if you will. Once this is done, problems should be addressed and resolved by the responsible officer.

■ **Note**
If all decisions have to flow up to the fire chief, why have a staff?

Discipline

discipline
a system of rules and
regulations

Before we conclude our discussion on the principles of management we need to mention one more word: **discipline.** Discipline is usually considered to be synonymous with punishment. We look at discipline/punishment as a leadership tool in Chapter 7. In this discussion we are looking at discipline as the organizational and individual responsibility to do the assigned job.

The Plain City Fire Department is now operating and has a mission statement. Officers are assigned specific responsibility in support of that statement. Companies are expected to respond to emergencies and take action in accordance with the department's policies and procedures. We expect this to happen. It happens because of discipline, not because we will be punished if we fail to perform but rather, because we accept the duty. We also expect others to accept their share of the duty.

Suppose that you, as a company officer, find that all of this stuff about training, preemergency planning, building surveys, physical fitness, and so forth are not really important. You might get away with that for awhile, but one day an alarm is transmitted for a working fire in your first-due area. The first-arriving companies are expected on the scene within 4 or 5 minutes, ready to quickly mount an aggressive fire attack. But because of your approach to company readiness, your people are not ready, your equipment is not ready, and you are not ready. By the time you get things sorted out, second due-companies are already at the scene, and so is the battalion chief. The fire puts the department to a quite a test. The fire is an occupied apartment building. Because of delays in the fire attack, the fire extends into several adjacent exposures, and what should have been a simple job becomes a major event. Several people are injured during the rescue operations and there is significant property loss. Organizational discipline helps keep this sort of thing from happening.

Every company in a fire department has a job and a place to perform that job, but it is also important that the company be competent in the performance of that job. We all know that some companies are better than others in certain tasks. Companies in larger cities often get more experience and are probably more proficient in those tasks they routinely encounter. But even within a city, some companies are better than others. Regardless of the size of the city, company preplanning and training activities are important. Well-disciplined companies do both and usually perform better during emergency activities.

ORGANIZING YOUR OWN COMPANY

Even within the company there are ways to organize the few people that work there. Suppose that you are the company officer. You have four firefighters working for you. One has been driving since the department had horses. Another is a respected firefighter who recently became a paramedic. One is a firefighter with

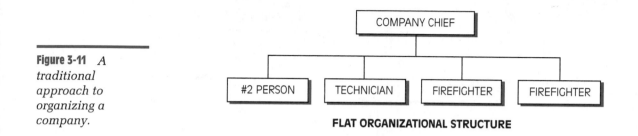

Figure 3-11 *A traditional approach to organizing a company.*

FLAT ORGANIZATIONAL STRUCTURE

5 years on the job. The fourth is a firefighter with 5 weeks on the job. Using the traditional approach, you might organize as shown in Figure 3-11.

This organization certainly follows some of the rules about flat and lean organizations and allows for good communications between you, the officer, and all of the firefighters. It reinforces the rules about unity of command and span of control.

You could also organize along the lines shown in Figure 3-12. This diagram illustrates the scaler organizational structure. Here we have a tall vertical structure. You can sort the four personnel into the slots as you feel appropriate. This organization reinforces the chain of command.

Another alternative might be to organize along the lines of a circular organization (see Figure 3-13). In this environment everyone is at an equal level in the organization and everyone has the opportunity to communicate with all of the others. As King Arthur explained to Lancelot as the latter was shown the round table at Camelot for the first time, "We are all equal here."

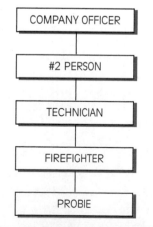

Figure 3-12 *This organizational approach reinforces the scaler principle.*

SCALAR COMPANY STRUCTURE

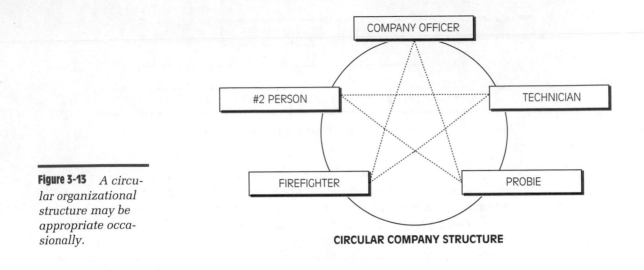

CIRCULAR COMPANY STRUCTURE

Figure 3-13 *A circular organizational structure may be appropriate occasionally.*

Each of these organizational structures has appropriate uses within the fire service. As an officer, you should understand the advantages and the limitations of each and use them appropriately.

Summary

The company officer has two roles in the organization. First, as a company officer, you represent the fire department to the average customer. The company officer is often the first on the scene and often the last to leave and, for many events, is the only person the customer talks to.

At larger events, other units of the department may be represented. Other companies may be present, and other parts of the organization are supporting the operation, whether or not they are at the scene of the emergency. Company officers must understand how the entire organization works, how they (as a company) fit into the total organization, what resources are available and appropriate for any type of incident, and if appropriate, help the customer connect to the appropriate part to get the service needed.

Second, as a company officer, you are a vital part of the organization that you represent. You should understand your role, how you represent your company and its needs to the department's management team, and how you represent the rest of the organization to the firefighters within your company.

Understanding organizations is important. In the past, people worshiped organizational structure, paid homage to the chain of command, and measured organizations by the number of rules they had. Those are good things, and we should understand their value, both in today's organizations and for those who grew up in this culture. But today we are learning that there may be better ways to run an organization. We are seeing more organizations move away from rules and structure. They are focusing more on values, especially where the employees are concerned.

Modern management concepts are discussed in Chapter 4. We do not want to leave you with an impression that all of this information about organizational principles is the end of the story. Far from it—it only is the beginning!

Review Questions

1. Define organization.
2. Why do we need structure in organizations?
3. How do organizational charts show lines of authority?
4. How do organizational charts show lines of communication?
5. Define division of labor.
6. Define span of control.
7. Define unity of command.
8. What is meant by delegation?
9. Distinguish between "line" and "staff" in an organization.
10. What is the role of the company officer in the organization?

Additional Reading

Klinoff, Robert, *Introduction to Fire Protection* (Albany, NY: Delmar Publishers, 1996).

National Fire Protection Association, Standard 1201, *Developing Fire Protection Services for the Public* (Quincy, MA: National Fire Protection Association, Latest Edition).

Paulsgrove, Robin, "Fire Department Administration and Operations," *Fire Protection Handbook,* Eighteenth Edition (Quincy, MA: National Fire Protection Association, 1997).

Peters, Thomas J., and Robert H. Waterman, Jr., *In Search of Excellence* (New York: Harper and Row, 1982).

Notes

1. Although this is not a math text, remember that in math we learned a formula for determining the combinations possible from a certain number of objects taken two or more at a time. For example, with a group of 25 people, there are 300 separate two-way relationships possible. With a group of 100 people, nearly 5,000 such combinations are possible. As the group gets larger, the number of relationships increases dramatically.

2. Under this concept, fire prevention is a supporting role. We think that this approach understates the importance of this vital activity and as a result, the traditional definitions for line and staff may not fit the fire service very well.

Chapter

4

The Company Officer's Role in Management

Upon completion of this chapter, you should be able to describe:

- Management.
- The functions of management.
- Recent major contributions to management science.
- The role of ethics in management.

INTRODUCTION

Many words describe the company officer. The company officer is a supervisor, a leader, and a manager. Certainly company officers are expected to effectively manage the department's resources at the station level. To be effective as a manager, you should understand what management is and the basic management concepts used in modern organizations.

WHAT IS MANAGEMENT?

The traditional definition of management is the activity of getting things done through people. Updating this definition a little, we can say that **management** is the act of guiding the human and physical resources of an organization to attain the organization's objectives. Management includes both the determination of what needs to be done and the accomplishment of the task itself. From this definition it is clear that managers plan and make decisions about how others will

management
the accomplishment of the organization's goals by utilizing the resources available

■ Note
Management includes both the determination of what needs to be done and the accomplishment of the task itself.

THE BEGINNING OF MANAGEMENT

Possibly the first reference we see to management is contained in the Bible, in Chapter 18 of the book of Exodus:

Jethro noticed that Moses, his son-in-law, spent all of his day serving as judge for his people, while they stood in long lines seeking his advice. "It's not right! This job is too heavy a burden for you to try to handle all by yourself. You're going to wear yourself and your people out" Jethro said to Moses at Mount Sinai.

"Let me give you some advice. Teach the decrees and laws. Show your people the way to live and the duties they are to perform. But select capable men from all the people—men who fear God, trustworthy men who hate dishonest gain—and appoint them as officials over thousands, hundreds, fifties, and tens. Have them serve as judges for the people at all times, but have them bring every difficult case to you; the simple cases they shall decide for themselves. That will make your load lighter because they share it with you."

Moses followed Jethro's advice. He chose able men from all over Israel and made them judges. They were constantly able to administer justice over the people in groups: thousands, hundreds, fifties, and tens. They brought the hard cases to Moses but judged the smaller cases themselves.

use the organization's resources. Managers should also be committed to accomplishing these activities efficiently and effectively.

Management in the Twentieth Century

There is evidence throughout recorded history of the use of management concepts. During the past one hundred years, much attention has been focused on the science of management. One of the first contributors was Henry Fayol (1841–1925), a very successful manager of a French coal mine. In addition to his ongoing management activities, Fayol was interested in understanding the process by which work was accomplished. Fayol's writings on this topic first appeared in 1900.

By 1916, Fayol had enough material for a book. Though originally published in French, the work was soon translated into English under the title *General and Industrial Management*. Fayol proposed some fourteen principles of management. Some of his principles related to organizational structure and some related to the work process itself. Even at this early date, Fayol was concerned about the importance of the *human resources* in the organization. (We see an evolution of his ideas in several other writers who follow.)

As you look at Fayol's list you will recognize that a few of these items were addressed in Chapter 3. Some writers are critical of Fayol where his ideas cross between organizational structure and the management process. But let us be kind here and remember that Fayol's ideas were developed a century ago. To say that he was ahead of his time would be an understatement: Fayol is called the "Father of Professional Management." Many of Fayol's ideas are still valid and quite relevant to our discussion here on management in the fire service.

Fayol also noted that management activities increase as one moves up in rank in an organization (see Figure 4-1). That was certainly the case in his time, and it remains true in many organizations today. However, where modern management is practiced, there is a clear pattern to decentralize management activities as much as possible. Where these modern management ideas are accepted, organizations are putting increasing responsibilities upon a greater number of managers, especially on first-line supervisors. In the fire service, the first-line supervisors are the company officers.

FUNCTIONS OF MANAGEMENT

Many years ago, Fayol and the other pioneers of management science recognized that management activities could be broken down into several discrete components, namely, planning, organizing, commanding, coordinating, and controlling. Over the years various authors have renamed these activities and some have repackaged them a bit. However, Fayol's concept of five basic activities or functions fits our needs here. Let us look at Fayol's five basic functions.

■ **Note**
Fayol is called the "Father of Professional Management."

■ **Note**
Management activities can be divided into several discrete components, namely, planning, organizing, commanding, coordinating, and controlling.

FAYOL'S PRINCIPLES

1. *Division of labor.* Divide the work into manageable portions and specialize the activities to improve efficiency. This is what Jethro told Moses to do.

2. *Authority and responsibility.* These terms go hand in hand. Where authority arises, responsibility must follow. Managers must have the authority to make decisions and give directions.

3. *Discipline,* as a characteristic of authoritarian organizations, includes good leadership, and clear agreement between management and labor on the role of each.

4. *Unity of command* suggests that one should receive orders from only one supervisor.

5. *Unity of direction* suggests a commitment from all to focus on the same objectives.

6. *Subordination of individual interests* to general interests in an organization suggests that the general interest must take precedence over the individual interests of the members.

7. *Proper remuneration,* a fair wage for fair work.

8. *Centralization of authority,* like the division of work, brings order to the organization. It provides accountability and responsibility where necessary. However, decentralization (we might think of it more as delegating) usually increases the worker's importance and value.

9. *Scaler (continuous) chain of ranks* or layers in an organization, and the notion that authority flows from the highest to the lowest ranks in the organization's chain of command.

10. *Order,* or a place for everything and everything in its place. Fayol was interested in both material order and social order. Today we would say that order refers to having the right personnel and the right materials ready to do the job so that employees are effective and efficient.

11. *Equity and fairness* are important in the treatment of the employees of any organization.

12. *Initiative* is important to the organization and its members. The ability to think out a plan and see it through to completion is one of man's greatest satisfactions.

13. *Stability of our personnel,* recognizing that there is value in a long-term relationship with our employees.

14. *Esprit de corps,* an organizational spirit, is essential to the survival of any organization.

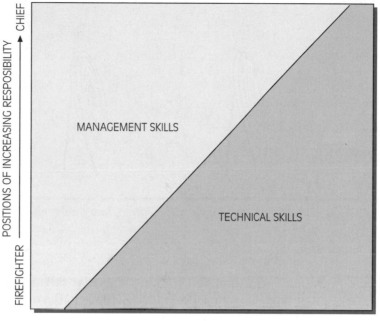

Figure 4-1
Management activity increases as you move up the organizational ladder.

THE NEED FOR MANAGEMENT SKILLS INCREASES AS ONE MOVES UP THE ORGANIZATIONAL LADDER

Planning

Any activity should start with **planning**. Many find planning to be a difficult and frustrating experience. Planning is largely a mental process and often requires the ability to envision things that have not yet happened. When planning is done properly, we challenge these visions with "what if" situations that make us think through the consequences of our proposals. This envisioning can be difficult. We are frustrated when things do not turn out exactly as we planned or our ideas are not approved. As a result of such experiences, managers may see planning as a waste of time and shy away from good planning activity.

Planning covers everything from the next hour to the next decade. We usually divide planning into three areas (see Figure 4-2):

Short-range planning looks at the rest of today, tomorrow, and the rest of the year. Many planning activities are driven by annual events. As a result, budgets, training, certification, and such are usually scheduled on a one-year cycle. For our purposes, we arbitrarily limit short-range planning to one year.

Mid-range planning reaches out from one year to five years. During this time we can usually convert goals into definitive plans. Most budget, procure-

planning
the first step in the management process, planning, involves looking into the future and determining objectives

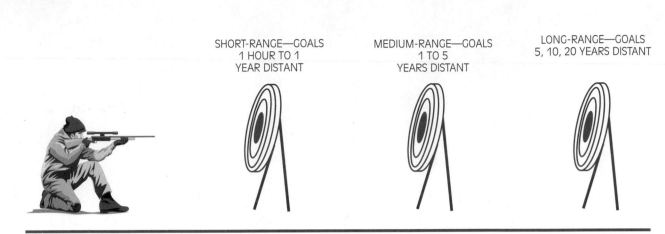

SHORT-RANGE—GOALS
1 HOUR TO 1
YEAR DISTANT

MEDIUM-RANGE—GOALS
1 TO 5
YEARS DISTANT

LONG-RANGE—GOALS
5, 10, 20 YEARS DISTANT

Figure 4-2   *Planning activity is generally divided into short, medium, and long range.*

ment, recruiting, and training activities fall into the realm of mid-range planning. Most departments have plans in the mid-range category, although some do this planning better than others. All departments should be looking well ahead at the needs for personnel, equipment, and facilities.

Long-range planning looks at the needs of the organization beyond five years. Many departments do not get involved with long-range planning, but those that do see the benefits of keeping ahead of the community's needs. When communities are changing, long-range planning is essential if the department is to change and provide service as the community changes. Planners must look at present growth and anticipate future growth. They have to increase (or relocate) resources to serve the expanding community. Managers should consider fire station location and access, the impact of major highways, and the increasing concern for the environment.

When planning precedes the action, the action is easier and is often accomplished faster.

Organizing

The second function of management is called organizing. We have already discussed organizations, but in this case we are looking at the activity of organizing itself. The organization breaks down the department's activities into manageable tasks. Likewise the process of **organizing** is the breaking down of large tasks into manageable activities.

In a sense, the development of the organizational structure of the department is the result of organizing. We indicated the benefits of this process when we

■ **Note**

When planning precedes the action, the action is easier and is often accomplished faster.

organizing
the second step in the management process, organizing, involves bringing together and arranging the essential resources to get a job done

talked about organizations in Chapter 3. When the Plain City Fire Department is established, the development of a structure and the assignment of tasks is the start of the organizing process.

Organizing keeps the department's functions in mind, and, when done properly, focuses energy on the department's goals and objectives. Organizing includes managing the resources available to the department. At the company level, these resources typically include human resources, physical resources, time, and money.

Commanding

Commanding is the third management function. Because others become involved in this area, the job of commanding can be difficult. Here we see the application of leadership skills and other human relations activities. We discuss the details of leadership in Chapters 6 and 7.

Commanding does not mean that you stand in the hall or in the street and shout orders. **Commanding** means that you are responsible for getting something accomplished, and because you control the resources of an organization, it means that you are using the talents of others to attain those goals. In this sense, commanding is leading, motivating, and building teamwork among your subordinates.

commanding
the third step in the management process, commanding, involves using the talents of others, giving them directions and setting them to work

Coordinating

The fourth management function involves the coordination of the available resources. These resources can include money, equipment, and personnel (see Figure 4-3). We should always consider time as a resource as well. To some extent these resources can be substituted for each other. For example, if time and money are available, and personnel or equipment are scarce, we might consider contracting for a particular service we need.

Managers should be constantly looking for the proper mix of resources to best meet the needs of the organization. Effective managers are constantly finding ways to do more for less. In the private sector, this is important if the company is to remain competitive. In the public sector, it is important because the citizens and their elected representatives are expecting good management practices from everyone in government.

Although most fire departments are nonprofit organizations, they operate in a competitive environment. Public fire departments are constantly competing with other government organizations, such as the police, libraries, and schools, for funds and support. With privatization, several organizations can offer the same service offered by fire departments, and usually do it for less money.

coordinating
the fourth step in the management process, coordinating, involves the manager's controlling the efforts of others

Coordinating is putting the various functions together into a smooth and well-operating organization. For the manager who is effective in planning, organizing, and directing, it would appear that coordinating would be easy. Indeed,

Figure 4-3 *For company officers, the greatest resource is your teammates. Effective coordination of their efforts is essential for company excellence.*

if these three activities have been well accomplished, coordination will be a lot easier.

Any one of several factors may interfere with good coordination. Poor communications, poor timing, resistance to change, lack of clear-cut objectives or at least the lack of understanding or acceptance of these objectives, and ineffective policies, practices, and procedures all can hinder a manager's efforts to coordinate activities. The group's acceptance of the manager as a leader plays an important part too. If the team accepts the leader, the leader will have a greater opportunity to coordinate their efforts.

THINGS THAT CAN BE COORDINATED

The resources of a station
The resources of a department
The resources of a community

---

controlling
the fifth step in the management process, controlling, involves monitoring the process to ensure that the work is accomplishing the intended goals and objectives, and taking corrective action when it is not

Controlling

Controlling is the fifth and final item on Fayol's list. **Controlling** is monitoring our progress. If we say that planning was looking ahead to see what we are going to

Figure 4-4   *Managers should always be watching performance, measuring effectiveness, and looking for ways to improve.*

do, then we can also say that controlling is looking back to see how we have done. Controlling is similar to the feedback we talked about in Chapter 2; it tells us how we are doing (see Figure 4-4).

CONTROLLING HELPS US GET TO THE RIGHT PLACE AT THE RIGHT TIME

When aircraft fly to a destination of any distance, they file a flight plan. Ships file a similar document when they leave port. There are several reasons for this information, some associated with safety, but for the crew involved in navigating the trip, it gives them a plan to follow during that trip. If they find that they are off schedule or off course, they can take appropriate corrective action early in the trip so that they arrive in the right place at the right time. Controlling is similar for managers; we want the organization's management actions to arrive at the right place and at the right time.

Controlling allows us to measure the effectiveness of our effort, to help us maintain our goals, to seek new ways to improve, to increase production, and to help plan for future undertakings. It should also give us an opportunity to identify those who should be recognized for their accomplishments.

CONTRIBUTORS TO MODERN MANAGEMENT

During the twentieth century, industry sought ways to increase production. The assembly line was established and machines were invented to increase the individual worker's productivity. As the pace increased, workers became subordinate to the machines they operated. Whereas Fayol and other management scientists had set out to create more pleasant work conditions, management had created situations that reduced the work to simple repetitive tasks. Workers became detached from both the product and from the work itself. Worker satisfaction decreased and with it, the quality of work decreased.

Many studies of management took place during the first half of the twentieth century. They make interesting reading to the serious student of management and you will find a good discussion of these events in any good college text on management. We briefly cover several of the significant contributors to this evolution for they help us understand where organizations are headed today.

Frederick Taylor

Frederick W. Taylor (1856–1915) was an American engineer working as a superintendent in a steel mill. Like Fayol, Taylor was concerned with increasing the productivity of both man and machine. However, Taylor's focus was on the first-line manager and the efficiency of the worker. He reasoned that if efficiency could be improved, the workers' productively could be improved, and with increased productivity, workers would have a good, steady job and would be well paid for their efforts. Taylor is called the "Father of Scientific Management."

Application of Taylor's work is evident in the fire service. We have division of labor, of course, but for many departments, the tasks have been narrowly defined so that little flexibility is allowed. Taylor would be pleased to know that many departments design their apparatus with standardized hose-bed and compartment arrangements "to increase productivity."

Abraham Maslow

Abraham H. Maslow (1908–1970) concerned himself with the full range of human relationships. In 1954 he published his *Hierarchy of Needs*. We discuss the application of Maslow's theory in Chapter 6, but we introduce it here because of the role he played in the evolution of management theory. Maslow recognized that most of us have five levels of needs, and that the very first of these are the basic physiological needs, the need for air, food, and water, and for clothing and shelter to protect us from the elements. Maslow said that these basic needs must be satisfied before any worker can be productive.

Peter Drucker

Peter Drucker (1909–) is credited with inventing a concept known as management by objectives (MBO). Drucker was a professor who had real world experience as a consultant in a variety of settings in the United States and elsewhere. He wrote several books including *The Practice of Management,* published in 1954.

The basis for MBO lies in managers defining areas of responsibility for their subordinates. Measurable goals are set for these areas of responsibility and these goals are used as a standard to evaluate the results obtained. MBO requires planning and thus reduces some of the surprises that face managers. It encourages innovation and improves performance because people are working together to meet the unit's goal. MBO has been challenged because it is labor intensive for managers; it places a heavy burden on them to plan, set goals, and provide a frank appraisal of the results. Many managers find that it is not worth the effort.

MANAGEMENT BY EXCEPTION

Management by exception (MBE) suggests that the manager gets involved only when something unusual occurs. This approach allows the manager to focus on complex and urgent matters. The idea is to define the normal parameters. When something gets outside these boundaries, attention is called to the problem. Budget tracking is a good example. If expenses are tracking according to plans, there is little need to spend time going over the numbers. A lot of things are like this. For example, most cars no longer have an engine temperature gauge on the dashboard. Most of us really do not need to know how hot the engine is, as long as it is not too hot! A red light will warn of this condition. We only need to take action when the situation is outside the normal.

---

management by exception
management approach whereby attention is focused only on the exceptional situations when performance expectations are not being met

Theory X
management style in which the manager believes that people dislike work and cannot be trusted

Theory Y
management style in which the manager believes that people like work and can be trusted

Douglas McGregor

Douglas McGregor (1906–1964) was also a professor. In 1960, McGregor wrote a book entitled *Human Side of Enterprise.* In that work McGregor introduced the now well-known concept of Theory X and Theory Y. McGregor said that many managers feel that they need to have tight controls over their organizations because workers are lazy and resist the demands of their managers. McGregor called this traditional style of management **Theory X** and said that it was not a good way to manage.

McGregor proposed an alternative style, something he called **Theory Y.** He felt that people would accept work and responsibility if given a chance, and that management should provide suitable working conditions so that people could achieve personal goals while satisfying the organization's goals. We will return to McGregor in Chapter 6.

William Ouchi

Theory Z
management style in which the manager believes that people not only like to work and can be trusted, but that they want to be collectively involved in the management process and recognized when successful

In the early 1980s yet another professor, William G. Ouchi, developed a concept he called **Theory Z.** Professor Ouchi focused on the worker's needs. Building on Maslow's and McGregor's ideas, Ouchi said that workers really want more than just the satisfaction of working productively; workers want to be involved in their own destiny. He also found that workers wanted to be trusted and to be entrusted. He found that many Japanese companies using this approach were realizing higher productivity and often making better products.

With Theory Z, the focus is on the employees rather than the work itself. The company encourages employee involvement at every level of decision making and accepts these ideas. Teams, rather than individuals, assemble products, rather than parts. As a result, workers share in the satisfaction that comes in seeing the results of their work and share in contributing to make the product even better.

In Theory Z, the relationship goes beyond the worker—family needs are recognized too. The workers reciprocate by being totally involved with the company over a long period of time.

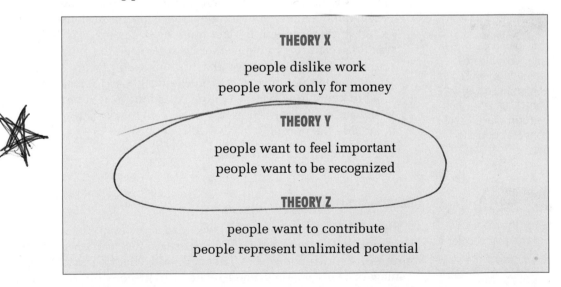

THEORY X

people dislike work

people work only for money

THEORY Y

people want to feel important

people want to be recognized

THEORY Z

people want to contribute

people represent unlimited potential

Edwards Deming

We would be remiss if we failed to mention the work of W. Edwards Deming (1900–1995). Deming wrote a book entitled *Quality, Productivity, and Competitive Position.* Deming's ideas were used by the Japanese in their manufacturing

total quality management
focus of the organization on continuous improvement geared to customer satisfaction

process (similar to theory Z) and he eventually brought the idea to the American workplace. The acceptance of his ideas is still ongoing. In some circles, Deming's ideas are referred to as **total quality management** or TQM.

Deming's Fourteen Points (altered a bit for the fire service)

1. Understand and accept the organization's mission, goals, and objectives.
2. Seek ways to improve the process and train the people to do it better. All improvements, big and small, are important. ~~—Train people Better~~
3. Use inspections as an improvement tool, not a threat.
4. Think beyond the bottom line. Consider quality and personal value along with the cost in every decision.
5. Demand quality, not better numbers, in everything you do.
6. Accept the idea of quality. Do every job right, the first time, every time.
7. Use training time effectively; some organizations train just to meet quotas.
8. Promote pride in the job. As supervisors, be sure that your employees have the equipment and materials they need, and recognize their good work.
9. Promote leadership. Work on giving supervisors the best tools possible to deal with the organization's most valuable resource—its people.
10. Create an atmosphere of trust and open communication.
11. Get rid of the organizational barriers that inhibit trust and effective communications, and force people to cooperate rather than compete.
12. Use analytical tools to help better understand and monitor the job.
13. Work as a team.
14. Use TQM all the time.

■ **Note**
Consistency of these purposes must be defined and maintained.

■ **Note**
TQM is based on employee participation and the concept that all must work together to achieve quality goals.

■ **Note**
TQM focuses on a continuing process of improvement.

To be successful, TQM must put the focus on the customer's needs. Long-term TQM commitments include establishing a vision, defining a mission, and providing guiding principles for the organization. Consistency of these purposes must be defined and maintained. The TQM process involves everyone in the organization, but, because of traditional values (we expect the leaders to lead), it is usually necessary for top management to sell TQM to the other members of the organization.

But it takes more than just words of inspiration from the leaders. TQM is based on employee participation and the concept that all must work together to achieve quality goals. Teamwork is essential. Effective communication, both vertical and horizontal, is critical to the process. TQM uses analytical techniques to provide the data necessary for timely and accurate decisions. Data can be used to define the process, identify problems, indicate solutions, and monitor improvements. TQM is not driven by analytical data; it is merely supported by such information.

Total quality management focuses on a continuing process of improvement. This approach combines all that we have learned about human resource management and adds the ability to use modern quantitative analytical techniques to provide timely data for making accurate decisions.

SOME COMPARISONS

Traditional Values	Total Quality Management
Looks for quick answers	Takes new management approach
Maintains old ways	Uses small innovations to improve
Focuses on short term	Focuses on long term
Looks for errors	Prevents errors
Decides using opinions	Decides using facts
Motivated by cost factors	Motivated by customer satisfaction

Kenneth Blanchard

The One Minute Manager, written by Kenneth Blanchard and Spencer Johnson in 1981, caused quite a stir when published. One of the key items in the book was the suggestion that we should try to find more ways to make people feel good about themselves. The authors suggest that people who feel good about themselves do good work. The lesson is that managers should focus on what is going well. The key action is: find someone doing something right and give them recognition.

Alan Brunacini

customer service
service to people in the community

internal customers
the members of the organization

Management evolution has led managers in many organizations to focus on a concept called **customer service.** Many companies are focusing on serving the customer better and starting that process with attention to the employees, often called the **internal customers.** The logic here is that if you make the internal customers (the employees) happy, they will make the paying customers (the citizens) happy. The goal is to create an environment where every employee wants to *delight* the customer. For businesses, it means they want the customer to come back. The employee handbook of a major restaurant chain illustrates the point beautifully. The job description for waiters is simply this: "Do whatever it takes to convince each of our guests to come back here for their next meal."

Alan Brunacini, the chief of the Phoenix (Arizona) Fire Department, has played a key role in introducing the concept of customer service to the fire service. He suggests that you look closely at what you are doing and think about how you treat our customers, both those inside the organization as well as the public you serve. While you are not looking for repeat business, you do want your customers to be *delighted* with your service. After all, they are paying the bill, and maybe your salary.

His department's mission statement reads something like this: "Deliver the best possible service to our customers." Clearly, turning this concept into a reality is not easy nor is it quickly accomplished; it takes time and energy to change attitudes. But being nice and taking care of the customers should be an essential part of every organization, especially for a public-service organization. We explore this topic further in Chapter 5.

MANAGEMENT IN MODERN ORGANIZATIONS

Management is the active part of making an organization run and includes planning, organizing, commanding, coordinating, and controlling. We usually think of management as something done by the senior staff of an organization, and as a result, we usually relate management to the administrative or nonoperational activities essential to the organization. For many organizations, this includes the support activities, such as human resources management, training, purchasing, and communications. But management activities are found at all levels and functions of modern organizations. Most fire chiefs spend more time working at their desks than commanding emergency activities (see Figure 4-5).

Figure 4-5 *Fire chiefs do most of their managing from their desks, not from the middle of the street.*

Most senior officers spend more time managing the organization than managing emergency incidents. Most company officers spend more time managing than fighting fires.

Management can be enhanced through good policy, procedures, and even by personal observation. **Policy** is usually published by fire department headquarters. Such policy prescribes what should be done in response to various situations. Policies are broad in nature and should be clearly understood. Personnel need to know where they stand with regard to the department's directions. The department's statement regarding harassment in the workplace is an example of a policy.

Procedure may be published by the headquarters, but a good part of procedure is initiated at the company level. Established procedures, like established policy, simplify the daily routine for firefighers and supervisory personnel alike. Instead of each problem having to be resolved each day it comes up, established procedures make these activities routine.

Procedure has a lot to do with organization and management. Procedure is how tasks are assigned. Procedure deals with questions regarding responsibilities and the order in which tasks are accomplished. Established procedure provides standardized solutions to similar problems and usually simplifies life at work. People like having solutions; employees tend to be more productive when they have standard solutions to recurring events. Some procedures are essential to the mission of the organization and to the safety of those who work there. The procedure for raising an extension ladder must always be the same. It takes practice and teamwork to do it well. Practice and teamwork do not bring rewards every day but they increase the likelihood that the ladder will be raised quickly and correctly when it is needed most.

Some procedures may not be so firmly fixed. For example, items dealing with the fire station maintenance program and meal preparation may vary, depending on the management style of the company officer. Some company officers may treat these procedures the same as that dealing with ladder raising, and never allow for variations. Others will allow their members to try alternative approaches to routine tasks as long as the job gets done. It is important that the members be able to separate those essential tasks that have very fixed procedures from those where some discretionary latitude is appropriate.

Another management technique is observation of the work and the workers. When company officers sit in their offices doing necessary paperwork, they may think that they are managing. When this situation occurs to excess, these company officers are overlooking the most important resource they manage—the human resources under their command. Company officers who dedicate their energies to their personnel will find that the overall productivity of the company is improved.

Much has been written recently about the art of managing by walking around and observing the work and the workers firsthand. Where practiced, it is

policy
formal statement that defines a course or method of action

procedure
a defined course of action

■ **Note**
Company officers who dedicate their energies to their personnel will find that the overall productivity of the company improves.

important to remember the respective role of everyone in the organization and not violate the rules we have covered. Personal observation allows one to see what is going on, engage subordinates in polite conversation, and make inquiries about the job as well as personnel issues. But be careful about saying things that might be inferred as direction or as responding to an organizational or personal need. You may be bypassing that chain of command.

Keep the Organization in Mind As You Manage

Situation #1. Suppose that while touring the station as a new company officer, you encounter one of your newest firefighters. You stop and visit briefly. At the end of the conversation the firefighter lets you know that he needs a few minutes off from work to go to the bank before it closes. It would be easy to deal with that request and approve it right there. Doing so would help you bond with the new firefighter. But let us look at the slightly larger picture.

Suppose that you have assigned this new firefighter to your most experienced firefighter for the purpose of training and monitoring the new employee's progress during the probational period of his employment, a mentoring process. Suppose that the new firefighter was doing a great job. Who should grant the request, the mentor or the company officer?

Suppose on the other hand that the new firefighter has been a royal pain for the mentor, but the senior firefighter is seeing some progress. How would the ability to reward the new firefighter aid the senior firefighter in motivating the new employee?

How about others in the fire station? Most fire stations are unique little societies. Suppose that they had been watching this relationship between the new firefighter and the mentor. What is going to be their reaction when the company officer grants the firefighter's request without consulting with the senior firefighter?

This illustration has two points. One has to do with the power of rewarding the individual. More importantly is the point of remembering the organizational structure.

Situation #2. Suppose that the chief or another senior officer in the department bypasses you and goes directly to one of your personnel with a complaint or request. The solution is not so simple here. First, you should direct your personnel to respect such requests and to let you know about the situation as soon as possible. If the incident is an isolated occurrence, it might be best just to overlook the situation. But, if this action seems to be a habit, you may need to get to the heart of the matter. When circumstances permit, you should politely and privately make your feelings known to the senior officer, noting that such actions undermine both your authority and the good order of the organization. Most seniors officers will respond favorably to such comments.

FAYOL'S BRIDGE

If employees precisely followed the formal rules for organizational behavior, the only people they could talk to would be their own supervisor and their own subordinates. All other communications should flow through the chain of command.

Suppose the officer of Ladder Company 1 wants to use the training academy's facilities during the coming weekend. Following all the formal rules, the company officer would have to ask his boss who in turn would ask his boss up to a point where there is a person who has supervisory authority over both the department's field activities (the ladder company) and the academy. In many departments, that person might be the fire chief. Then the request flows down the organization until it gets to the academy. The answer would have to flow back the same way.

This arrangement is obviously impractical. While we usually think about delegation and decentralization as relatively modern concepts, we can look back to Henry Fayol for a solution to this basic organizational problem. Fayol realized a century ago that rigidly following the chain of command was an impractical arrangement. He said that for organizations to be effective, the manager in one section should be able to communicate with the manager in another section, without having to go all the way to the top of the organizational structure. The shorter path is called **Fayol's bridge** (see Figure 4-6).

If the fire chief authorizes the officer of Ladder Company 1 to call the training officer of the academy to use the academy's facilities, and if the fire chief authorizes the training officer to respond directly to such requests, the chief has created an environment where employees do not have to refer routine issues up the organizational chain of command for resolution. This speeds the process and leaves the senior officers free to deal with more important issues.

For those who work in organizations where this takes place daily, you may be wondering why we have taken half a page of text to describe the process. For those of you who work in organizations where this direct liaison never occurs, you may be wondering, "How can this work?" Our purpose here is to point out that this practice is common in well-run organizations, yet there are a lot of places where using Fayol's bridge would be unthinkable. How about your department?

Modern management suggests that we should do away with as many of the obstacles to getting our business done as we can. Certainly we need to have organizations and understand organizational structure. Certainly we need to have some rules of conduct for those who work in these organizations. But at the same time, we need to provide a workplace where business can be done effectively.

Fayol's bridge
organizational principle that recognizes the practical necessity for horizontal as well as vertical communications within an organization

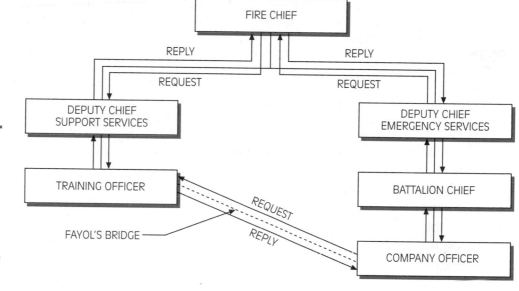

Figure 4-6 *Information relationships and effective delegation within organizations facilitate crossing the organizational lines to get things done. This concept is known as Fayol's bridge.*

> There will always be exceptional situations that are outside the parameters of the normal day-to-day business that should be referred to upper management. No problem here; we want that. But resolving the routine issue at the lowest possible level adds to the quality of life of the organization and for all who work within it.

THE ROLE OF ETHICS IN MANAGEMENT

This discussion on management would not be complete without some mention of ethics.

The fire service has a long and rich history of providing service to the citizens of our communities and generally enjoys an excellent reputation, a reputation that has been well earned and that we do not want to lose. The ethical standards of a fire department are influenced by what society expects and what the local community believes is the function of the department. Recent polls indicate that more people have trust in their fire department than almost any other public institution.

Anyone in public office faces a challenge of having a greater ethical obligation than the average citizen. Everyone in the fire service is serving in a public office. For officers, there is an increased obligation, because officers are expected to lead by personal, positive example and take appropriate corrective action when things are wrong.

What are ethics? **Ethics** can be described as a system of conduct, principles of honor and morality, or guidelines for human actions. For the most part, these definitions are rather vague. Even where a professional organization sets standards for its individual members, there is always some degree of variance in how these "rules" are to be interpreted, and there are always a few who will try to beat the system.

Ethics are not new. We usually have a feeling of guilt when we do something wrong. This feeling is manifested in many ways, including physical reactions on our skin and in our stomach. The history of ethics goes back thousands of years. You can find evidence of ethics in the earliest recorded history. Most of our religions have strong ethical principles. These rules cover our actions with others: honesty, courage, justice, tolerance, and the full use of our talents.

Many civic, fraternal, professional, and even religious organizations have a set of ethical values. These organizations provide guidelines for their members and encourage members to follow these practices at all times. Doing so improves the relationships within the group, and the group's image also enhances our relations with others.

Ethics have a direct impact on the management of the fire service, from the chief to the company officer. At the company level, the company officer must often make decisions for which there are no clear-cut guidelines. At the same time, the company officer, through personal example and positive leadership qualities, is expected to set an example for others in the department.

There are laws and regulations that address ethical issues, but these have little influence on those who wish to be unethical. What is effective is a well-defined culture or a value system within the department that acts as a regulator of conduct. The laws and regulations that impact upon us as citizens change. Ethics often begin where laws leave off. In the area of employee relations, issues surrounding diversity and harassment are very much in the news these days. These are ethical issues. We read stories of allegations being brought against individuals regarding their personal conduct on these matters. Not too many years ago, such conduct was accepted, though we might argue that it was improper then, too.

With a change in public attitude, increased litigation, and the attention usually brought by the press, what used to be okay is no longer acceptable. These changes in attitude impact upon our delivery of services as well as the way we operate within our own organization.

Laws and regulations and even an organizational code of ethics can only serve as guidelines. Our personal system of values is what really determines our behavior. This system is influenced by many factors, including family values and culture, community attitudes, the influence of laws, our own set of values, and life's experiences. We all have ethical values. Where we get into trouble is when our values do not coincide with those of our organization or society.

---

ethics
a system of values; a standard of conduct

■ **Note**
We usually have a feeling of guilt when we do something wrong. This feeling is manifested in many ways, including physical reactions to our skin and in our stomach.

■ **Note**
Ethics often begin where laws leave off.

THE INTERNATIONAL ASSOCIATION OF FIRE CHIEFS' CODE OF ETHICS

Maintain the highest standards of personal integrity, be honest and straight-forward in dealings with others, and avoid conflicts of interest.

Place the public's safety and welfare and the safety of firefighters above all other concerns; be supportive of training and education that promote safer living in occupational conduct and habits.

Ensure that the lifesaving services offered under the member's direction are provided fairly and equitably to all without regard to other considerations.

Be mindful of the needs of peers and subordinates and assist them freely in developing their skills, abilities, and talents to the fullest extent; offer encouragement to those trying to better themselves and the fire service.

Foster creativity and be open to consider innovations that may better enable the performance of our duties and responsibilities.[1]

The decisions faced by officers in the fire service do not fit neatly into a tidy framework of values. Often operating without supervision, the company officer is responsible not only for the conduct of the company members but also for their productivity. It is up to the company officer to see that the company remains ready to deal with emergencies, and that the many activities needed to maintain readiness are accomplished.

If company personnel are expected to have daily physical activity, conduct fire safety inspections, and prepare preplans for their first due area, the zeal and effectiveness with which these activities are conducted will probably go unnoticed by the public and to a large extent by the department. It becomes a question of doing what is right. If the company officer fails to properly accomplish these tasks, who will step forward to see that they are done?

■ **Note**

The standards for ethics change.

The standards for ethics change. Today, there seems to be increased attention on ethics, especially for those in public office. Nearly every day's news contains at least one story of alleged unethical conduct on the part of someone in public office. Although our Constitution guarantees that all citizens are presumed innocent until proven guilty, public opinion can quickly condemn an individual or group without the benefit of a trial. Once an individual or organization is tainted by allegations of unethical conduct, restitution is difficult.

Officers have an obligation to act in an ethical manner at all times, to demonstrate integrity, honesty, courage, and faithful productivity. Some will say, "That's not the way we have been doing things here." We acknowledge that, and note that few of us can make major changes in history. However, all of us certainly can control the present and maybe even shape the future in some small positive way.

Every person in the fire service should live by this.

Code of Ethics

Be loyal to the highest moral principles and place allegiance to country above loyalty to any person, party, or government department.

Uphold the Constitution, laws, and regulations of the United States and all of the governments therein.

Give a full day's labor for a full day's pay, giving earnest effort and best thought to the performance of duties.

Seek to find and employ more efficient and economical ways to accomplish tasks.

Never discriminate by dispensing special favors or privileges to anyone, and never accept favors or benefits under circumstances that might be construed by reasonable people as influencing the performance of duties.

Make no private promises of any kind that are binding on the duties of your office because your private word cannot be considered as binding on your public duty.

Engage in no business that is inconsistent with the conscientious performance of duties.

Never use any information gained confidentially in the performance of duties as a means to make a private profit.

Maintain high standards of personal integrity, honesty, and straightforwardness.

Provide fair and equal lifesaving service to all.

Expose corruption wherever you discover it.

Uphold these principles, ever conscious that a public position is a public trust.[2]

■ **Note**

Ethics are an important management issue.

Ethics is an issue today and must be faced by all organizations. Managers must accept the fact that they are responsible for their actions as well as those of their subordinates. Managers must define their ethical values and follow these values through personal example.

Summary

As a company officer you are part of the department's management team. In keeping with contemporary management trends, many departments are passing much of the management process down to the company officer, and beyond. When permitted to do so, company officers can have a great impact on the effectiveness and efficiency of the entire organization.

The information in this chapter can be used at the company level. Company officers should be aware of and should be using all the functions of management. Modern officers should be aware of the models of management style that have evolved through management science and use them as needed.

As with other tools you use on the job, no one uses all of the tools all of the time. However, it nice to know which tool is the right tool for the job at hand, and to know how to use it effectively.

Ethics are important for the company officer too. The best way to have good, fair, and honest people in the station environment is for the company officer to set a good personal example and to expect the same level of performance and behavior from all others. We further explore practical applications of management in Chapter 5.

Review Questions

1. Define management.
2. What are the functions of management?
3. What resources are available to manage at the company level?
4. Characterize workers according to McGregor's Theory X and Theory Y.
5. Describe Theory Z.
6. What is TQM?
7. How can TQM be used in today's fire service?
8. Who are the customers of a public service organization?
9. Why are ethics important in management?
10. What is the company officer's role in management?

Additional Reading

Blanchard, Kenneth, and Spencer Johnson, *The One Minute Manager* (New York: William Morrow, 1981).

Blanchard, Kenneth, and Norman Vincent Peale, *The Power of Ethical Management* (New York: Fawcett Crest, 1988).

Carter, Harry, and Erwin Rausch, *Management in the Fire Service,* Second Edition (Quincy, MA: National Fire Protection Association, 1989).

McGregor, Douglas, *The Human Side of Enterprise* (New York: McGraw-Hill, 1960).

Peter, Lawrence J., and Raymond Hull, *The Peter Principle* (New York: William Morrow, 1969).

Peters, Tom, *Thriving on Chaos* (New York: Alfred A. Knopf, 1987).

Notes

1. Code of Ethics of the International Association of Fire Chiefs. Used with permission.
2. From "A Code of Honor" by Robert D. Carnahan, *Fire-Rescue Magazine,* January 1998, p. 78. Reprinted with permission.

Chapter

5

The Company Officer's Role in Managing Resources

Objectives

Upon completion of this chapter, you should be able to describe:

- The importance of organizational vision and mission statements.
- How goals and objectives support the department's mission.
- How to effectively manage human resources, time, and the department's financial resources.
- How to introduce and manage change.
- A process approach when solving problems.
- The role of the customer.

INTRODUCTION

This chapter brings together everything we have covered so far. We look at the company officer's role in the organization, how the organization works, how management helps the organization do its assigned mission, and finally how we communicate effectively within the organization. Some of this material may seem like a repeat of the previous chapter, but the focus here is on the company officer's practical application of the material presented in previous chapters.

VISION AND MISSION STATEMENTS

vision
an imaginary concept, usually favorable, of the result of an effort

Let us start with the big picture, a **vision** of what you are and where you are going. With a clear understanding of your organization's vision, you can set goals and plan your own destiny. In this chapter we look at how we do this for your work position, and to some extent, for your personal life.

Organizations should have a vision too. As employees, we should be part of that organizational vision and realize that if we want to be comfortable and effective in our activities at work, we need to share at least some of the same vision that has been adopted in our workplace.

mission statement
a formal document indicating the focus and values for an organization

Mission statements declare the vision for the organization. A mission statement is like putting up a sign that says what you are and what you do. Mission statements tell your employees and your customers the services you provide. Many mission statements define the priorities and values of the organization.

In the past, a fire department may have had it a little easier in writing its mission statement. In those days, a completely adequate mission statement may have read something like this:

Our job is to prevent fires, but when fire occurs, to protect life and property and confine and extinguish the fire.

In recent years the job of the fire service has become more complicated (see Figure 5-1). Many departments have worked hard to define their mission in the face of these changes. Here are two examples:

Our mission is to provide emergency and nonemergency services for the protection of the lives and property of the citizens entrusted to our care.

Our mission is to save lives, preserve property and the environment, and ensure the health and safety of ourselves and the community.

Expanding the basic list of services a bit we get:

The mission of The Plain City Fire Department is to provide a range of programs to protect the lives and property of the inhabitants of Plain City from the adverse effects of fire, sudden medical emergencies, or the exposure to dangerous conditions created by man or nature.

Some mission statements go beyond defining the scope of services that are provided to the community to say something about the organization's philosophy or to suggest the way that these services will be provided. The following is an example:

This department will provide emergency services to the citizens and businesses of the city at fires and other disasters. We will also attempt to prevent such emergencies through safety education programs and code enforcement. These services will be provided in a prompt, courteous, and professional manner.

A mission statement may also suggest that there is a continuing effort to improve on what the organization does as well as a comment about management style. The mission statement should also suggest everyone is committed to the mission. This commitment must be a voluntary all-hands everyday action. Here is an example:

We serve with honor and integrity. We treat everyone with fairness and respect. We provide friendly and efficient service. We value the public's confidence and trust. We are responsible for our actions. We pursue excellence. We work as a team. We make our community even better.

Figure 5-1 *As fire departments take on new missions, their personnel are expected to take on new skills. Photo courtesy of Michael Regan.*

And also consider the following example from a leading retailer. With just a little work, this could be used in a public sector organization:

> Our mission is to provide the customer with quality apparel and service in a responsive, honest, and ethical manner. As we strive to accomplish this we feel that it is important to base our actions on core values that we share. These values represent our mission for being the best we can be, both as a profitable company and as individuals. We are all responsible for making this vision a reality.

■ **Note**
Mission statements should go beyond nice words and a sign.

Mission statements should go beyond nice words and a sign: They should accurately reflect the focus and the philosophy of the organization and should be the result of a collaborative effort to gather ideas from representatives of the entire department, working together in a committee process. The mission statement should be accepted and supported from the top down, not just with words but with actions that clearly indicate that these words are really what we are all about. Most importantly, the words in the statement should be understood and practiced by everyone in the organization. We will see more about management style, employee commitment, and a focus on improvement in the following sections.

GOALS AND OBJECTIVES

goal
a target or other object by which achievement can be measured; in the context of management, a goal helps define purpose and mission

Following the mission statement, goals and objectives must be established to bring the vision of the mission statement to life. **Goals** are the larger, strategic statements that may take several years to accomplish. Goals help you identify where you are going. You might ask, if you do not know where you are going, how will you ever know when you get there?

Objectives are usually an element within a goal, and represent the efforts of a shorter period. Goals and objectives should be consistent with the department's mission, and in fact, they should support it. Goals and objectives help us identify where we are going and help us measure progress as we make the trip.

objective
something that one's efforts are intended to accomplish

To illustrate, let us suppose that you are undertaking a significant personal trip, one that will require several days of driving. Because the purpose of the trip is to attend an event that is occurring at your destination, let us say your grandmother's surprise birthday party next Saturday, you must be there at a certain time, or the whole effort is a waste of time. As you plan the trip, your *goal* is to arrive at your destination on Friday afternoon. Your *objective* may be to arrive at specific intermediate points by suppertime of each day while en route.

■ **Note**
Goals and objectives help us identify where we are going and help us measure progress as we make the trip.

You might question the value of all this goal-setting activity, and say that it appears to be a waste of time and energy. Most successful organizations have mission statements and set goals. They train their employees on the importance of these visions and provide training to prepare them to do their duties consistent with these goals.

You may think that all of this goal-setting activity is done by the department's top management, but it is better if everyone has an opportunity to contribute to the setting of goals and objectives for the organization. The more participation in the process, the better the results. Participation also enhances acceptance by those upon whom these goals and objectives will impact.

USING THE FUNCTIONS OF MANAGEMENT

In Chapter 4, we listed the five functions of management as <u>planning, organizing, commanding, coordinating, and controlling</u>. Let us see how planning and organizing relate to goal-setting at the company level.

Planning

Planning is a systematic process and should be ongoing. Planning should always start with a clear understanding of the goals and objectives. A good plan has clearly defined goals, a logical set of objectives, a list of resources needed, and a method to measure progress.

> **■ Note**
> The more participation in the process, the better the results. Participation also enhances acceptance by those upon whom these goals and objectives will impact.

> **■ Note**
> Planning should always start with a clear understanding of the goals and objectives.

CHARACTERISTICS OF A GOOD PLAN

A good plan has

- A clearly defined goal
- A logical set of objectives
- A list of resources needed
- A method to measure progress

GOAL: Do something each day to make this place better.
OBJECTIVE: Make each day productive.

Questions you should ask yourself:
What can I do at the company level

- to define work that needs to be done?
- to provide the training, if needed?
- to provide the necessary materials?
- to plan and schedule the activity?
- to check the progress of our efforts?
- to recognize the accomplishments of individuals?

Organizing

Organizing involves the integration of human and physical resources, money, and time.

Managing Human Resources As company officers you are assigned personnel to staff your company. How effectively you use these personnel is a function of your leadership style. We discuss leadership in Chapters 6 and 7, but let us look here at how you *manage* these human resources.

In Chapter 4 we listed Fayol's Principles of Management. In larger organizations, these principles cover the way we manage in a organization where a management hierarchy is inevitable. As an organization gets larger, increasing numbers of mangers are needed to supervise the work. The consequences are additional layers of managers. One way to slow this process is through delegation.

Delegation is a process of entrusting some of our work to others. Delegation happens in large organizations and in small ones. A mom-and-pop ice cream stand has some degree of delegation, especially if mom and pop hire their kids to work there during the summers.

In the delegation process, two things happen:

The supervisor assigns duties to the subordinate, and

The subordinate takes on an obligation or responsibility to perform these duties.

authority

the right and power to command

The supervisor should also assign **authority** to do the job. To be effective, delegation of the duty and acceptance of the obligation for the job should be coupled with the authority to do the job (see Figure 5-2). Limits are usually set upon how much authority is granted, but the authority should be sufficient to allow the subordinate to be productive in the assigned task.

Delegation is often referred to as the granting of authority to another to carry out an assigned task. Delegating requires that the task be clearly identified and that the duties and limits of the delegation be spelled out. Effective delegation does not mean that the task is eliminated from the supervisor's responsibilities. Rather it implies that a partnership has been established in which the actual accomplishment of the task may be done by another. However, the supervisor ultimately remains responsible for the work being accomplished. As such, the supervisor is expected to continue to manage the activity and provide appropriate guidelines.

With the exception of a few specifically identified tasks, most of the duties of the company officer can be delegated. How well this delegation occurs is a clear reflection of how well you understand yourself and your subordinates. In both cases, it is important to understand the strengths and weaknesses involved and the experience and motivation of all concerned.

Once delegation has taken place you must continue to manage. Hopefully planning and organization have already taken place. But you should continue to coordinate and control the activity, monitoring the progress. Provide help and assistance where needed. Give encouragement and recognition when appropriate.

Figure 5-2 *Effective management of any organization involves the ability to delegate certain tasks.*

There are many benefits of delegating. Certainly it will take some of the work off the supervisor. It will also make the supervisor look good—the company is a team, not the work of individuals. The troops are part of that team and will be motivated to do more by their own accomplishments.

Delegation is difficult for some people, especially for new first-time managers. And that is quite understandable, for in many cases they have never had an opportunity to delegate work before, and are hesitant, either due to a lack of knowledge or experience in the process.

BARRIERS TO DELEGATION

I can do it better myself.

It will take time to train someone.

I am not sure that anyone else can do this.

No one else can do this as well as I can.

I really cannot take a chance on this project.

I think my boss would want me to do this myself.

Delegation is like other management activity: It should follow a regular pattern. You start by planning, giving some thought to the delegation process. Maybe you are bringing a new employee into the team or maybe you are reallocating some of the work among existing employees. Some planning must precede the actual handoff of work. You should define the work and establish procedures for its accomplishment. At some point you should actually bring the subordinate into the picture and tell him or her what you want. When delegating to trained and qualified employees, define the purpose of the work and leave the details to the subordinate.

In many cases some training is needed to prepare the new person for the duties at hand. But eventually, it will become time for the new employee to start the task. Usually they work with someone else, then gradually they take on the task alone. Finally, some sort of feedback is necessary to make sure that our expectations as supervisors, are being met. You might ask for a regular meeting time or some sort of reporting system so that you can effectively monitor the employee's progress. You have probably been through this process many times, as the subordinate.

Managing Human Resources on a Larger Scale Most employers are fair; they hire employees, treat them fairly, and pay them a reasonable wage. Some employers are not fair; they exploit their employees and treat them unfairly. Past practices of exploitation and unfair treatment of the workforce led to the development of the labor unions.[1]

Employees usually join unions as a result of frustration with management. In many cases these frustrations grow out of the employee's dissatisfaction with working conditions. Employees take their concerns to their supervisors, but the problems are not fixed. Some of the problems lie in the lack of good communications and some from management's inattention to the employees. Eventually the employees feel that they not only have a bad situation but that they cannot get the situation fixed.

Labor relations refers to a formal relationship between a group of employees and an employer. The relationship includes the negotiation of a formal contract concerning such things as wages, hours, and conditions of employment as well as the interpretation and administration of the contract over the period of the contract.

The administration of the contract lies with both the personnel specialists of the organization and the organization's managers. Personnel specialists deal with the details of administering the union contract. Hopefully these specialists provide some guidelines to the managers on how to make the contract work effectively.

As with most everything else in an organization, the attitude of top management toward a particular issue filters down throughout the organization and permeates the relationship at every level. It is important for the fire chief and other top officials to look upon labor relations as a positive opportunity. Likewise,

■ **Note**

When delegating to trained and qualified employees, define the purpose of the work and leave the details to the subordinate.

■ **Note**

The attitude of top management toward a particular issue filters down throughout the organization and permeates the relationship at every level.

union leaders should look at their relationship with the department's management as a positive opportunity.

When you think about the benefits that unions have advocated for their members, you should think of several major issues: improved wages and benefits and formal rules for hiring, promotions, pay, discipline, and remuneration. Unions also take a strong stand on having an opportunity to be a part of the decision-making process on issues that impact upon employees. One could argue that these should have been offered by management in the first place.

The question of unionization and the selection of the union to represent the workers is the subject of an election among the employees. Collective agreements are of fixed duration and provide a binding contract between the workers and management. Collective bargaining allows the workers to be represented by the union.

CHARACTERISTICS OF UNIONS

A characteristic of unions is that they have exclusive representation—only one union is authorized in a given job territory. The questions of unionization and the selection of the union to represent the employees are subject to an election among the employees. Labor contracts are of fixed duration and provide a binding contract between the employees and management. Collective bargaining allows the workers to be represented by the union.

In the private sector, union membership has been declining and today, less than 20% of all nonfarm workers are union members. There are indications of a recent reversal in this pattern and some employment sectors are seeing growth in union membership. Whatever their status, unions remain a strong social, political, and organizational force. Where unions exist, management must work with the union rather than directly with the employees on many issues.

Union representation among public sector employees is a significant part of total union membership. These employees work for federal, state, and local governments. At the federal level, labor relations are regulated by the president. At the state and local level, labor relations are determined by state law. Most states have passed legislation pertaining to labor relations issues for state and local government employees. The laws differ in the states and you would be prudent to understand the laws in your particular state. Most firefighters are represented by the International Association of Fire Fighters (IAFF), a union with 225,000 members.

For most paid fire and emergency medical services (EMS) personnel, the right to form, join, or assist a labor organization is governed by state law. These laws are generally modeled after the federal laws. The most conspicuous law in this area is the Labor Management Relations Act. Provisions of this act protect the interests of both sides. For example the act protects the rights of individuals to

■ Note
Most firefighters are represented by the International Association of Fire Fighters (IAFF), a union with 225,000 members.

form, join, or assist in the activities of the union. The act also protects the unions from interfering with an employee's rights as a citizen and as an employee.[2]

LABOR RELATIONS ARRANGEMENTS

A **closed shop** usually results from an agreement between their union and the employer that requires prospective employees to become members of the union prior to being hired. Such agreements are generally unlawful.

A **union shop** usually results from an agreement between the employer and the union that requires membership within a specified period of being hired. This practice is lawful.

In an **agency shop** the agreement provides that union membership is not required, but payment of union dues is mandatory for all employees.

Finally, there is the **open shop.** From organized labor's perspective this form of union security is the weakest. The employees are free to decide if they will join the union. If they do join the union, they must pay union dues. If they do not join, they do not have to pay dues. Usually employees can terminate membership only during certain periods, such as when the contract is up for renewal. In those states where right-to-work laws have been adopted, only the open shop is authorized.

---

closed shop
a term in labor relations denoting that union membership is a condition of employment

union shop
a term used to describe the situation in which an employee must agree to join the union after a specified period of time, usually 30 days after employment

agency shop
an arrangement whereby employees are not required to join a union, but are required to pay a service charge for representation

open shop
a labor arrangement in which there is no requirement for the employee to join a union

Most labor relations contracts spell out a grievance procedure. Grievances occur when there is a differing interpretation of the contract between representatives of labor and management; when a violation of the contract has occurred; when there is a violation of a law, regulation, or procedure; or even when there is just a perception of unfair treatment. The employee grievance process is a systematic process for resolving these differences. Most union contracts define how the grievance process should work. The idea is to bring the problem to resolution as quickly and as fairly as possible. There are generally three steps in the grievance process. The first step involves filing a complaint with the supervisor. Most grievances are resolved at this level.

Other provisions of labor relations contracts usually address several major issues as follows:

Compensation and working conditions. Many of these provisions cover personnel administration issues—hiring, pay, benefits. However, many also address items related to working conditions, items that are under the purview of the supervisor.

Employee security. Issues addressed here include such things as protecting an employee's position during times of promotions and layoff.

■ **Note**

Typical labor–manage-

ment issues include:

discharge, discipline,

discrimination, griev-

ance process, health

and safety, hiring,

insurance, leave, pro-

motion, and shift

duties.

mediation

the process by which
contract disputes are
resolved with a
facilitator

fact finding

a collective bargaining
process; the fact
finder gathers infor-
mation and makes
recommendations

arbitration

the process by which
contract disputes are
resolved with a
decision by a third
party

■ **Note**

To achieve long-term

success, labor and

management must

learn to accommodate

each other rather than

confront one another.

Union security. These provisions pertain to what the union can do to encourage employees to become members.

Management security. In this area, the contract defines those issues that are the exclusive domain of management.

Contract duration. The length of the contract. Contacts usually run from 1 to 5 years.

The role of the federal government in union contracts is relatively passive. The National Labor Relations Act provided employees with the right to engage in union activities. That act also established the National Labor Relations Board to supervise representation elections and to investigate allegations of unfair labor practices.

The power of strike has been reduced in recent times. Growing antiunion sentiment, court decisions, and a pool of workers ready to replace the strikers have generally reduced the unions' power through strike action. We frequently hear about strikes in the news, but in reality a very small percentage of employee time is lost through labor conflict.

Typical labor–management issues include: discharge, discipline, discrimination, grievance process, health and safety, hiring, insurance, leave, promotion, and shift duties. When the union and management cannot agree on the issues, they may agree to a third party coming in to help them settle their differences. There are three levels of involvement: mediation, fact finding, and interest arbitration.

Mediation involves a third party acting as a facilitator. The mediator opens communications channels, persuades the two sides to meet, helps the parties readjust their positions on issues, and controls the process by scheduling the meeting time and agenda. The Federal Mediation and Conciliation Service may be invited to participate in the negotiations. There is no specific time in the process at which it enters, and in most cases, it does not enter at all. Most labor contracts are settled without any assistance.

The next level of involvement is **fact finding.** The third-party fact finder investigates the truthfulness of facts that are relevant to the impasse. The fact finder can also make a recommendation to the parties regarding the reasonable settlement of the issue.

Arbitration is the final step. Arbitration allows the third party to settle the issue. The arbitrator hears both sides and determines the results. Arbitration is a last resort: It provides a solution, but neither party may like the results.

To achieve long-term success, labor and management must learn to accommodate each other rather than confront one another. By so doing, labor and management can learn to increase productivity and improve the quality of work life. Unions and management groups depend upon one another to achieve the goals, yet frequently confront one another at every turn.

Research has shown that good labor relations occur when there is mutual trust and respect, honest and open communications, mutual problem solving, good (consistent) contract administration, and timely resolution of grievances.

budget

a financial plan for an individual or organization

Managing Financial Resources Money is the second resource you manage. The principal tool for managing money is called a **budget**. Budgets tend to be unpopular topics, yet budgets are necessary. In their simplest terms, budgets are a management tool, and in fact, they represent nearly all of the five functions of management: Budgets allow us to plan, organize, coordinate, and control our financial activities.

Budgets usually encompass one year of activity of an organization. Budgets usually are tied to the goals and objectives, and they identify the resources needed to accomplish these goals and objectives. If you want to provide basic or advanced EMS capabilities on every engine and truck company in your department, you have to think about the training and equipment needed to support this program. You should plan for the procurement of these resources, determine the costs involved, and calculate these costs into the budget for the coming years.

capital budget

a financial plan to purchase high-dollar items that have a life expectancy of more than one year

Budgets are usually broken down into two categories, operating and capital. **Capital budget** items are big, expensive items that last more than one year. For the fire service, classic examples of capital budget items are fire stations and apparatus (see Figure 5-3). Local jurisdictions usually have a definition for capital budget items in terms of cost and expected life span. Because of their cost, capital budget items get a lot of attention and are usually clearly identified and carefully defended in the overall budget process. The **operating budget** includes nearly everything not covered in the capital budget. Most of the department's expenses are in the operating budget. For paid fire departments, most of the operating budget covers salaries and related personnel costs. Other obvious operating expenses include fuel, utilities, supplies, and equipment.

operating budget

a financial plan to acquire the goods and services needed to run an organization for a specific period of time, usually one year

Like the management process, developing the budget starts with planning. The planning should focus on what resources are needed to turn the organization's goals and objectives into reality. Depending upon the size of the department

Figure 5-3 *A new fire truck represents a significant long-term investment. Photo courtesy of Jack K. McElfish, chief, Richmond Fire Department.*

and the complexity of local government, this may take anywhere from a month to a year to accomplish. Once the budget is approved the plan become the guideline for the coming year's activity.

When the new budget year starts, the budget moves from the planning phase to the action phase. Funds are transferred to the accounts and the department can start to purchase the items identified in the budget.

As the weeks and months unfold, it is appropriate to check the level of activity in the budget. The budget can be set up in a variety of ways to facilitate this check.

line-item budgeting
collecting similar items into a single account and presenting them on one line in a budget document

Today most organizations use some form of **line-item budgeting.** With line-item budgeting and accounting, like items are collected for purposes of keeping track of the expenditures during the year. Suppose we consider the cost of electrical utility service as a line item. We can plan for a certain expenditure for electricity, but extreme weather conditions may cause this cost to increase. A comparison of weather data and the electric bills will show that the expenditures were unpredicted, but appropriate. On the other hand, suppose the department planned on spending money on urgent replacements for turnout gear. Six months into the year, the money is still in the account. Managers should be asking the personnel responsible for this purchase some questions about their "urgent" procurement needs.

You might note that the *controlling* function of management is necessary as the budget year passes. Suppose that the extreme cold weather added significantly to the total cost of electricity and heating fuel. With the cold weather, it is likely that during this same season there was an increased amount of alarm activity further adding to departmental costs. These added costs may not be apparent right away, but will show up in time. These obligations may require some adjustments to the budget to keep the department operating as the end of the year approaches. Purchases may be deferred. Travel and training may be cancelled. Other discretionary costs may be delayed or eliminated to keep the department in business.

program budget
the expenses and possible income related to the delivery of a specific program within an organization

Many organizations use a **program budget.** A program budget gathers expenses that support a particular program. For example, if the department provides EMS services, a program budget would identify all of the costs associated with the EMS program.

The Role of the Company Officer in Budgeting The level of participation of the company officer in budgeting varies. Some departments control budget planning and administration at the very top whereas other departments maintain control at the top, but keep everyone informed through a variety of timely reports. This approach allows everyone to at least see what is going on and have some understanding of why thing are or are not happening.

As the control of the budget is more decentralized we see that the company officer has increasing involvement with the procurement process. Many departments allow the company officer to identify needs for the coming year. These are forwarded up the chain of command and then ranked in some order of priority. As money becomes available, procurement is authorized.

SAVING MONEY HELPS EVERYONE

Even where traditional budget controls are in place, there should be incentives for saving money. Turning out lights that are not being used, setting the thermostat at a reasonable level, easing off the "gas" pedal, and shutting down engines that may idle for long periods may seem like small gestures, but collectively, they can have a significant impact on the department's finances. These small savings can be converted into purchases that are really needed.

For routine supplies, many departments allow the company officer to draw supplies against an account number. Some departments take care of this supply activity by providing a stocking service. Each fire station is checked once a week for needed supplies by a contractor. The contractor then stocks the needed items with no effort from the station personnel.

If the department makes it a practice to look at its spending activities regularly, it can determine if the spending is following the planned budget. Deviation reports, monthly position reports, and other methods are available to help managers keep track of the budget.

When the expenditures are racing ahead of the budget plan, it may become necessary to slow things down a bit through procurement delays and curtailment of travel and training.

Empowering the Company Officer with Money　Some departments have taken the rather radical approach of empowering the company officer with funds, based on historical data, to run their station. If the officer can plan and manage well, he will likely be able to save some funds to buy discretionary items that the station needs. On the other hand if he makes no effort to reduce costs, it is unlikely that the station will see any rewards from the process.

GOOD FINANCIAL MANAGEMENT

The company officer should:

- Keep track of the station's needs
- Identify items that would improve efficiency
- Involve everyone in the budget process
- Talk with other company officers to learn how to effectively manage the station's resources

---

Managing Time Of the three resources on this short list, time may be the most precious for everyone. At work, some employees are content to come in, put in their hours, and go home. Given their choice, they watch television or sleep. They occasionally eat. They are wasting their life! What they do on their days off is their business. What they do when they are on duty is your business. Wasting the day away may have been okay at one time in the fire service, but it is no longer acceptable. Wasting time is clearly unacceptable for a company officer.

PARKINSON'S LAW

Have you ever noticed that activities take about as long as the time allowed? This is called Parkinson's Law, a concept named for C. Northcote Parkinson and first published in 1957. According to Parkinson, work expands to fill the time available for its completion.

You should manage your time just as you manage any other resource. You should find new ways of doing things so that you are not just putting in your time, but you are making good things happen, all the time.

A lot of attention has been given to time management in recent years. The popularity of this topic suggests that many people have problems managing their time. Like management in general, time management starts by setting goals and objectives. In this case we are probably setting personal goals and objectives.

Next comes personal planning and scheduling. This activity is more than just making a list of things that need to be done. Good time management correlates the management of our time with our previously set goals and objectives. Planning allows us to prepare for events, rather than react to them. When you are in a reaction mode, you are inefficient and unproductive.

TIME MANAGEMENT

Plan

Delegate as appropriate

Establish goals for yourself

Be organized and reduce clutter

Make a schedule and stick to it

Fit other things around these events

Identify and work to eliminate time wasters

Give first attention to difficult and complex tasks

Announce meetings, start them on time, and have an agenda

Some people seem to enjoy this mode. They operate in a crisis mode, and run from one crisis to another, never being able to break out of the rat race and get ahead of the events of their life. People who wait until the last moment to do their Christmas shopping or prepare their income tax reports are good examples. Most would benefit on focusing their time and energy on fewer things and concentrate on doing them well.

Of course, there are many things in your life that you cannot control. In the fire station, you should be available to your personnel, to your boss, and to the public. You also need to be available to respond to alarms. But there is still a lot of time you can manage. Set up a program that makes the day become a routine, and where your precious time is used effectively (see Figure 5-4). Here are several suggestions:

- If meetings are needed, have them at a regular time and have an agenda. Make the meeting brief but effective.

- If you want time to work in the office, let your people know when that is. Ask your people not to disturb you unless they have an urgent need.

- Avoid putting things off. Taking time to decide to put something off simply wastes time and gets nothing done.

- Distinguish between *important* and *urgent* matters. Important things need to be done, but not right away. We know about most important events well in advance of any deadline. Most important events take time to plan and execute. A training program for the coming year is an important event.

- Urgent events usually need to be done right away. Some urgent events are unexpected. You cannot anticipate emergency responses; you have to deal with them when they arise. Likewise a call from your supervisor asking for some information that is needed right away is an urgent event. A call from home that indicates that your spouse or child is seriously ill is also an urgent event.

SAVE TIME BY PLANNING

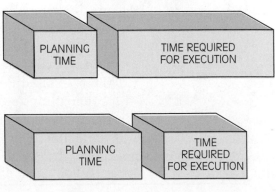

Figure 5-4 *Good planning will save time in the long run.*

People allow important events to become urgent events through the lack of good time management techniques. Putting off the work on the training program until the last moment makes it become an urgent event. Avoid urgent events as much as possible. Your goal should be to manage time, not let time manage you.

TEN PROVEN TIME EATERS

1. *Lack of personal goals and objectives.* Where are you headed?

2. *Lack of planning.* Set priorities, set a schedule to meet your goals, and follow through on your priorities. Manage your time; do not let time manage you.

3. *Procrastination* is putting off things that need to be done. It is poor time management. Making the decision to put off the activity takes time. It is often more effective to just bite the bullet and do it.

4. *Reacting to urgent events*, often the result of procrastination. These events force you to rearrange your schedule to meet the need of the organization or other individuals.

5. *Telephone interruptions.* They are frequently unwanted and frequently unnecessary. Learn to conduct business in a polite, businesslike way and get off the phone.

6. *Drop in visitors.* Same comments as for the phone item above. Be polite, do your business, and move along.

7. *Trying to do too much yourself.* Learn to focus on what is important and learn to delegate what you can.

8. *Ineffective delegation.* When delegating, learn the steps for doing it effectively.

9. *Personal disorganization.* In its milder forms disorganization presents problems due to scheduling conflicts, the inability to find things when needed, and so forth. In a more serious form, it means that you have not set goals and objectives and that you lack focus.

10. *Inability to say "no."* For some, this item may be the hardest on the list. Tie your time to your goals and focus on priorities. If the request supports your goals, go for it. If it does not, politely say that you are sorry, but you are unable to take on that activity. Obviously this may not set well with your supervisor, but even here, supervisors sometimes can overload your plate. Show your supervisor a list of what you are working on and ask for help in setting priorities. He will give you the guidance that you need and may also find (possibly to his surprise) that you really are overloaded.

Protect your time by setting goals and objectives. Starting with goals and objectives helps you decide what really must be done by you, what can be delegated to others, and what can be omitted. Without a goal, you will not know your destination.

Time management usually requires some method of record keeping, usually accomplished by using some sort of planner. Many planning tools are available for computer applications, but using some kind of portable planner (see Figure 5-5) permits you to have it with you all the time. Whatever form is used, it is important to keep the planner up to date. Use it to help set priorities and to identify items that can be delegated to others.

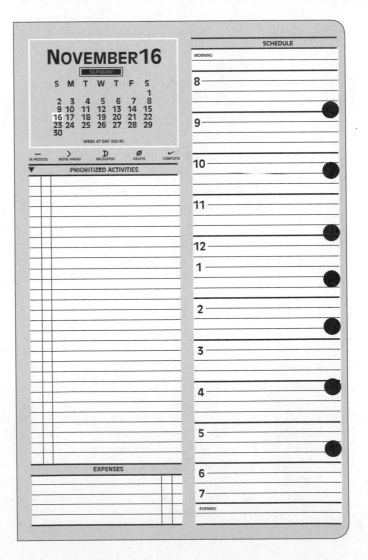

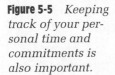

Figure 5-5 *Keeping track of your personal time and commitments is also important.*

Of those items that remain on your list, number them to identify their priority. Try to avoid the temptation to clear up the little things first. Do the important things first. The little things can be fitted in later or held over until another day.

Planning your time helps you to control your time. When you are managing your time, you are effective and productive (see Figure 5-6). When you let time control you, you will become frustrated and stressed as work piles up around you.

Planning can be done at the start of the day and at the end of the day. At the start of the day, your mind is fresh and you probably have a pretty good idea of what the day will bring. On the other hand, planning at the end of the day allows you to focus on the accomplishments of the day and helps you identify what remains to be accomplished. A listing of these items is the starting point for the next day of work. An advantage of this practice is that you and your subordinates can subconsciously think about the work, even while you are off duty.

Give attention to the vital few tasks that are really important (see Figure 5-7). Research has shown that we often spend inappropriate amounts of time dealing with trivial issues, thus saving little of our time for the really important items. Give priority to the important management tasks.

Most fire officers have come up through the ranks, and some spend an inappropriate amount of the time doing firefighter stuff rather than managing firefighter activities (Figure 5-8). Many feel more comfortable doing the job themselves. They are proficient (that is how they got promoted) and enjoy these activities. These officers may also have some anxieties about delegating. They

Figure 5-6 *Effective time management increases productivity.*

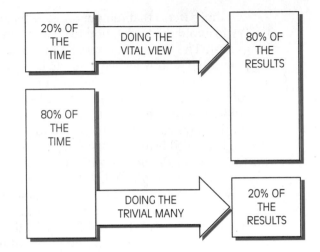

Figure 5-7 *Effective management involves giving appropriate attention to the vital few tasks that are really important.*

Figure 5-8 *Planning the day involves all affected personnel.*

may have some concern about their work being evaluated and are unwilling to let that evaluation rest on the accomplishments of a subordinate. And many would rather work with their hands than their minds. As we indicated earlier, the management functions, such as planning and organizing, require considerable mental effort.

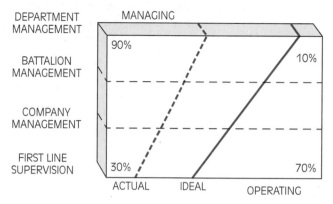

GIVE PRIORITY TO MANAGEMENT FUNCTIONS

Figure 5-9   *Give proper attention to management functions.*

As you rise through the ranks you will find that you will become increasingly concerned with management activities. At the lower levels in our organization, technical skills are important. In the middle ranks, interpersonal skills are critical. In the upper ranks, management skills become most important (see Figure 5-9).

We defer commanding and coordinating for now. That leaves us with controlling. When controlling, you are looking for problems. Solving problems is one of the manager's jobs.

SOLVING PROBLEMS

From time to time you will face problems that will require a bit of thinking for their solution. When you have a problem, you should follow a proven, logical approach to find the solution. The first step may be simply recognizing that you have a problem. Many managers do not see problems occurring or, if they do, they fail to take any corrective action. Here are the generally accepted steps for solving problems:

1. *Accurately define the problem.* That may seem simple, but in many cases it the most difficult step. And in many cases, it is a critical step, for if you do not correctly identify the problem, the rest of the process is doomed.

2. *Gather information.* Review the laws, policy statements, and regulations from federal, state, and local sources that may address the issue.

3. *Analyze the information.* See if you have everything you need.

4. *Look for alternative solutions* and list all possible solutions. At this point in the process, you want to be open minded. Invite others to offer ideas.

5. *Select one or more of the best solutions.* At this point you want to start ranking the solutions in terms of expected outcomes. Some solutions may cost too much or take too long. There are always compromises. If the answer were easy, you would not be in this process.

6. *Take action.* Usually at this point you must get approval to implement your solution. You have done all the work—identified the problem, gathered data, considered alternative solutions, and listed one or more solutions you are recommending. This information is usually conveyed to the supervisor in a short memo. Your supervisor should be able to quickly review what you have done and give approval.

7. *Monitor the results.* This step is also important but often overlooked. You want to be sure that you follow the process and see what happens. Most changes bring about some negative consequences. You want to be sure that the good consequences outweigh any bad items you introduce.

8. *Take corrective action if necessary.* At this point you may have to reboot the process and start all over.

■ **Note**

When the answer to the problem comes too easily, make sure that you correctly identified the real problem and that you found the correct solution.

Notice how the five management functions show up here: In steps 1 and 2 you are *planning,* in step 3 you are *organizing* your information, in step 6 you are *commanding* and *coordinating,* and in steps 7 and 8 you are *controlling.*

When the answer to the problem comes too easily, make sure that you correctly identified the real problem and that you found the correct solution. When you solve your problem, you may have to change a procedure or policy within your organization. Changes should be carefully managed.

MANAGING CHANGE

As a manager, you should make change a requirement in your organization. Identify the desired outcomes, sell the benefits, involve your people, and reward the participants. Some individuals and organizations deal with change quite well. They carefully analyze their problems, rapidly adapt new ideas, and go on with business. Others tend to be more resistive. In these cases, change often passes them by.

■ **Note**

With change, conflict may become more apparent.

Change breeds resistance. With change, conflict may become more apparent. We should convert the energy of conflict into positive values: Recognize that competition is healthy and that we can use the results of the conflict to further positive outcomes. With conflict we tend to take a position, and as a result, there are winners and losers.

Change that is easily accomplished, or has the highest visibility, or that makes you look good to your boss, may not be what is really needed for the organization.

We discuss how to introduce change in Chapter 7.

TECHNOLOGY ADAPTATION LIFE CYCLE

When change is ongoing, some accept it more rapidly than others. In today's fast-paced world, change often comes with the development of new technology. Employees adapt to new technology in various ways (see Figure 5-10). These individuals can usually be identified and placed into one of five categories:

Techies or Innovators (5%). This small minority is the first to the store to buy any new product. New technology is their main interest. They like the challenge of being in the forefront. Their opinions are important as they often influence those who follow.

Visionaries or Early Adopters (10%). Like the pioneers before them, these individuals are ahead of the pack. They are interested in using technology to improve productivity. They are more challenging and more demanding of new technology for they are looking at the results, not the means.

Pragmatists (35%). Pragmatists are not pioneers. They are pushed into the adoption of technology by the need to increase productivity, rather than being pulled by technology itself. They need support, understanding, and interaction with peers to survive.

Conservatives (35%). Conservatives are not comfortable with technology. Tradition is strong with them, and they join only when there is no real alternative. When they do adapt to new technology, they tend to adopt popular concepts and proven products.

Skeptics (15%). Skeptics resist new technology. They will cost the organization more than the results are worth. Best leave them alone.

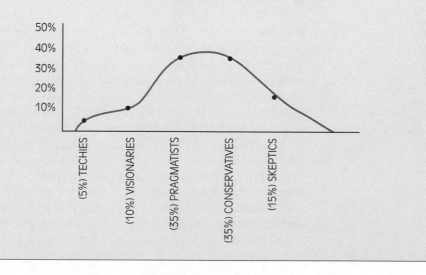

Figure 5-10   *The Technology Adaptation Curve indicates an approximate distribution of employees' ability to deal with the changes brought by new technology.*

MANAGING CUSTOMER SERVICE

You work in a unique environment. Your workplace may be the middle of a busy highway or someone's bedroom. You are likely to be called to work where fire is burning or chemicals are leaking. Wherever you go, you have been called because people are having a problem. They call us when something hurts or is broken. They call us when something is burning or leaking. These people are our customers.

Unlike some businesses, we do not have to look for customers; they come to us. But just like a business, our job is to serve them. Not just serve them but help them to the very best of our ability, every time, every day! This is what customer service is all about.

Three Major Components of Customer Service

First: Always be nice. Treat people with respect and consideration. Show patience and consideration, even if it is three o'clock in the morning. Your customers may be impressed with the rapid arrival of your big fire truck, but what really matters, and what they will remember the longest is what you do after you get there. Fire trucks do not treat medical problems or put out fires; people do. People also show compassion. Remember to treat the customer, every customer, the way you would want your mother to be treated. Provide the best service, and show compassion.

Second: Regard everyone as a customer. Many times the problem is straightforward and simple for us, but for the victim's family and friends, things are very complicated. Factor these folks into the customer equation as well. What you do and say has a lasting impression on everyone present. Regardless of whether you do well, or not so well, they will tell others what happened. Which story would you want them to tell?

There are others who wander into our workplace. Most of what you do attracts spectators. While you must control them and keep them out of harm's way, you should be aware that they have a natural and genuine concern about what is going on. You, as a provider, have the advantage of a front row seat at every show. But that does not mean that they cannot stay and watch from a safe distance. Good Samaritans also stop to help and must sometimes be relieved with an informal and polite transfer of command. Imagine finding that the apparently uninformed foreign-born female civilian who is helping at the scene of an accident involving a child with a broken leg is actually an orthopedic trauma surgeon.

Our customers are everywhere. They walk into fire stations looking for directions to the post office across the street, or directions to a street that is located across town. They are school kids and scouts who come to visit, and high school students working on a science project. Firefighters are heroes to the citizens, but do not forget all those folks behind the scenes who seldom get the glory—the mechanics, payroll clerks, and dozens of others who make our job a little easier. They are customers too.

Part of your customer service is how you *look*. Uniforms are a big deal for most members of the fire service. They indicate pride and identity. They connect us to our local government and the long and glorious tradition of the fire service. The uniform should be complete, neat, and worn properly.

Your vehicles are another highly visible clue that you are out and about (see Figure 5-11). They should be overmaintained with a personal and professional labor of love. The tools you use are also impressive. They range from primitive to space-age, but they all have a function, and all should work when needed. Anyone watching you will be quickly able to tell if your vehicle, your tools, and yes, even you, are sharp and ready for the task at hand. A well-maintained crew and a well-maintained vehicle send a positive action-packed message to the world: You are ready, willing, and able to handle your customer's needs.

Third: Constantly raise the bar, raise the standard, and look for ways to improve, to offer more and better service to your customers. You should be constantly working to improve yourself and your organization (see Figure 5-12). We have to do that just to keep up with the not-so-perfect but constantly changing world around us. Remember, you want your customers to be *delighted* with your work.

Figure 5-11 *Your vehicles are another highly visible clue that you are out and about. Over-maintenance should be performed as a personal and professional labor of love.*

Figure 5-12 *The citizens paid for this equipment. They let you manage it for them. Show pride in the way it is maintained and operated. Photo courtesy of Fairfax County Fire and Rescue Department.*

ASSESS THE NEEDS OF YOUR CUSTOMERS

- Seek feedback from customers about their needs.
- Seek feedback from customers about their degree of satisfaction.
- Define requirements of the workplace to meet these needs.
- Establish goals.
- Define a management philosophy.
- Develop a self-directed team.
- Provide feedback to the team regarding performance.
- Incorporate new ideas and technologies.
- Seek to make continuing improvements to the process and the product.
- Fine tune the mission.

BUILDING A MANAGEMENT TEAM

The following are characteristics of good management in a committee or group activity, including your company:[3]

1. Provide an informal, comfortable, relaxed environment, a working atmosphere in which people are involved and interested (see Figure 5-13).

Figure 5-13   *Successful company officers create an environment where there is open discussion and agreement by consensus.*

2. Provide an environment that encourages discussion in which virtually everyone participates.

3. Provide an environment in which the objective is well understood and accepted by the members.

4. Provide an environment in which the members listen to each other, in which the discussion does not jump from one idea to another. Every idea is given a hearing and people are not afraid of being thought foolish by putting forth a creative idea, even if it seems extreme.

5. Provide an environment in which disagreement is allowed. The group is comfortable with this concept and shows no signs of having to avoid conflict.

6. Provide an environment in which agreements are not suppressed or overridden by premature group action. On the other hand, there is no tyranny in the minority. Individuals who disagree do not appear to be trying to dominate the group or to express hostility. Their disagreement is an expression of a genuine difference of opinion, and they can expect a hearing in order that a solution may be found.

7. Provide an environment in which most decisions are reached by a kind of consensus, in which it is clear that everybody is in general agreement and willing to go along.

8. Provide an environment in which criticism is allowed but in which there is little evidence of personal attack, either openly or in a hidden fashion. The criticism has a constructive flavor in that it is oriented toward removing an obstacle that faces the group and prevents it from getting the job done.

9. Provide an environment in which people are free to express their feelings as well as their ideas on a problem and on the group's operation.

10. Provide an environment in which clear assignments are made and accepted.

11. As the leader, provide an environment in which you lead but do not dominate. In fact, allow the leadership to shift from time to time, depending on the circumstances. Different members, because of their knowledge or experience, are in a position at various times to act as resources for the group. The members utilize themselves in this fashion and they occupy leadership roles while they are thus being used.

To be a good manager, help make these conditions happen at your workplace.

Summary

This chapter contains practical tools to assist you in managing your resources. Learn to use your resources well, especially your own time. As you get older, you will find that it is the most precious resource you have.

Review Questions

1. What is a mission statement?
2. What is meant by labor relations?
3. What are some of the typical issues covered by union contracts?
4. What is a budget?
5. What is the difference between a capital budget and an operating budget?
6. How can company officers improve the use of the department's financial resources?
7. How can company officers improve the use of the department's human resources?
8. How can company officers improve the use of the department's time?
9. How can company officers improve the use of their personal time?
10. What is the company officer's role in managing customer service?

Additional Reading

Carlson, Richard, *Don't Sweat the Small Stuff . . . and It's All Small Stuff* (New York: Hyperion, 1997).

Cascio, Wayne F., *Managing Human Resources,* Third Edition, (New York: McGraw-Hill, 1992).

Covey, Stephen R., *First Things First* (New York: Simon and Schuster, 1995).

McGregor, Douglas, *The Human Side of Enterprise* (New York: McGraw-Hill, 1960).

Paulsgrove, Robin, "Fire Department Administration and Operations," *Fire Protection Handbook,* Eighteenth Edition (Quincy, MA: National Fire Protection Association, 1997).

Peters, Tom, *Liberation Management* (New York: Alfred A. Knopf, 1992).

Ritvo, Roger A., and others, *Managing in the Age of Change* (Burr Ridge, Illinois: Irwin Professional Publishing, 1995).

Notes

1. Unions are discussed here for several reasons, even though we realize that not all firefighters are union members, or that they even benefit from union representation. This section discusses a process that, while formalized by a labor–management relationship, can bring benefits to the employees of any organization.

2. Unfair labor practices include interference with employees in their right to organize, interference with a labor organization, interference with employees who do not want to participate in union activities, discrimination in employment because of union participation, refusing to bargain in good faith, asking the employer to discriminate against an individual, and featherbedding provisions in the contract.

3. Adapted from "The Managerial Team," in *The Human Side of Enterprise* by Douglas McGregor.

Chapter

6

The Company Officer's Role: Principles of Leadership

Objectives

Upon completion of this chapter, you should be able to describe:

- Leadership.
- The difference between leadership and management.
- The leader's role and responsibilities in the organization.
- Issues facing today's leaders.

INTRODUCTION

We discuss organizations in Chapter 3. In Chapters 4 and 5 we discuss management in organizations. We now look at the human side of our organizations, the people who work in the organization.

To be effective, the people in those organizations need leadership. Leadership focuses on people—people working together in organizations. In this chapter, we look at some principles of leadership. In Chapter 7, we look at the applications of these principles in your workplace.

ORGANIZATIONS AND GROUPS

organization

a group of people working together to accomplish a task

An **organization** is a group of people working together to accomplish a task. There are many types of organizations. Fire departments are formal organizations. A fire department bowling league represents an informal organization. Both meet the definition: a group of people working together for a common purpose. In spite of the varied nature of organizations, they share several common characteristics:

All organizations have a purpose. The purpose may be to make a profit, protect the environment, provide an essential service to the community, or just have fun.

Second, all organizations are made up of people. Although a purpose statement is essential to provide the organization with goals and objectives, it takes *people* to make these things happen.

Third, these people will arrange themselves into some type of organizational structure. They will define the duties and responsibilities of the organization's members and establish rules and procedures.

By their very nature, groups represent people who share common interests and goals. They may share a vision of what their group can do. For many groups, there is also a sense of permanence, a desire to see that the group flourishes and survives. As a result of these common interests and visions, the members work together to satisfy these interests. These shared values are what bind the members together as a group.

As we have already indicated, most of us like to be affiliated with groups. The members of a fire department are a group, and they are part of at least one group at work. There are also informal groups at work, groups of individuals who share common interests. Their interests may be bowling, softball, golf, fishing, photography, or other pursuits. Even here, we see some variation in how groups work. Softball requires a group just to field a team, and at least two teams are needed to play a game. Clearly, there is organizational structure here, for the team members must be able to play the various positions and must be able to work together to be successful. On the other hand, individuals in a group that goes fishing together may be more autonomous and unstructured. In these informal situations, we have the right to select what we want to do.

We may also belong to groups away from work. We may be active in a church, civic association, or neighborhood resident association. We may be involved with activities that support the development of our children, such as scouting or soccer leagues. All of these are groups. Within all groups there are positions to be filled. Someone has to be the leader. That person may be appointed or elected. In most informal groups, that person may be the one who comes forward with an idea. The group accepts the idea and responds to the requests for help.

WE ALL BELONG TO INFORMAL GROUPS

Suppose that one of your neighbors suggests a party. If the weather is nice it can be held outdoors. Someone starts with an idea, others add their ideas and agree to participate. The group comes to a consensus on the time and place that works best. Everyone agrees to make great food and make purchases to support the gathering. The weather cooperates and everyone has fun.

Is it work? Yes, of course it is work. Does someone have to take the lead? Yes, at least initially. Will there be some who do not participate? Of course. But for most, the benefits far outweigh any efforts. All who attend go home feeling great and agreeing to do this again next year.

Who elected the leaders? Who made the rules? Clearly they were there, but they were very informal. All agreed to the idea and in so doing accepted some responsibility to the others for the overall success of the event. A group of people with a purpose. In this case, the purpose was just to have fun.

WHAT IS LEADERSHIP?

leadership
the personal actions of managers and supervisors to get subordinates to carry out certain actions

What is leadership? **Leadership** is defined as the personal actions needed by managers to get employees to carry out certain activities. Leadership is achieving the organization's goals through others. Simply put, leadership is getting other people to do what you want done. Management, leadership, supervision, and administration; we hear these words often used and we may confuse them. Indeed, they are often used interchangeably.

Most management texts contain a section on leadership. Leadership is one's influence over others. This definition infers that leadership is a management tool. If you think of the management of the resources of the organization, you must include the human resources. Managing human resources deals with personnel administration, the administrative activities associated with staffing an organi-

■ Note
Leadership is one's influence over others.

zation. These activities include hiring, promoting, and so forth, which are usually the job of the human resource or personnel department.

For the supervisors in the organization, managing human resources is more about leadership, the human relationships in the organization. Effective leadership deals with changing the personal conduct of others. Leadership skills are based on the feelings and attitudes that have grown out of both the relationship with a subordinate and the sum of all of the experiences that have occurred during the supervisor's life.

> President Harry Truman said, "A leader is a (person) who has the ability to get other people to do what they don't want to do, and like it."

■ Note
We let our leaders use their experiences to attempt to control our actions.

We let our leaders use their experiences to attempt to control our actions. How much tolerance we have for this control over us is determined both by our personality and the personality of the leader. Some are more tolerant than others. We have all known people who were tolerant of their organizational situation, and who seemed to stick around, regardless of the conditions. We have also met people who frequently changed jobs for one reason or another. In some cases, they may have lacked the tolerance needed to survive in a difficult work environment.

Several things generally help make the work environment a better place:

- Sound organizational objectives
- Clear policies and guidelines
- Consistent management
- Clear definition of duties
- Open lines of communications
- Individuals well matched with jobs
- Recognition of good work

We have discussed several of these important issues in previous chapters. Where these characteristics are present, it is easier to be a good leader. When employees can see their place in the overall organization, when they are given information and allowed to contribute their ideas, and when their services are appreciated, they will be satisfied and productive. Textbooks, magazine articles, and professional development seminars all address the positive values of these important concepts. Yet we still see organizations where these basic qualities are missing. In these cases, effective leadership is more challenging. Many of the subordinates become frustrated and some of the leaders do too. The organization is likely to be operating at less than full potential.

When poor work conditions occur in a fire department, everyone loses. Most firefighters join the fire service because of a love for the job. Most look for opportunities for professional growth and for serving their community; they like working with others and they are generally willing to take on any reasonable task to help the citizens.

When the essential qualities of effective organization listed previously are missing, these employees become disillusioned. Some employees quit, while others just hang around, doing as little as they can. In these cases, everyone loses and the department is probably not really serving the community as well as it could.

When we looked at simple organizational structure in Chapter 3, represented by a triangle (as shown in Figure 6-1), we saw that there were four major categories of employees: employees, supervisors, middle managers, and top managers. Applying these generic labels to the fire service, we usually call these same layers firefighters, company officers, senior staff, and the fire chief. In this chapter we are focusing on the supervisory role of the company officer.

Company officers supervise firefighters. Because they are the first supervisory rank in the organization, and because they usually do not supervise other managers, we call them **first-level supervisors.** The leadership component of the supervisor's job is a large part of the overall responsibility.

The perception of the supervisor's job has changed over the last few years. Before World War II, and even for a time afterward, the supervisor's role was simply to see that the work got done. This was where management interfaced with the employees. The supervisor had the authority to define the job, monitor the workers' performance, discipline those who needed it, and fire those who did not respond.

Today, the role of the supervisor is much expanded. You might think of some other titles that help define the leader's job. Today's team leader is also a coach, foreman, and supervisor. Regardless of the exact title or the perception it implies,

■ Note
When the essential qualities of effective organization are missing, employees become disillusioned.

first-level supervisors
the first supervisory rank in an organization

■ Note
Today's team leader is also a coach, foreman, and supervisor.

Figure 6-1 *Company officers represent the first supervisory rank in the organizational structure of the fire service.*

CHIEF

SENIOR STAFF

COMPANY OFFICER

FIREFIGHTERS

■ **Note**
An important part of
the leader's job is to
develop and maintain
the organization's
human resources to
their fullest potential.

■ **Note**
Increasing services
while holding the line
on budgets means that
fire departments must
become more efficient.

the job is to lead the organization's human resources. An important part of that job is to develop and maintain the organization's human resources to their fullest potential. The role of the supervisor is important today in any organization and it appears that the importance of the supervisor's role will continue to grow.

Increasing Responsibilities for the Company Officer

In many communities, the cost of providing government services is under close scrutiny. At the same time, fire departments are expected to provide increasing services. Increasing services while holding the line on budgets means that fire departments must become more efficient. In organizations of every kind, part of this cost reduction is being accomplished through the streamlining of the organization. In particular, middle management is a common target. For the fire service, this means that there are fewer layers of supervision between the fire chief and the station offers. More responsibility for the day-to-day operation of the organization is being passed to the company officer.

For many fire service organizations, the greatest opportunities for further improvements in efficiency will come at the company level. While the organization is getting smaller, the job at the company level is becoming more complicated. We just noted that more work is being passed down to the company officers and at the same time, most fire departments are responding to an increasing variety of emergency situations. Company officers must personally remain proficient in the skills needed to respond to the myriad of emergency situations that can be expected, and at the same time they must see that their personnel are proficient and ready to perform their respective duties (see Figure 6-2).

Figure 6-2 *Company officers are expected to provide leadership in the fire station as well as at the scene of emergency events.*

In addition to the duties required at the emergency scene, many other supervisory tasks crowd the day. Preplanning, company inspections, and physical training all compete for the company's time. Other issues, such as diversity training and improved customer service beg for attention. All of these require good leadership.

Where Do We Get Good Leaders?

It has been said that there are only two positions in a fire department to which outsiders can apply. One is the fire chief, and the other is as an entry-level firefighter. With few exceptions, all other positions are filled by promoting people from within the organization.

There are many benefits of such a promotion process. Certainly, the company officer needs to understand the firefighter's job, and will even be a firefighter during certain emergency activities. Company officers, by the nature of their experience, also know the organization, understand the rules, and in most cases know the people with whom they work.

We usually promote our personnel based on some combination of past performance, test scores, and evidence of relevant training and education. But new supervisors soon find that their technical competence does not mean as much at the supervisory level; they find greater challenges in people issues, administrative duties, and the responsibility for managing others.

MODELING HUMAN BEHAVIOR

To be a good leader, you should understand a little about human behavior. Most of us in the fire service have had thousands of hours of training on the technical aspects of the job. You have been trained in fire suppression, emergency medical procedures, and how to deal with hazardous materials. Most fire companies spend about 10% of their time dealing with emergency activities. They spend most of the rest of the time they are together, relating together as humans. Few of us have had much formal training in human relations. Most of us would work better together if we could better understand how others react in various situations.

Working together is all about group behavior. Studying group behavior is a part of the science of sociology. Sociology includes the study of groups ranging from families and small workplace groups to entire communities and ethnic groups. We look briefly at several of the significant contributions to this field.

It is hard take a photograph of human behavior. It would be even harder to show a picture of an attitude. Yet, we all have some concept of these terms. To help us understand these ideas, many of the writers on human behavior have conceived **models** to represent some aspect of human behavior. A diagram or chart usually represents these models.

Models are used to represent real things. When we model human behavior we attempt to represent real human behavior. We use diagrams or drawings to

■ Note
Most of us would work better together if we could better understand how others react in various situations.

model
representation or
example of something

help convey our thoughts in books. Unfortunately, the models do not always give a complete or precise picture.

Using models of human behavior helps us understand people. We do not need to be behavioral psychologists to realize that not all people are alike. But we can all benefit from the available information that enables us to better understand people. Understanding helps us get along better with others. These basic concepts are important to everyone, especially leaders.

Maslow's Hierarchy of Needs

Maslow's Hierarchy of Needs
a five-tiered representation of human needs developed by Abraham Maslow

We briefly mentioned Maslow in Chapter 4. Maslow conceived a model to represent a hierarchy of human needs. For example, in the model of **Maslow's Hierarchy of Needs** (see Figure 6-3), definite steps represent each level or layer. The following list defines each of Maslow's five needs:

Survival needs are primarily related to personal survival, starting with air, water and food, and rest. In addition, clothing and shelter may be needed to protect us from the elements.

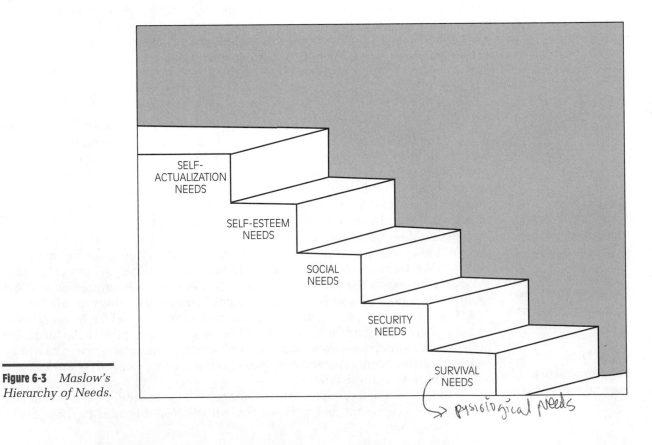

Figure 6-3 *Maslow's Hierarchy of Needs.*

Security needs focus on protection from harm. The concern is for safety for self and family, now, and in the future. In a more modern context, this also means a steady income, for without income, the other security and survival needs are in jeopardy.

Social needs include the desire to have friends, to be a part of a group, and to be accepted and have the respect of others. Social needs should be met in the workplace, as well as away from it.

Self-esteem needs take the desire for respect beyond just a desire; it suggests that we work on it and work for it. We also see here the needs for recognition and status in our group.

Self-actualization needs are used by Maslow to describe the highest level. At this level all of the other levels have been satisfied and we have about reached our full potential.

Life is not as easy as Maslow's model might suggest. One does not suddenly meet the needs on one level and step up to the next level; we move in small increments up and down life's road, gradually moving from one level to another. There are no road signs along the way; in fact, it is unlikely we will know when we have moved from one level to another.

To some, Maslow's concept suggests one-way movement, that we are always moving *up* the steps in the hierarchy. Unfortunately, that is not true, because we move *up and down* the steps of the hierarchy during life's journey. We have good days and we have setbacks: We lose our job, our home is destroyed by fire or flood, our spouse or a child is injured. We are all at motion in this system, sometimes moving so slowly that we cannot perceive the motion. At other times, we move through the system rather quickly, especially when descending.

Although you may think of human behavior theory as being removed from our purpose here, today's fire officers should understand and apply Maslow's ideas. They should understand where their employees are on the scale and be able to help them move up the scale when they are interested in doing so. Company officers should realize that as they attempt to motivate people, they must find motivational tools that are consistent with their employee's needs. Fame and glory are nice, but for the young firefighter who has a sick child or who is struggling to pay the rent, the basic needs are more important.

In the case of Maslow's Hierarchy of Needs, we are looking at a useful model or tool to help us understand and motivate others. As a company officer, you should recognize that your workers have needs on one or more levels and you should try to determine which level they are on. You should also realize that once those needs have been satisfied, your subordinates will have greater needs. In other words, as subordinates' needs change, so do the factors that are effective in motivating them. Understanding human needs helps you better motivate others to improve themselves and the organization.

■ Note
As subordinates' needs change, so do the factors that are effective in motivating them.

Herzberg's Perspective on Motivation

Many consider money to be the ultimate motivator in the workplace. But experience has shown that when you provide employees with a raise, the euphoria lasts about one payday; after that, most people settle back into their same old level of attitude and performance.

Frederick Herzberg described two factors that act on employees at work. He called the first set **hygiene factors**, factors that are needed just to get people to come to work and to prevent *dissatisfaction*. Hygiene factors include company policies and administration, supervision, salary, interpersonal relations, and working conditions. Herzberg suggested that if these are done well, the employees will be satisfied. If they are done poorly or not at all, the employees will be dissatisfied.

hygiene factors
as used by Frederick Herzberg, hygiene factors keep people satisfied with their work environment

Just as good physical hygiene helps prevent dissatisfaction with our health and fitness, so do the hygiene factors at work help prevent worker dissatisfaction. Good hygiene does not cure disease but it does help prevent it. By the same token, Herzberg's hygiene factors do not motivate employees; they prevent employees from having bad feelings about the workplace.

Dr. Herzberg called his second set of factors **motivators**. Motivators encourage employees to rise above the satisfactory level and do excellent work. As he defined them, motivators include achievement, recognition, the work itself, responsibility, and advancement. Keep Herzberg's ideas in mind when you are motivating others.

motivators
factors that are regarded as work incentives such as recognition and the opportunity to achieve personal goals

---

HERZBERG'S MODEL

Motivators

responsibility
job with a purpose
recognition of good work
opportunity for promotion
opportunity for achievement
opportunity for personal growth

---

threshold of satisfaction

Hygiene Factors

good working relationships
considerate supervisors
good working conditions
good pay and benefits
job security

When you really stop to think about it, both Maslow and Herzberg suggest a similar hierarchy of needs. Maslow had five layers; Herzberg has only two. In the case of Herzberg's model, remember that everything below the line will just keep employees from feeling bad. As supervisors, we have to add the items above the line to get positive results.

Needs Theory[1]

There is yet another way of looking at motivational factors. We have already seen that affiliation and recognition show up consistently as motivators. Research has also shown that some individuals like to have some influence over the group as well. So we add status or **power** to our list of motivators as we look at the way we can best motivate others.

Consider persons with high status or power needs. They like to be in charge and to have influence over others. They like the structure of organizations and they like to get work done. Remember these three important characteristics: persons with high power needs like to be in charge, persons with high achievement needs like challenges, and persons with high affiliation needs like to work in groups.

People who have **high-achievement needs** typically take personal responsibility for their efforts, set their own goals, and take on new and demanding challenges. They tend to be creative, and for many with achievement needs, the strongest characteristic may be their need for feedback. These individuals usually set high goals for themselves, and they look for ways to get feedback on their performance.

The third group are those who have **high-affiliation needs**. We have already discussed group dynamics to some extent. These people thrive on it. They desire to belong and be accepted by the group. Although they like to work, they prefer to work as part of a group. In many ways these individuals are on Maslow's third layer—they have a need to belong.

These three characteristics, the need to have status, to achieve, and to affiliate with others, are present in all of us. In many people, one of these needs characteristics may be far more dominant. When that is the case, you may be able to find work that fits their needs well. When you can match their needs with the task or the environment, they will be more than just satisfied, they will be motivated to work well. However, you should realize that you cannot always put people into the roles that they would most like. Frequently the organization's needs must also be considered.

DETERMINING THE BEST LEADERSHIP STYLE FOR MOTIVATING OTHERS

We have been discussing what motivates employees and how understanding their motivations will make us more effective in motivating others. Let us look at other factors that might determine leadership style. These factors are more a function of the leader's needs and feelings, rather than those of the employee.

power
the command or control over others, status

■ Note
Persons with high power needs like to be in charge, persons with high achievement needs like challenges, and persons with high affiliation needs like to work in groups.

high-achievement needs
according to the needs theory of motivation, individuals with high-achievement needs accept challenges and work diligently

high-affiliation needs
according to the needs theory of motivation, individuals with high-affiliation needs desire to be accepted by others

Leaders have various styles based on their own needs. We mentioned in Chapter 4 that according to Douglas McGregor, there were two generally held management styles, determined by the leader's belief about the workers. McGregor's Theory X and Theory Y are called management style but they have a great deal to do with the way you supervise. How you supervise is how you motivate others to do what you want to get done.

While having concern for people is admirable, in the real world, we are all expected to get the job done. Some might see the needs of the people and the need for getting the work done as a conflict. Being nice, as suggested by the Theory Y style of leadership, might not give you the productivity you want. On the other hand, being a taskmaster, as possibly represented by the Theory X style of leader, will not be well accepted in most situations by your employees. It would appear therefore that these two forces, namely a concern for people and a concern for production, are in opposition, or at least in competition with one another, as you determine your own leadership style.

The Managerial Grid

The grid shown in Figure 6-4 represents a field in which the leader's concern for people and production are plotted. The vertical axis represents the leader's concern for people, and the horizontal axis represents the leader's concern for pro-

> ■ **Note**
>
> While having concern for people is admirable, in the real world we are all expected to get the job done.

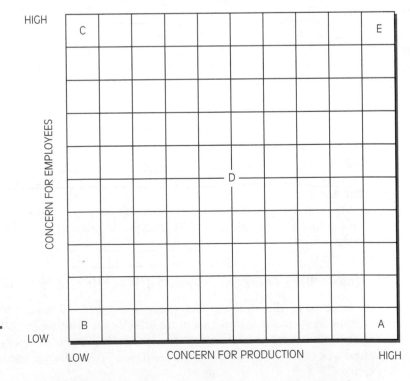

Figure 6-4   *The Management Grid.*

duction. The Managerial Grid was introduced in 1964 by Drs. Robert R. Blake and Jane S. Mounton. Their model of managerial behavior in an organizational setting suggested that morale (the people factor) and productivity are independent of one another. Thus, either one could flourish, or fail, in spite of the other.

There are infinite positions on the grid, but five positions are noted here for purposes of explanation:

A. In the lower right-hand corner we see position A. Position A represents a leader with a passion for production while having a relatively low concern for people. Position A may be similar in some ways to Theory X. Persons having this style of leadership use their power to control people, telling them what to do and how it is to be done.

B. At position B we see a minimum regard for both people and production. Here the leader provides the minimum amount of supervision needed to survive within the system.

C. In this corner the leader has a strong bias for interaction with people, while having a relatively mild concern about production. In some ways, this person is a strong advocate of McGregor's Theory Y.

D. In the center of the grid we have a leadership style that is represented by a middle-of-the-road approach on both scales. While this position may be somewhat better than the person who operates from position A, B, or C, we see that neither people nor production gets full attention and that either can be compromised to accomplish the other.

E. At position E we have the supervisor who has a high regard for both the employees and for production. These leaders use their own talents to integrate all the positive leadership qualities to bring out the full potential of all of their employees.

Understanding Power

The word *power* has been used several times in this chapter. How do you define power in the context of leadership? We said that leadership is achieving the organization's goals through others. To have this influence, the leader must have some sort of power over subordinate employees. Power is the ability to influence others. You get leadership power in several ways.

legitimate power
a recognition of authority derived from the government or other appointing agency

For fire service personnel, the first type of leadership power comes with the badge of the office. Some refer to this as **legitimate power**, the power that is bestowed upon you as an officer in the organization. As an officer you wear insignia that indicates that you are a representative of local government. The name of the governing agency is usually prominently displayed on that insignia. So, some power comes with the position.

reward power
a recognition of authority by virtue of the supervisor's ability to give recognition

With that legitimate power, two additional types of power are implied. The first of these is the power to reward people, or **reward power**. As a supervisor, you have the power to approve requests, recommend individuals for special assignments, and write recommendations and evaluations that will help the employees in your company attain their personal goals.

Reward power is more than giving someone a medal or some time off. Reward power may be nothing more than name recognition, a smile, or an acknowledgment of effort. Reward power can be taking a moment to help someone who is having a bad day. Reward power is not taking your bad day out on someone else.

The other implied power that goes with the job is the power to punish, or **punishment power**. We usually think of punishment power as the authority to administer discipline, although, fortunately, this is not really an issue for most employees.

Far more often we punish people by simply failing to give recognition for a job well done, or withholding information that might be useful. This is not done on purpose, it is just a failure on the part of the supervisor to do things that are important in the job of being a good leader. Legitimate, reward, and punishment power come with the badge.

There are additional forms of power that one *earns*. While the legitimate, reward, and punishment power come with the position, this added power is earned by individuals through their personal actions. It may be influenced by the way they treat people, by their ability to communicate, and by their knowledge of the job. These qualities can be illustrated with terms like charisma and knowledge. Clearly these qualities are much more subjective than the legitimate power evidenced by the gold badge, but they are just as real. Terms such as **identification power** and **expert power** are often used to describe the qualities of role models and knowledgeable individuals.

punishment power
a recognition of authority by virtue of the supervisor's ability to administer punishment

■ **Note**
Legitimate, reward, and punishment power come with the badge.

identification power
a recognition of authority by virtue of the other individual's character or trust

expert power
a recognition of authority by virtue of an individual's skill or knowledge

HOW TO DETERMINE THE LEADERSHIP STYLE THAT IS RIGHT FOR YOU

Effective leaders work at achieving the organization's goals through the efficient labor of others. They strive to do this as well as possible, and at the same time, they are helping subordinates reach their full potential.

But these goals are seldom easily attained in real life. We are frequently faced with the old debate: Which is more important—the task or the people? Sometimes the task is critical. At other times, the people are far more important. Sometimes both are important; sometimes neither requires much attention.

Three factors can help determine your own leadership style: the employee, the leader, and the situation. Let us look at how each of these impacts on a leader's style.

Elements That Determine Leadership Style

The Employee

The employee's experience

The employee's maturity

The employee's motivation

The Leader

The leader's self-confidence

The leader's confidence in the employees

The leader's feeling of security in the organization

The leader's perception of the organization's value system

The Situation

Risk factors involved —— *Value, Time, Size*

Time constraints imposed

Nature of the particular problem

The organizational risk climate

The ability of the individuals to work as a team

■ **Note**

Many leadership styles can be used. Dynamic and effective leaders make their style fit the situation.

directing
controlling a course of action; as a leadership style, it is character-ized by an authoritar-ian style

consulting
seeking advice or getting information from another; as a leadership style it implies that the leader seeks ideas and allows contributions to the decision-making process

supporting
as a leadership style, the supporting process involves open and continuous communications and a sharing in the decision-making process

Many leadership styles can be used. Dynamic and effective leaders make their style fit the situation. Certainly the factors we have been discussing here should be integrated into the process. Four representative styles follow:

(Telling) **Directing.** This style is characterized by lots of direction and mostly one-way communications. The supervisor tells what has to be done, provides direction, and monitors the results. The task gets more attention than the people. The employees have little input into the decision-making process.

(Selling) **Consulting.** Here there is some discussion in which the supervisor seeks ideas, explains the needs and decisions, and "sells" the idea. The supervisor still gives lots of direction but maintains close presence, providing encouragement and reassurance. As the supervisor is available to provide support, both the task and the people get lots of attention.

(Participating) **Supporting.** The leader encourages participation at all levels and shares responsibility for the process. Two-way communications are encouraged on a continuing basis. There is a sharing of the decision-making process. The leader facilitates growth by sharing information and asking questions that will enhance the employee's understanding of the situation. In this case the leader has taken a supporting role, supporting the employees. There is little direction, but the leader continues to provide support, encouragement, and recognition.

(Delegating) **Delegating.** The supervisor essentially turns the management of the task over to subordinates. Direction is limited to setting the goal and defining the parameters. Communication may be limited but when it occurs, it will be a cordial two-way process.

In these four definitions, you see a gradual change from the strong directive-type of leadership behavior to one in which the leader is more of a supporting player (see Figure 6-5). The first type is mostly supervisor-centered leadership and communications are mostly one way. As the process moves toward a more subordinate-centered leadership style, you see two-way communications that are more effective, more freedom of thought and expression for the subordinates, and employees that are more involved with the decisions and the outcome.

Based on Relationship/confidence w/employee

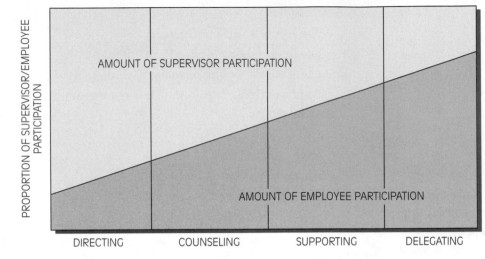

Figure 6-5
Leadership style varies depending upon the amount of employee input allowed.

George S. Patton, Jr. said, "Never tell people how to do things. Tell them *what* to do and they will surprise you with their ingenuity." Many of us tend to be too directive. Ease up a bit and learn how and when to use each of the four leadership styles effectively. During a normal day, there are situations in which each is appropriate. Use the one that is most productive.

GOOD LEADERSHIP IN TODAY'S WORKPLACE

Several major social issues have confronted our society in recent times. These issues deal with diversity and harassment in the workplace. Diversity and harassment in the workplace are generally considered to be management issues, but we consider them here in the context of good leadership. Indeed, the organizational tone for policies regarding these important issues must come from the top of the organization. But this is not a book for fire chiefs nor is it a book about what fire chiefs should do. The focus of this text is the company officer, and in that regard, the leadership issues that company officers face.

The fire service has had significant problems in the area of diversity and harassment. We still see headlines in the trade journals and local papers suggesting that personal prejudice and a lack of good leadership on the part of a few have cast a cloud over many. Although much progress as been made, there is still plenty of evidence to suggest that the fire service needs to continue to train officers and other personnel in dealing effectively with diversity and harassment issues.

Both diversity and harassment deal with personal attitudes toward others. Those attitudes may be the product of a lifetime of family and community values over which we have little control. Regardless of our past culture and present

■ **Note**
"Never tell people how to do things. Tell them what to do and they will surprise you with their ingenuity."
–George S. Patton, Jr.

■ **Note**
The fire service has had significant problems in the area of diversity and harassment.

■ **Note**
Both diversity and harassment deal with personal attitudes toward others.

beliefs, we must realize that within the workplace, laws and regulations determine the legal boundaries our actions. To avoid problems, we should understand these laws and regulations.

Many federal laws prohibit discrimination. Starting with the U. S. Constitution, the Fifth Amendment provides that "No person shall . . . be deprived of life, liberty, or property without due process of law." The Fourteenth Amendment provides that "No state shall . . . deny any persons within its jurisdiction the equal protection of the law."

Given these constitutional guarantees, one has to wonder how these problems have occurred. Some suggest that the government failed to address the sociological problems in the communities and workplace with quick and positive leadership action, and that if it had, these problems would not have happened. Others suggest that employers were not fair in providing an equitable workplace for their employees; that if their action had been more effective, we would not see the headlines in the newspapers nearly every day regarding diversity and harassment issues in our government, our military, and our major corporations.

If we lived in a perfect world, and if all government officials and all employers were good and considerate human beings, these things would not happen. But our world is not perfect and people sometimes fail to understand the problem or the solution. Where good judgment and morality failed, the laws now provide incentives for proper action.

In 1964 the Congress passed the Civil Rights Act. Title VI of the act deals with programs and activities that receive federal funding and prohibits discrimination on the basis of race, color, national origin, or sex. Title VII of the Civil Rights Act deals with employment and prohibits discrimination by employers of the basis of race, color, religion, sex, age, or national origin. The act also authorized the establishment of the **Equal Employment Opportunity Commission (EEOC)** to enforce the act. Of these two, Title VII has the greater impact on our workplace.

Equal Employment Opportunity Commission (EEOC)
federal government agency charged with administering laws related to nondiscrimination on the basis of race, color, religion, sex, age, or national origin

---

EQUAL EMPLOYMENT OPPORTUNITY

A condition in which all employees, and potential employees are treated fairly. Equal employment opportunity is more than a legal right; it is a human right.

Several other significant laws have bolstered the power of the Civil Rights Act. The Equal Pay Act of 1963 prohibits discrimination in compensation on the basis of sex. The Equal Employment Opportunity Act of 1972 strengthened the authority of the Civil Rights Act of 1964, and expanded the power of the EEOC. The Civil Rights Act of 1991 provided that individuals may be *personally liable* (as well as their organizations) for discrimination in the workplace. It authorized jury trials and increased the limits on financial settlements. As a result, we have seen increasingly larger punishments, especially in cases involving sexual harassment.

Finally, the Americans with Disabilities Act (ADA) of 1990 prohibits discrimination on the basis of disabilities, and covers employment, public accommodations, transportation, and telecommunications. The ADA is a complex law that applies to employment issues, our 911 systems, and providing proper access for citizens and employees in public places such as fire stations. Questions regarding local application of the ADA should be referred to higher management levels of your department or local government.

diversity
a quality of being diverse, different, or not all alike

■ **Note**
In the very near future, the white male will be a minority member of the workforce!

Diversity

Our communities are becoming more diversified. Likewise, most government and private business activities are, or at least should be, becoming more diversified as well. Until recently, white males dominated the workplace in nearly every industry, but this is changing, and in the very near future, the white male will be a minority member of the workforce!

Changes are taking place in the fire service as well (see Figure 6-6). For the most part, the fire service has long been dominated by white males. However, that is changing, and women and minorities are entering the fire service in increasing numbers.

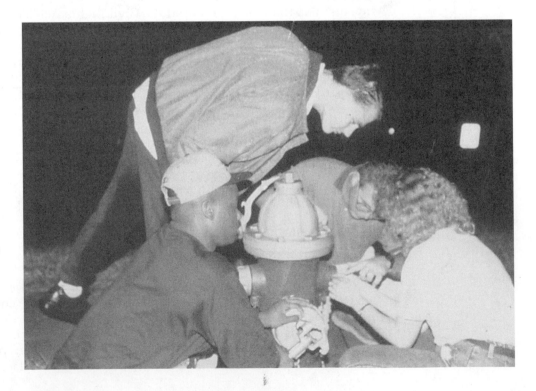

Figure 6-6 *Today's students are tomorrow's firefighters. Tomorrow's firefighters will be more diverse than those in the fire service today.*

Women deserve special mention here. They represent slightly over half the total population, but they represent a very small percentage of the fire service. Most of those who have entered the fire service have proved themselves to be capable of performing all the tasks associated with fire fighting. In spite of some formative barriers that they had to overcome, many of the females who entered the fire service have done well. Some of those pioneers have proved themselves over and over and are serving capably as chiefs and senior officers in large, progressive fire departments.

Think about America's great history. Nearly all of our ancestors were immigrants, and at one time, they too were minorities. Half of them were female. Yet our parents and their parents were able to become successful. Many have given great service in public office, academia, and in industry. Today's generation and tomorrow's generation should have the opportunity to accomplish their personal goals.

Public and private organizations are responding to these changes by seeking ways to hire, train, promote, and retain the best talent available, regardless of the gender, race, or ethnic background. As a result we are seeing a more diverse workforce. Effectively supervising a diverse workforce is what good leadership is all about.

Benefits of Diversity We should all be committed to making our organization the best it can be. We should be looking at the total community and all of its citizens as a source of new members for our organization. As new employees enter our organizations, we must be aware that they may have different values. As standing members of the organization, we should look on these new values as an opportunity, rather than a problem. The new employees may have new perspectives on existing social problems, may be able to speak a second language, and may provide many other benefits that we lack. But all too often we close the door on these opportunities because these employees are different.

■ **Note**

All of us want highly motivated individuals working for us.

All of us would want highly motivated individuals working for us. Opening the doors of the fire station to groups that have been traditionally excluded allows a greater number of interested and qualified persons to apply to our organization. A greater applicant pool means that departments can be more selective about the people they hire. Many large metropolitan fire departments have the luxury of selecting one candidate from 20 or more applicants. If done carefully, the selection process provides highly qualified candidates who are likely to survive the recruit training process and become well-motivated, long-term employees of the organization. Their gender, race, or ethnic background should not be an issue.

Maintaining good working relationships for this increasingly diverse group of employees is another matter. The new employees enter a culture that is traditionally a while male organization. They may be uncomfortable. The more senior members of the fire service have had little experience in dealing with diversity issues, and some do not know how to deal with these diverse groups or how to handle the problems that arise.

For many women and minorities, the path to acceptance and the opportunities for advancement are a little harder than for the traditional employee. In some cases these are merely perceptions, and in some cases, there are real barriers that prevent them from having an equal opportunity.

Many of the organization's rules are unwritten. We must be sure that the same information is presented to all people, regardless of our personal feelings about them. We must be sure that the same information is presented to all of our employees.

How to Supervise Diversity Understanding and effectively managing diversity must start at the top of any organization. But we quickly see that top management can have little impact without the understanding and cooperation of everyone in the organization. We all need to proactively embrace the positive values of diversity and work hard to give all of our subordinates the very best opportunity to grow to their full potential and satisfy their own goals.

Supervisors should take the time to learn about all new subordinates and their personal values. The fact that these values may not be identical to our own should not be considered as a barrier, but rather as an opportunity for learning more about ourselves and our employees.

Supervisors should also be sure that females and minorities are fairly represented and involved with all of the department's activities. Put them on committees. Listen to their opinions with an open mind and give their ideas a fair try. Give recognition and support when appropriate. If these actions are not completely comfortable and natural, supervisors should be honestly assessing their own weaknesses in this area and looking for opportunities to improve their dealings with others. Many of the problems you are facing today are the result of inappropriate actions in the past.

Harassment

To **harass** means to annoy, tease, or torment. Harassment of any kind should not be allowed in the workplace. This ban includes harassment of new employees, or any other efforts to make any individual feel uncomfortable. Where this harassment has sexual overtones, it is a clear violation of federal law. The law deals with both those who harass and those who permit harassing.

Title VII of the Civil Rights Act of 1964 covers sexual harassment in the workplace.[2] The act states that the behavior must be unwelcome sexually oriented conduct of a verbal or physical nature. There are two situations in which sexual harassment generally occurs. The first of these is considered a hostile work environment. A hostile work environment occurs when there are pictures, comments, and other offensive acts that inhibit a worker's performance. A second situation occurs when one asks for sexual favors as a condition of employment, promotion, or transfer. This implies an act as power by one over another, in which a favor is requested in exchange for some personal action. Such harassment is referred to as quid pro quo, a Latin phase meaning "this for that."

We usually think of harassment situations in which there is a male–female relationship between two employees, with the male being the aggressor. In this stereotype male–female harassment situation, the female is often the victim. Because of physical size and organizational position, men are generally thought to be more powerful. Sexual harassment is often about power.

While such circumstances do occur, you should realize that many other gender combinations can occur as well. The roles might be reversed—the female may be the aggressor. And it possible that both parties could be of the same sex. Likewise you should realize that not all of the parties have to be employees. Harassment may come from a visitor, a vendor, a contractor, or a service provider in your workplace. Finally, sexual harassment need not be about sex; it may simply be a general derogatory comment about a person because of his or her gender.

■ **Note**

Supervisors have a special obligation to protect employees from the consequences of sexual harassment.

What to Do When Harassment Occurs Supervisors have a special obligation to protect employees from the consequences of sexual harassment. When an employee comes to you, as a supervisor, with a complaint about sexual harassment, you must take action promptly. Encourage the complainant to provide you with specific information. You may need to help the process by asking open-ended questions like:

- Where did this occur?
- When did this occur?
- Who was involved?
- Were there any witnesses?
- Have you talked to anyone else about this?
- Has this happened before?
- How long has this been happening?
- Did you try to stop (this conduct) on your own?
- What was the reaction?
- What do you want me to do?

Notice especially the last question in the series. In some cases, no corrective action is desired. The complainant may just want you to know that the event occurred, to sort of "get it on the record." Be sure that you follow your organization's policy. Many organizations require at least some notification be made, even in these situations.

In some cases you will have to take some action. Be sure to follow your department's policies here. Some departments direct the company officer to make the initial investigation, while others expressly preclude such action. If the policy directs you to conduct an investigation, you should follow the initial interview by checking with the witnesses, if any. Find out what they might have seen or heard. If talking to witnesses, start by saying that you are investigating a complaint of sexual harassment and that they have been named as witnesses to the act. Do not provide them with specific information regarding either the name of the accused or the accuser; just focus on their observations.

Finally, after you have a good understanding of what has happened, talk to the accused. Again, do not reveal the names of the accuser or any witnesses you have interviewed. Be serious and to the point, but at the same time, be open minded and fair. Here are some guidelines on asking questions:

When talking to witnesses start by saying, "I am investigating a complaint of sexual harassment. I would like to ask you a few questions regarding what you may have seen or heard earlier today. . . ."

When talking to the accused start by saying "I am investigating a complaint of sexual harassment by you. I would like to ask you a few questions regarding your actions (or conversations) this morning. . . ."

This approach makes the process a little easier for both you and the person you are questioning.

In talking to the accused, ask the individual to respond to the allegation(s). In most cases, the accused will acknowledge his or her action, and offer an excuse that the actions were misunderstood. Instruct the person on the seriousness of the accusation and advise him or her regarding what is and is not acceptable workplace conduct. Advise that there are laws and regulations that deal with these matters, and that violations can have serious consequences.

GENERAL COMMENTS REGARDING ANY HARASSMENT ISSUE

- Move quickly.
- Let everyone you talk to understand that you are concerned, that you take these matters seriously, and that you will make every effort to be fair and just.
- Meet in private.
- Make notes of the meetings and safeguard them.
- Provide feedback to the accuser as to what you found.

■ **Note**

Most organizations have established policies for dealing with harassment. Know and follow your organization's policies.

■ **Note**

The supervisor is a key player in the prevention, recognition, and the resolution of any harassment-related problems in the workplace.

The laws provide clear guidelines for organizations to follow in these matters. Most organizations have established policies for dealing with these issues. Know and follow your organization's policies. Supervisors may be personally at risk if they violate the law or their organization's policies in this important area. The supervisor is a key player in the prevention, recognition, and resolution of any harassment-related problems in the workplace.

■ **Note**
Volunteer organizations are expected to provide a fair and equitable workplace for their members and for the public they serve.

Although a general statement elsewhere in this book indicates that the terms "employees" and "subordinates" include volunteer members of the fire and rescue services, for purposes of this text, let us be very clear that the laws regard members of volunteer organizations in much the same way as they do employees and employers. Volunteer organizations are expected to provide a fair and equitable workplace for their members and for the public they serve. Although the threat of liability may be present, the risk of adverse publicity should be an even greater motivator to do the right thing. The loss of the reputation of a volunteer organization could lead to the loss of community support; most volunteer organizations need strong community support to exist.

Good leaders can prevent sexual harassment in their workplace. Take a proactive stance on sexual harassment. Should you become aware of any sexual harassment, take quick, firm action to let people know that the action is inappropriate. Failure to do so could be embarrassing and expensive for both you and your organization.

Summary

One of the most important roles of the company officer is leading others. Leading involves working with others and trying to influence others to accomplish the organization's goals. This can be done in many ways, but when done effectively, productivity is increased and both supervisors and workers seem happier and more satisfied. Part of working with others involves accepting proper organizational behavior in which everyone is treated fairly and with respect. We continue this discussion on leadership and provide some practical tools for effective leadership in Chapter 7.

Review Questions

1. What is a group? What are some of the characteristics of a group?

2. Why do people like to work in groups?

3. What is leadership?

4. What is leadership power?

5. How does Maslow's theory apply to the leader's job in the fire service?

6. Why should the demographics of the fire department mirror those of the community it serves?

7. How does diversity help the fire service better serve its community?

8. Name and briefly describe the principal laws that provide guidelines to encourage diversity in the workplace.

9. Name and briefly describe the federal law that prohibits sexual harassment in the workplace.

10. What action should a company officer take when an employee comes to him or her with an allegation of sexual harassment?

Additional Reading

Carter, Harry R., and Erwin Rausch, *Management in the Fire Service,* Second Edition (Quincy, MA: National Fire Protection Association, 1989).

Conger, Jay A., *Learning to Lead* (San Francisco: Jossey-Bass, 1992).

Covey, Stephen R., *Principle-Centered Leadership* (New York: Simon and Schuster, 1990).

Effective Supervisory Practices, Second Edition (Washington, DC: International City Management Association, 1984).

Herzberg, Frederick, "One More Time: How Do You Motivate Employees?" *Harvard Business Review,* Jan–Feb 1968, pp. 53–62.

Hogan, Lawrence J., *Legal Aspects of the Fire Service* (Frederick, MD: Amlex, Inc., 1995).

Morehead, Gregory, and Ricky W. Griffin, *Organizational Behavior,* Third Edition (Boston: Houghton Mifflin, 1992).

Plunkett, W. Richard, *Supervision,* Eighth Edition (Upper Saddle River, NJ: Prentice Hall, 1996).

Robbins, Stephen P., and David A. DeCenzo, *Supervision Today!* Second Edition (Upper Saddle River, NJ: Prentice Hall, 1998).

Schneid, Thomas D., *Fire and Emergency Law Casebook* (Albany: Delmar Publishers, Inc., 1997).

Notes

1. Not to be confused with "needs" as used earlier by Maslow.

2. The U.S. Code defines sexual harassment as follows: "Unwelcome sexual advances, requests for sexual favors, and other verbal or physical conduct of a sexual nature constitute sexual harassment when (1) submission to such conduct is made either explicitly or implicitly a term or condition of an individual's employment, (2) submission to or rejection of such conduct by an individual is used as the basis for employment decisions affecting such individuals, or (3) such conduct has the purpose or effect of unreasonably interfering with an individual's work performance or creating an intimidating, hostile, or offensive working environment.

Chapter

7

The Company Officer's Role in Leading Others

Objectives

Upon completion of this chapter, you should be able to describe:

■ How to help new subordinates integrate into the organization.

■ How to help all subordinates rise to their full potential.

■ How to understand and reduce conflict in your organization.

■ How to introduce change into your organization.

INTRODUCTION

This chapter continues our discussion on leadership. In Chapter 6 we discussed the basic concepts of leadership. Here we fit them into the reality of the fire service. Remember that the focus of this book is the company officer, the first line supervisor in the fire service.

BEING A SUPERVISOR IS DIFFERENT

■ Note

When people are promoted to supervisors, they quickly find that the work is different and in some ways more difficult than when they were one of the workers.

When people are promoted to supervisors, they quickly find that the work is different and in some ways more difficult than when they were one of the workers. The typical firefighter has developed an impressive array of technical skills, knowledge, and abilities that make him or her quite competent in that role. The new job as a supervisor is quite different.

Administrative duties, understanding the organization's rules and regulations, and supervising others are demanding tasks. Communications skills are vital. As a supervisor you have to communicate with your subordinates. As a company officer you are expected to be able to communicate with the rest of the organization and the public.

Few organizations, including most fire departments, prepare prospective supervisors for the position. Fire departments promote good firefighters and expect them to become good officers. Some fire departments provide excellent training as new supervisors enter the officer ranks, whereas others provide a continuing program of professional development to provide the supervisor, leader, and manager with the leadership tools he or she needs for the increasing responsibilities.

■ Note

Continue the development process: Read every article and take every class on leadership you can.

No one can anticipate every situation that may occur in the role as a new supervisor. The fact that you are reading a book on the topic is a positive step, and if you are using this book as a text in an officer development program, so much the better. But even here, what is learned from the book or in a classroom will only begin the process of personal professional development. You are encouraged to continue the development process: Read every article and take every class on leadership you can.

The selection of new firefighters into your organization and the new firefighter's entry-level training are normally not the province of the company officer. However, once these new firefighters have survived the selection and initial training process, they are usually handed to the company officer. They will likely work in that environment until they too become company officers.

■ Note

For most of those entering the fire service, the person who has the greatest impact on them in their entire career is probably their first company officer.

As supervisors and as company officers, it is your job to develop the new firefighter into a true professional. This is a critical time for both the officer and the new firefighter. For most of those entering the fire service, the person who has the greatest impact on them in their entire career is probably their first company officer. Company officers have an awesome opportunity and responsibility here!

You might look upon new firefighters as simply a resource. Indeed, they are now a part of the company team. At the same time you should look on new firefighters as part of your team and your profession, and help them move along in a continuing process of development, just as you have done. In fact, a significant part of any supervisor's time should be devoted to the development of each of his or her subordinates. The task of developing our subordinates is one of the most challenging tasks a company officer faces, and at the same time is one that can provide great personal satisfaction. Just like watching your children grow, there is nothing more satisfying than watching a subordinate develop to full potential.

BUILDING A RELATIONSHIP WITH YOUR EMPLOYEES

The First Meeting

The first meeting with a new firefighter should define the job, establish positive relationships, and build a friendly atmosphere. While new firefighters are qualified to serve in that position, they may have not have actually worked in this particular environment.

You may have to show them some pretty basic things, like where their locker and bunk are located, where to hang their turnout gear, and where you want them to sit on the apparatus. You will have to pick up where the recruit academy left off in terms of training. Many fire departments have a structured program that allows the new firefighter to review what was covered in recruit school while learning new material that is needed on the job. And the new firefighter should be expected to become familiar with the buildings and streets in the company's first-due response area. Much of this can be done using a senior firefighter as a mentor.

■ **Note**

Many fire departments have a structured program that allows the new firefighter to review what was covered in recruit school while learning new material that is needed on the job.

> ### THINGS THAT YOU CAN DO FOR YOUR SUBORDINATES
>
> - Share information about the job and the organization.
> - Share knowledge about the role of the department and how they fit in.
> - Share knowledge about trends in growth, promotions, and opportunities.
> - Identify alternative career plans and the implications of each.

■ **Note**

The supervisor should establish a clear definition of the job.

The supervisor should establish a clear definition of the job. This seems rather elementary to the established firefighter, but think back to the first days when you were on the job: Wouldn't it have been nice if someone had taken a few minutes a day to explain the job in some detail to you?

When you are talking to a new subordinate, try to avoid information overload. Take just a few minutes a day to cover the needed items, rather than trying to do it all at once. When we get too much information, it is difficult to sort out all that we have heard, and it is certainly difficult to remember all the details, especially when we are entering a new profession. When you are explaining the tasks that need to be done, clearly define the subordinates' role and show them how they fit into the overall organization. Give the subordinate a chance to talk and to ask questions.

■ **Note**
Give the subordinate a chance to talk and to ask questions.

Teaching individual small activities may seem trivial to the seasoned professional, but we all know that proficiency in the little details is what keeps us prepared for the big emergencies. For example, if the new firefighter is responsible for maintaining some piece of equipment on the apparatus, explain what the tool does, and why the maintenance is important, both in terms of being able to use the tool effectively and the personal safety of the firefighter using the tool.

■ **Note**
Set well-defined job standards with new subordinates.

Set well-defined job standards with new subordinates. There should be plenty of information available to help you determine what these job standards are. Your organization should have a detailed position description of the firefighter's duties. That position description may refer to additional requirements, such as those listed in NFPA 1001, *Standard for Firefighter Professional Qualifications*. Make sure that your subordinates understand the way the job *should* be done, which may not necessarily be the way that it is done now.

For most new employees, sorting through all of this information will be a bit overwhelming. As a supervisor, you should help subordinates by assigning some priorities. If appropriate, put some deadlines on the first few items on the priority list to give the new subordinate some specific goals to attain in the first few days and weeks on the job.

■ **Note**
An important part of setting standards should include a regular review of the new employee's progress.

An important part of this process should include a regular review of the new employee's progress. Although a formal review will occur after some specific time of service, typically six months or a year, you should not wait that long with a new employee. Hold a minireview to help track the new subordinate's accomplishments, provide positive reinforcement for the good work he or she is doing, make adjustments where necessary, and set new goals for the short-term future. Such meetings should be held once or twice a month.

Determining the Employees' Competency and Commitment

The employees' performance is determined by their competence and their commitment to the work. Their competence is determined by their knowledge, skills, and abilities; their commitment is determined more by their attitude, motivation, and confidence in themselves. Together these two dimensions present four possible situations:

Type 1: High competency/high commitment = good performer. This employee has the skill and self-confidence needed and is ready to move to areas of greater responsibility. Keep her challenged.

Type 2: Low competency/high commitment = good student. This characterizes many of us while we are on the learning curve. The novice employee is learning new skills but now realizes that this process is a lot more complicated than initially thought. As a supervisor, your role is to provide assurance and help her gain self-confidence.

Type 3: High competency/low commitment = poor attitude. In this case the employee is learning but lacks the self-confidence or motivation needed to go it alone. Work on attitude and self-confidence.

Type 4: Low competency/low commitment = unwilling and unable. May need to review the individual's goals. Look for learning disabilities. Take shorter steps and have patience. Provide positive motivation and evidence of your confidence.

THREE EFFECTIVE TOOLS FOR EMPLOYEE DEVELOPMENT

We have discussed at some length the leadership role of the supervisor in the company environment. Although the role is critical to the collective productivity and safety of the members assigned to that company, we should also look at the role of the supervisor as it relates to each member as an individual. Effective supervision promotes an environment that encourages each member to develop to their full potential, which not only benefits the department, but also provides the very kind of intangible rewards that separate the *good* workplaces from the *great* workplaces. This interaction is accomplished in three ways: coaching, counseling, and evaluation.

Coaching

coach
a person who helps
another develop a skill

■ Note
**When coaching, you are
helping a team member
improve his knowledge,
skills, and abilities.**

Coaching is a term we borrow from the sports world but it fits well here. When coaching, you are helping a team member improve his knowledge, skills, and abilities. Coaching usually focuses on one aspect of the job at a time, and the coach works with the student until the desired level of competence is demonstrated (see Figure 7-1). We do this with our children. We should do it with our subordinates too. It is an easy and satisfying process.

Coaching is usually initiated by the leader. It is an informal process that requires little preparation. The leader already knows the technical aspects of the task. Teaching the task and having the patience to work with the subordinate may require some self-control, but the process is really quite easy. Take time to describe the steps that make up the job, demonstrate the proper method, and ask the subordinate to demonstrate the same sequence. Ask him to talk his way through the task, just as you did when you were demonstrating the task. This does two things: First, it reinforces the learning process, and second, it verifies that the subordinate's thinking, as well as his actions, are correct. While watching and listening to your subordinate, give positive feedback and reassurance.

Figure 7-1 *The traditional coach helps people develop physical skills.*

Coaching is an informational process that helps subordinates improve their skills and abilities. Coaching implies a one-on-one relationship that treats subordinates as full partners, that communicates a sense of mutual respect and trust, a willingness to listen, and even a willingness to compromise. An effective coach has self-confidence and the ability to win the confidence of others. A good coach is flexible enough to recognize the vision of success and the fact that there may be several paths to getting to that goal.

An effective coach has that vision, the self-image of success, and the ability to communicate that vision effectively to others. An effective coach can see beyond the obvious.

For a new person entering a workplace, this means that the coach or supervisor (see Figure 7-2) can see beyond the present and envision the full potential of the new employee. There is much to be learned from the way coaches treat their players and from how the players react to that treatment.

Performance often is a function of how subordinates are treated and the expectations that are indicated. Where success is expected, success will likely follow. Where failure is expected, it will surely follow. Successful leaders have the ability to transmit high expectations to their subordinates. Subordinates, more often than not, will perform to these expectations. High expectations lead to high performance. This, in turn, reinforces the high expectations, and even higher performance is expected next time. The converse is also true.

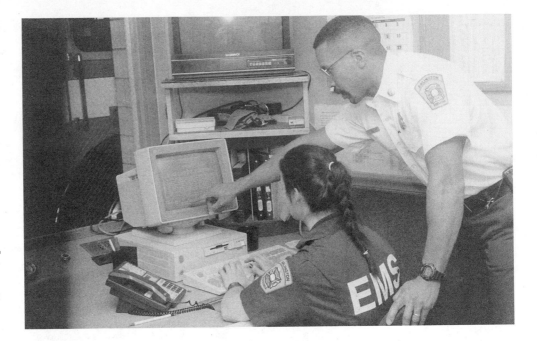

Figure 7-2 *Company officers coach people to help them develop professional skills.*

Think about what happens when you have a good working relationship with good subordinates. You tend to be comfortable with them and are likely to spend more time with them. You find favor with their work and, therefore, you find it easier to talk to them and compliment them on their work. As a result, they will probably continue to get better. But what happens when you do not have a good working relationship with a subordinate? You probably are not comfortable with that person. Therefore, you do not spend time with her and you do not see her work, good or bad. As a result, she probably does not improve. Whose fault is that?

AN EXAMPLE OF COACHING

Let us use teaching knot tying as an example of coaching. Knot tying is a basic skill, but some firefighters have a hard time remembering the basic knots, especially when they are not often used. The officer or a more senior firefighter can take the new firefighter to a quiet corner of the station and work on just one knot. Help the student and then watch him tie it a few times. A few hours later, ask him to do it again. Next day, have him tie the knot again and then try another knot. Within a few days the firefighter will have the skill and self-confidence to tie every knot on the list. This same idea works with just about any lesson being taught. Coaching takes time, but the satisfaction for both parties is huge!

counseling
one of several
leadership tools that
focuses on improving
employee
performance

■ **Note**
Counseling is the sec-
ond leadership tool
available to help subor-
dinates. Counseling
may focus on some
specific aspect of the
job, but more likely the
focus is on general
attitude or behavior.

■ **Note**
To be effective, coun-
seling should be accom-
plished shortly after
the unsatisfactory
behavior was observed.

Counseling

Counseling is the second leadership tool available to help subordinates. **Counseling** may focus on some specific aspect of the job, but more likely the focus is on general attitude or behavior. Counseling should always be done in private. Keep the counseling session businesslike but friendly. In fact, counseling can be very informal. After all, the purpose is to help the subordinate improve. To be effective it should be accomplished shortly after the unsatisfactory behavior was observed.

For example, a new firefighter while on an emergency medical services (EMS) call, shows some rather obvious signs of displeasure at having to treat a patient at the scene of the incident. While no one said anything, other firefighters and bystanders saw what was happening. The time to correct this problem is back at the fire station (see Figure 7-3). As the company officer, you should call the firefighter into your office and ask her to discuss her feelings. Counsel the individual that while these feelings are understandable, they should be suppressed, at least in public. If this is the first time this situation has been noted, just point out to the subordinate what she did, why it is unacceptable, and define what is acceptable. Most subordinates will get the message.

The counseling session should be as constructive as possible under the circumstances. Good leaders are able to counsel subordinates without either one becoming emotional. If the supervisor and the subordinate have a good relationship, and if the communication process is working well, the information can be quickly and effectively conveyed without having to make a major production of the event. Ridiculing, threatening, or verbally attacking the subordinate are generally ineffective.

Figure 7-3
Counseling focuses on improving some specific aspect of employee performance, attitude, or behavior.

Counseling is a tool for employee improvement, but it also is a problem-solving process. The problem should be clearly identified and an explanation should be offered as to why it is a problem. Have agreement with the subordinate about the problem and the corrective action that must be taken. Give the subordinate an opportunity to explain her actions or express her feelings on the matter. Close the session by agreeing on the next step. A follow-up meeting may be needed in some cases. Pick a specific time and date that are mutually convenient for the next meeting, and make a point to log that meeting in your personal activity planner.

One final comment: Remember that the subordinate will naturally be feeling down after the discussion. Once the counseling session is over, consider the option of rebuilding the relationship. Go and get a cup of coffee. Together!

THE FOUR CAREER DEVELOPMENT ROLES OF A LEADER

coach	helps employees with self-assessment
appraiser	provides employees with a reality check
adviser	helps employees identify their goals
referral agent	connects employees with their own future goals

---

Performance Evaluations

■ Note

Performance evaluations are another important employee development tool.

Performance evaluations are another important employee development tool. Many subordinates and even some supervisors look upon performance evaluation as a necessary but undesirable part of the job. If we look upon evaluations as a way of *improving* performance, things might be a lot easier.

Compared to coaching and counseling, evaluations are usually quite formal. Evaluations are structured in that they are usually done on a regular schedule, and they follow a standard procedure, usually dictated by the reporting form.

If coaching and counseling have been taking place as they should, there should be no surprises for subordinates in the evaluation interview. In fact, the subordinates should be able to write the evaluation themselves. Although the written record is important, the performance evaluation interview (see Figure 7-4) is important too, especially for newer members of the organization.

In most organizations, formal personnel evaluations occur once or twice a year but in reality, the evaluation procedure should be a continuous cycle. The cycle should start with goal setting, then move to encouragement, self-assessment, watching performance and providing feedback during the period, and a formal performance review and goal setting. The cycle is continuous (see Figure 7-5).

Figure 7-4
Performance evaluations are an important employee development tool.

CYCLE OF PERFORMANCE MANAGEMENT

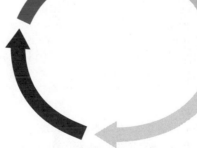

DOCUMENTING PERFORMANCE
PROVIDING ONGOING FEEDBACK

PERFORMANCE PLANNING
SETTING EXPECTATIONS

MID-YEAR REVIEW
PROCESS

PERFORMANCE EVALUATION
INTERVIEW

Figure 7-5   *The cycle of performance management represents the continuous process of goal setting, observation, and performance evaluation.*

■ **Note**

A good evaluation interview starts with a review of the significant accomplishments of the period.

■ **Note**

Every counseling session, even for the best employees, should identify areas where improvement is desired.

A good evaluation interview starts with a review of the significant accomplishments of the period. It should focus on the positive accomplishments. Good news is always easier to deliver and receive. Every counseling session, even for the best employees, should identify areas where improvement is desired. It may be a matter of improving performance levels in existing skills, or it may be a matter of motivating the subordinate to move on to new areas. Give your subordinates a specific target to aim for. Your expectations become their goals for the following reporting period.

When the interview is over the subordinate should have a clear understanding of the completion to the following sentences:

- My leader wants me to do more _____
- My leader wants me to do less _____
- The part of my job that matters most is _____
- Excellent performance in my job requires _____
- I should learn to take care of _____
- I think that my leader would say that my performance is _____

The subordinate should be motivated to set some goals too. Examples might include:

- I want to do more _____
- I want to do less _____

When the subordinate and the supervisor concur in this process, they are almost certain to see some progress. It may be necessary to compromise on a few of the items, but remember that your job as supervisor is to motivate your subordinates to their full potential.

During a good evaluation interview, it is important to let the subordinate talk too. Many subordinates, especially younger ones, are reticent during these interviews, but they should be encouraged to express their feelings and views on matters pertaining to their own performance and goals as well. Frequently subordinates are their own severest critic; they overestimate their shortcomings while underestimating their own accomplishments.

At every counseling session or evaluation interview, you should be sure to reinforce your subordinate's contributions with compliments about what was done well. Show confidence in her abilities and express personal pleasure in her contributions. Always give the subordinate an opportunity to ask questions or speak freely before the meeting is ended (see Figure 7-6).

■ **Note**

Make a habit of noting something about the performance of every employee on a regular basis.

Writing Employee Evaluations Many find that writing the employee evaluation is an arduous task. Here are several suggestions:

First, make a habit of noting something about the performance of every employee on a regular basis (see Figure 7-7). During each work cycle or work week, a supervisor should see something good, or bad, about each of his subordinates.

Figure 7-6
Encourage your subordinates to openly discuss their performance.

LOG FOR OBSERVING PERFORMANCE

Employee's Name: _____

Supervisor's Name: _____

Date	Behavior Observed	Discussed (Yes or No)

Figure 7-7 *Good supervisors have an organized system that simplifies the recording of observations of employee performance.*

Remember the lessons of *The One Minute Manager*: Look for people doing something right. We often remember to record shortcomings; make it a habit to record accomplishments, too. And remember that when you see something that requires corrective action, you should counsel the individual right away. There should be no surprises at the evaluation session.

Through these recorded observations you will likely have plenty of quality material to use on the evaluation itself. In fact you will probably have too much. That is okay; you can select the best and most representative material and write great evaluations filled with specific details. And you will not have to try to remember six month's or a year's worth of accomplishments as you try to write the report. Once you get in the habit of doing this recording, the process becomes easy.

Several weeks before the date scheduled for the employee performance evaluation interview, let the employees know that the evaluation session is approaching. Tell them exactly when it will occur and ask them for input into the process. They should be able to identify their accomplishments, shortcomings, and personal goals as well as you can. This may not work for employees in their first year or two of service, but it usually works for members after they have been through the process several times.

Measure your employees against a fixed and known standard. Several tools are helpful in benchmarking performance levels. First, there are national performance standards for firefighters, emergency medical technicians (EMTs), and the various technical specialties in the fire service. Use these as a measurement gauge in determining competence. Most departments add some additional performance requirements. Keep these in mind and let your subordinates know that you are using these as the criteria for performance measurement.

In addition, many departments have career development programs for their employees. These programs usually involve both training and educational requirements to help employees increase their skills and abilities in the workplace. Successful accomplishment and application of these career development requirements should be encouraged and formally recognized.

■ **Note**
Measure your employees against a fixed and known standard.

■ **Note**
Let your subordinates know what you expect, and when they give a little bit more, give them some recognition for their effort.

GIVING RECOGNITION

We may differ in the way we want to be recognized for our work, but recognition is one of our basic needs. Maslow, Herzberg, and others have identified the need for recognition. If you only talk to your employees when their performance is below acceptable standards, you lose the value that comes from positive recognition. Let your subordinates know what you expect, and when they give a little bit more, give them some recognition for their efforts (see Figure 7-8).

Giving recognition is easy, however there are several pitfalls. Do not be overly zealous in giving praise. If you compliment every act, you eventually dilute the recognition process. On the other hand, it is also easy to fail to give recognition when it is due, or to acknowledge that the work is worthy of special recognition.

PHOENIX FIRE DEPARTMENT
Phoenix, Arizona

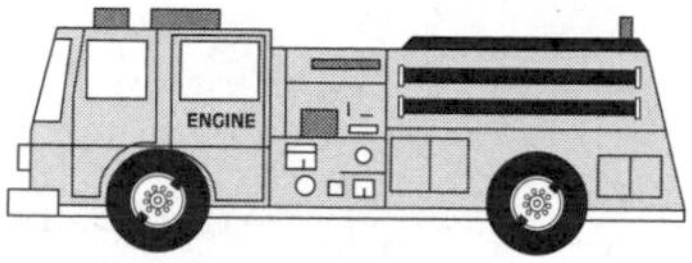

RECORD OF EXCEPTIONAL PERFORMANCE

1. Employee	2. Division/Section
3. Employee's Classification	4. Date Prepared

5. Initiator of Commendation

6. Description and Date of Exceptional Performance

7. Supervisor's Comments

8. Employee's Comments

9. Supervisor's Signature	10. Employee's Signature

Original: Personnel File
Copy: Employee Chief
           Fire Chief
           Initiator

90-1 D
615-8215-0027
Rev. 9/95

Figure 7-8 *Awarding certificates of accomplishment or placing notes of exceptional performance are an important part of the recognition process.*

There will be times when exceptional performance should be recognized with an award from the department. The award may be for valor at the scene of an emergency, or it may be for exceptionally meritorious service during nonemergency activities. The recognition may come in the form of a medal, citation, or other formal presentation (see Figure 7-9). As the company officer, you should remember that the awards process starts with you; if you do not take the initiative to recommend the award, it is unlikely that anyone else will. There will be plenty of opportunity for the department's senior staff members to pass judgment on the merits of the award. They will never have that opportunity if the employee's act or performance is not brought to their attention by the company officer.

Recognition does not have to be a medal or pay raise. It might be just a comment, a word of appreciation, or only a nod. At times, such gestures are as good as any tangible reward you could provide. Whatever form the recognition may be in, make sure that it is sincere. One of the basic tenants of the Dale Carnegie program is, "Show sincere, genuine appreciation."

Recognition should be personal. It should identify the individual, the act, and that you are personally grateful for the work. We usually teach students to try to avoid starting a sentence with "I." This is one of the exceptions. It means a lot when you say something like, "I was very pleased with the way that you. . . ." Think about how your subordinate will feel!

As we have already stated, recognition fulfills a basic human need. Recognition motivates employees to want to do even more and to be even more effective.

Figure 7-9 *A formal award provides highly visible recognition of an employee's accomplishments. Photo courtesy of Fairfax County Fire and Rescue Department.*

TAKING DISCIPLINARY ACTION

disciplinary action
an administrative process whereby an employee is punished for not conforming to the organizational rules or regulations

Having talked with a subordinate about a performance problem and having seen no change, your next step may be to take some form of **disciplinary action**. If you do not take such action, the problem will likely continue. In addition, if you do not take action, you will lose part of your credibility and part of your authority as a leader.

When taking disciplinary action, think of it as another way of improving employee performance and behavior. This is usually far more comfortable for both you and the subordinate than thinking about disciplinary action as a form of punishment.

When counseling or administering punishment, remember to: *golden rule:*

- Focus on the behavior, not the individual

■ Note

When taking disciplinary action, think of it as another process for improving employee performance and behavior.

- Help the subordinate maintain self-esteem

- Work to maintain a constructive relationship

Make an effort to be consistent and fair in the treatment of subordinates over a period of time. This is often difficult. No two employees are the same and no two offenses are quite the same. However, you should consider several factors when deciding whether to use discipline. First, the discipline should have some relation to the offense. Stealing and dishonesty are far more serious than arriving 5 minutes late for duty. One should also take into consideration the employee's recent work, his past disciplinary record, and any mitigating circumstances. Finally, some organizations require that the company officers inform their supervisors of the action taken. Find out what is appropriate.

■ Note

Make an effort to be consistent and fair in the treatment of subordinates over a period of time.

Discipline Procedures

Most organizations have an established discipline procedure. In most cases there is a progressive system that generally follows the steps of oral reprimand, written reprimand, transfer, suspension, demotion, and finally, termination. Let us briefly look at each of the steps and see how they might be effectively used to help improve a subordinate's performance.

oral reprimand
the first step in a formal disciplinary process

Oral Reprimand An **oral reprimand** is much like a private counseling session between the supervisor and the subordinate, except that it is likely that the unsatisfactory conduct in question has been discussed previously. The company officer should let the subordinate know that the conversation constitutes an oral reprimand and that the next step in the progressive disciplinary process is a written reprimand. The nature of the unsatisfactory behavior problem and the corrective action that is desired should be clearly identified during this session. Try to get the subordinate to understand the nature and consequences of his actions. Ask him to help spell out the corrective action. Although no formal report is required, it is a good idea to make notes of the conversation.

written reprimand
documents unsatisfactory performance and specifies the corrective action expected; usually follows an oral reprimand

Written Reprimand This second step in the disciplinary process indicates that the unsatisfactory behavior has continued. The company officer should administer the **written reprimand**. Lacking any established procedure to the contrary, a brief memo to the subordinate that identifies the nature of the unsatisfactory behavior, the fact that previous discussion has been had, and the fact that the unsatisfactory behavior has continued should be spelled out, along with the desired corrective action. A sentence or two on each of these topics is adequate, often in memo form. The subordinate should be asked to sign a copy of the reprimand acknowledging its receipt. Give the original to the subordinate.

transfer
a step in the disciplinary process that provides the employee a fresh start in another venue

Transfer The next step might be a possible **transfer**. A transfer may provide a solution to the problem, not just shifting the problem to someone else. Transferred employees are entitled to the same fair treatment you would give any other employee, but at the same time, the new supervisor should be aware of the prior discipline problem.

suspension
disciplinary action in which the employee is relieved from duties, possibly with partial or complete loss of pay,

Suspension At this point we are well beyond the punishment authority of most company officers in the fire service. **Suspension** is a serious punishment and is usually administered by a senior fire department officer. Disciplinary procedures vary, but most departments provide for a formal hearing, usually in the more senior officer's office, with the company officer and the employee each having an opportunity to present their version of the story. Most of the real evidence will be in the form of written records showing that the unsatisfactory conduct occurred and that previous disciplinary action has not been effective.

■ **Note**
The purpose of the disciplinary process is to improve the subordinate's performance or conduct.

Again, we should emphasize that the purpose of the disciplinary process is to improve the subordinate's performance or conduct. If that occurs, the punishment is effective. There are plenty of senior fire officers on duty today who have been on the carpet at some time during the early part of their career. The punishment, coupled with a natural maturing process, helped them to correct the problems they may have had in their younger days.

demotion
reduction of an employee to a lower grade

Demotion The punishment of **demotion** has a lasting impact. It takes away both pay and power. Obviously demotion is not very effective in the case of firefighters, because they are already in an entry-level position. As with suspension, the disciplinary procedure is usually conducted by a senior fire department officer. Make a detailed record of the procedure.

termination
the final step in the disciplinary process or in the Incident Command process

Termination **Termination** is used when all other steps have failed. Termination is costly for both the employee and the organization. A terminated employee must be replaced. A replacement must be recruited, selected, trained, and brought into the organization. However, in spite of these considerations, termination may be appropriate in some situations.

For serious offenses, one or more of these steps may be omitted.

Punishment should:

- Be associated with a specific unsatisfactory conduct or behavior
- Occur soon after the offense
- Be firm and fast
- Be uniform among all persons
- Be consistent over a period of time
- Have informational value
- Clearly identify the problem and the desired corrective action
- When possible, be administered by the same person who can reward good performance, usually the company officer

TAKING CARE OF YOUR SUBORDINATES

Fairness for One and All

■ Note
Everyone should be treated fairly.

Throughout the process of developing your team, it is important to remember that everyone should be treated fairly. People have different needs and goals. Some have a high need for recognition. Others have a greater need for affiliation. Recognize these needs in your subordinates. Finding the best way to motivate each of your subordinates is one of the most important things you do in being an effective leader. Use the most effective approach on each employee.

We have discussed the importance of setting goals. Be sure that the goals are reasonable and fair for each employee. While all firefighters should have the same skills, some will be better in some things than in others. Some learn some skills faster than others. However, it should all balance out in the final analysis. Work with individuals to help them overcome their weaknesses, and recognize their strengths by letting them do that which they do best, at least some of the time.

Encourage all of your employees to participate in the goal-setting process for themselves and for their company. Get them involved in working together to help the company reach its full potential. Getting subordinates involved is called **empowerment**. Empowerment allows employees to have a feeling of ownership in the organization.

empowerment
to give authority or power to another

Maintaining fairness has taken on new dimensions with the increasingly diverse workforce we see in today's fire service. Diversity comes in many dimensions. Variations in race, gender, age, experience, and other factors tend to make each employee unique. Leading such a diverse group presents it own special challenges. The key to understanding and leading a diverse workforce lies in understanding and flexibility. Take time to understand each employee's needs. Try to be flexible in accommodating those needs.

Along with being fair, the supervisor must take the lead in making sure that the workplace is comfortable for all who work there. Set a personal example and encourage others to imitate your lead in the following:

- Avoid disrespectful slang terms and names.
- Eliminate offensive pictures, cartoons, and other printed matter from the workplace.
- Avoid ethnic jokes.
- Speak calmly and slowly.
- Listen equally to each subordinate.
- Spell out those unwritten rules.

your own
expectations (handwritten margin note)

Dealing with Conflicts

With all the demands upon them, your people are under considerable pressure. In the fire service, one may see more of life's raw edges in a day than most citizens see in a lifetime. We expect our firefighters to treat people and go places that most citizens would like to avoid. We expect our drivers to operate on busy streets, getting us safely and quickly to the scene, without the risk of an accident. And we ask our subordinates for increasing efforts around the fire station and in the community. It is no wonder that our people are under pressure.

conflict
a disagreement, quarrel, or struggle between two individuals or groups

Pressure can create **conflict**. When people work together under pressure, conflict is natural. Conflict can be healthy. It implies that people are working hard and trying to be successful in meeting their goals, both their personal goals as well as those of the organization. Conflict can be unhealthy when it involves personality clashes. When this type of conflict occurs, it needs to be quickly resolved—another job for the company officer.

■ **Note**
When people work together under pressure, conflict is natural.

As an effective company officer you should be able to detect a conflict arising while it is still in its incipient stages. Although it is not always easy, you should work with both parties to understand what is happening and resolve the problem. Here are some suggestions:

- It takes two (or more) to have a conflict. Get the parties together, tell them that you think "we have a problem," and ask the parties to help you find a solution.
- Try to get the parties to acknowledge that there is a problem and that the conflict is causing a problem in the workplace.
- Work with the involved parties to calm the situation and to solve the problem.

- When a solution is found, get a verbal commitment from each party on how they are going to support the solution. Make sure that everyone understands the other's position.

- Set a specific time to review progress on the situation.

Employee Complaints and Grievances

gripe
the least severe form of discontent

complaint
an expression of discontent

grievance
a formal dispute between employee and employer over some condition of employment

■ **Note**
Gripes and complaints should be channeled into positive action.

■ **Note**
Sometimes just letting the subordinate talk over the problem helps to resolve the issue.

grievance procedure
formal process for handling disputed issues between employee and employer; where a union contract is in place, the grievance procedure is part of the contract

It is natural for employees to gripe about things in the workplace, especially in fire stations. When you take four or five bright and active young people and lock them up together in a building away from their own homes for 24 hours, gripes and complaints are inevitable. Some of this is just ordinary grousing, but sometimes the issues and the people involved can get quite serious.

A first expression of dissatisfaction might be considered a **gripe**. Prolonged or repeated dissatisfaction over the same topic might lead to a **complaint.** The difference is not important. However, the next step, a **grievance**, is serious and can occur when the complaining becomes significant or does not bring relief. All three deserve the attention of the company officer.

If someone complains about the weather, there is not much you can do about it. If someone complains about the coffee pot not working, that may require some action. Not every gripe or complaint needs to be personally resolved by the supervisor, but the supervisor should provide a healthy atmosphere where problems can be raised and resolved.

Gripes and complaints should be channeled into positive action. The supervisor should expect adults to solve some of their own problems and to try to get help from others when this does not work. When neither of these work, employees may turn to their supervisor for help. Company officers should set some ground rules here. If a subordinate complains about some work-related item at the dinner table or while the crew is engaged in recreational activity, the officer can say, "I would like to hear more about that. Please come see me about that when I am 'at work'," which may mean, come talk to me about this in the office. The officer is politely saying, let's not mix our business with our pleasure. When subordinates have a problem, they should take the initiative to go see their officer when the officer is at work in his office.

Sometimes just letting the subordinate talk over the problem helps to resolve the issue. At other times, the officer may suggest some form of action to help a subordinate to help himself. And at times, the officer will have to take action to help resolve the problem. Listening to and acting on material complaints will improve the workplace environment. At the same time, we do not want to pamper our people by listening to and attempting to accommodate their every whim.

We have discussed **grievance procedures** in Chapter 5. Company officers must be familiar with their organization's grievance procedure. Where employees are represented in a bargaining unit by union, the grievance procedure is part of the union contract, and contracts usually allow the employee to have someone from the union with him during any action that may be taken.

If employee frustration has escalated to the grievance level, it indicates that there is a problem that they cannot resolve, that they are unhappy with the situation, and that they have decided to make a formal issue of their complaint. It may also indicate that there is a communications problem in the organization. Most organizations have a formal grievance procedure; the procedure forces the issue to be heard in a timely manner.

Dealing with the Disenchanted Firefighter

Some fire companies have a seasoned firefighter who has developed a negative attitude. This firefighter may be sullen or even openly hostile. When he does speak, he complains a lot, but offers few suggestions and seldom takes the initiative to solve problems. He also tries to do as little work as he can. Not only does he set a bad example, his negative attitude can become contagious if left unchecked.

As a supervisor, you should take the lead and try to gradually restore the disenchanted employee. You may want to start with improving his attitude. Give him a specific task that he can accomplish. When it is successfully accomplished, give him some recognition for his effort. Repeat the process. By taking the rehabilitation process in small increments, you may be able to turn the employee with a negative attitude into a positive performer.

INTRODUCING CHANGE

Some say, "Things are always changing here." That may be a bit of an exaggeration, but for many organizations, it does seem that there is always some new idea being implemented. It is important that an organization be able to change. If it cannot, it may not survive. Likewise, it is important for individuals to be able to change as well.

■ **Note**

Most change is good; it allows us to keep up to date and to use modern ideas and technology to get the job done.

Most change is good; it allows us to keep up to date, and to use modern ideas and technology to get the job done. As the speed with which we share information and technology increases, so will the rate of change. As one sage put it, "If you think we have had a lot of change in the last 10 years, wait until you see what happens in the next 10 years!" (See Figure 7-10.)

We all change during our lifetimes, but accepting change is an individual thing: Some do it better than others. For us as individuals, accepting change is something we can plan, control, and monitor on our own. When it become too painful, we can slow down or reverse our decision.

Changes in organizations are another matter. Although the change will impact upon us as individuals, just as with our own initiatives, with organizational change we probably do not have much opportunity to set the pace. We

Figure 7-10 *The next generation emergency response vehicle will be quite different than that seen in fire stations today. Photo courtesy of E-One.*

often have little or no knowledge of the change in advance of its implementation, and we have little opportunity to plan, control, and assess its impact. As a result, changes in organizations are much more difficult on the employees.

It is natural to resist change. Our resistance may come from the fact that we are comfortable with the present process. We may not have the training or education to deal with the change or the new process that comes as a result of the change. We are intimidated by the unknown or the rumors we have heard. All are valid reasons why people resist change. We all want to share the benefits of change, but few of us want to pay the price of changing from the present.

Although things may be running well at your fire station, there will come a time when you have to make a change. Make sure there is a valid reason for the change. Change that is easy to accomplish or that has the greatest visibility may not be what is really needed. Change may be the result of one of your our own ideas or it may be the result of an initiative that has come down the chain of command. Regardless of its source, it has to be implemented at the station level. Introducing change and managing its effective implementation is a leadership skill.

Good leadership provides a culture where change is not only welcomed but where it is required. With good leadership you should have open lines of communications (see Figure 7-11). You should be encouraging employees to offer new ideas about their work and the service they provide to the community they serve. As a supervisor, you should provide a climate where a subordinate's ideas are given serious consideration.

Let us look at several key ingredients for providing the environment where change is accepted. First, keep the employees involved in the process. Let your employees know that you want to hear about their ideas. Let your employees know that you are open to other ways to do things, even where the present system may be of your own creation. We are surrounded with hard working subor-

■ **Note**
It is natural to resist change.

■ **Note**
Introducing change and managing its effective implementation is a leadership skill.

Figure 7-11 *New ideas are not always welcomed.*

dinates. Some take the job more seriously than others, but all have the ability to offer ideas on how to do things better.

Work to overcome the obstacles of change. We usually fear the unknown. Change introduces uncertainty into our lives and often introduces new events over which we have no control. We often do not know all the details when changes occur. We do not know if or how it will affect us. We do not understand why the change has occurred and what future change may be about to occur. All of these anxieties can be overcome with good communications.

When introducing change, keep your employees informed of what is going on, and let them be a part of the process, rather than a victim of its passing. Let them know the time frame and the depth of the impact it will bring. Depending upon the time, circumstances, and the conditions involved, you may be able to use one of the leadership styles described in the last chapter to sell the idea, rather than ramming the idea down your subordinates' throats. Let them ask questions. Try to provide honest answers to their questions. Be positive and supportive of both them and the change.

You can expect to hear some negative comments. Let those comments be heard and try to respond with honest and positive answers if you can. If you need information from elsewhere in the organization, say so. One of the main components of successfully introducing change is to keep the communications open.

The greatest fear occurs when change involves personnel action. We all like to know what is happening in our lives, especially when we are at work. When people get moved around, when they have their working hours changed, and when they have to take on new duties, they want to know about these things in advance. Try to keep these events to a minimum, but at the same time, let your people know about the necessary events well in advance.

■ **Note**
Change introduces uncertainty into our lives and often introduces new events over which we have no control.

■ **Note**
One of the main components of successfully introducing change is to keep communications open.

Others you work with—the police, other fire departments, hospitals, and of course the public—may be affected by the change. Let them know what is going on. For most organizations, this is an excellent time to build some good public relations.

Once a change has been introduced, it is always a good idea to follow up, to remain in touch with what is happening in the organization. Review the change and the impact it has brought to the organization. Change seldom comes without a price. With each good event, there are usually some negative side effects. The goal is to have the positive effects outweigh the negative ones. Monitoring and fine tuning the results of the change will increase the net results (see Figure 7-12).

When change is successfully implemented, it is a good time to recognize those who made it happen. Recognize the individual and the team that led the change. Also give some recognition to those who were able to successfully accommodate the change. Some key steps for successfully introducing change:

- Create a climate where change is welcomed and not feared.
- Let employees be involved in finding new ways and in implementing new ideas from others.
- Reduce stress and anxiety by keeping people informed.
- When introducing change, be positive.
- Consider the impact of the timing of the change on your employees and introduce the changes as gradually as possible.
- Follow up and monitor the total impact of the change.

Figure 7-12 *New ideas should always be given fair consideration. Shown here is a new "all-steer" ladder truck.*

REFLECTIONS ON THE DEPARTURE OF OUR LEADER

Our leader is leaving us. I can recall when he arrived. We were concerned, as we always are when a new officer is assigned to the company. What will the new company officer be like? What things will he change? Well, we were to soon find out.

First, we noticed how personable he seemed; he seemed comfortable with everyone. As the days passed, we noticed that he made it a point to spend some quality time with everyone. Sure, he spoke with everyone, but more importantly he listened to everyone, too. He did not make promises that he could not keep and did not try to solve all of our problems by himself. But he always listened and offered encouragement.

We also noticed that the new officer was refreshingly unassuming. He was not here on some ego trip. He was here because he wanted to be, because he had earned it, because he was qualified for the job, and because he enjoyed being here. He was as sharp as anyone around, but he sometimes admitted he did not know the answer. But he always knew where to go to find it.

He enjoyed the job and maintained a sense of humor. He worked us hard, but he also made sure we had fun too. We recently won an award for being the best company in the department. At the award ceremony, our company officer thanked the chief for the award, but he also publicly thanked each of us and made sure that everyone present knew that we had done the work. We excelled because of his leadership; he did not mention that however.

We also performed well because of his teaching. I guess you might call it mentoring or coaching. He was always concerned about our professional growth and gave us room and motivation to grow. He let us make decisions and take responsibility for our actions. He praised us when we did well. When we did not do as well as he thought we could, he let us know that too, but always calmly, tactfully, and fairly.

Our boss is getting a promotion and a transfer. We will miss him. But we are far better off for his having been here with us these past several years. He helped each of us grow to be better than we were before, and collectively, we are now a far better team—the best one around, in fact! I guess some of his influence will still be here. In many ways he taught us to think and work like he does. And I think we are getting pretty good at it. Do you suppose that he was preparing us to take his place?

Summary

The following information summarizes what we have covered in Chapters 6 and 7 and should provide you with a list of practical tools to help you become an effective leader.

1. *Have an orientation session.* We usually have an orientation session for new firefighters. What do we do for experienced firefighters who are transferred to a new station? What do we do when the supervisor changes? In every case, there is a need for taking a few minutes to get acquainted. For the supervisors who do this regularly, it is a good idea to have a short checklist, not unlike a teaching outline, to help facilitate the transition and cover every issue with every employee. This session is a good time to explain personal values as well as any policies that may be unique to this workplace. This session is a good time to listen to the employees, hear about their goals and ambitions, and possibly their problems and needs. Remember—do not try to cover everything in one session. Give the new employee information in installments that can be easily absorbed.

2. *Have regular meetings and share information with employees.* One effective meeting that should happen regularly is a planning meeting. The planning meeting may be nothing more than getting together with everyone after the equipment checks at the start of the shift for a cup of coffee and a brief discussion about what will happen during the upcoming shift. Some planning meetings may be longer, such as planning the training schedule for the next three months. Plan for future events and let your employees share the information. Let employees be involved in the decision-making process whenever possible.

3. *Set goals* for the company and for every employee. Some goals will be collaborative efforts that require group acceptance, whereas other goals may be for individual accomplishment. In any case, get the employee(s) actively involved in goal setting. Use these goals as a benchmark for measuring progress during evaluations.

4. *Provide regular feedback on performance.* This does not require a formal evaluation with a complicated report. It may only take a minute or two, and usually focuses on some specific task or time frame. It is a lot easier for both the supervisor and the employee to deal with these performance issues when they are addressed in small increments. When corrective action is desired, it will come a lot sooner.

5. *Empower your people.* Offer opportunities for your employees to enhance themselves and provide opportunities for their growth and learning. Encourage them to seek training and education at every opportunity. Provide opportunity and incentive for individuals to increase their knowledge.

6. *Provide instruction and help,* be available, show interest and support, and recognize accomplishments. But do not supervise, unless necessary.

7. *Ask your employees for ideas and suggestions.* Listen to the employees' ideas and suggestions. Respond to all of them and, when appropriate, see that the ideas are forwarded up the chain of command for action. Be sure that employees get credit for their work and suggestions.

8. *Be a role model.* Do the right thing.

One final comment. While coaching and counseling those around you, take time to work on your own development needs. Steve Covey, author of the recent best-seller *Seven Habits of Highly Successful People,* calls this taking time off from your work to improve your skills as taking time to "sharpen the saw." Covey compares our learning process to working with tools, noting that more gets done with sharp tools, but that production is sacrificed in the process. He suggests that we also have to take time off to sharpen our thinking tools. His message: "Take time to keep your mental saw sharp!" There is a wealth of information on the topic of supervision and leadership. Every library and bookstore has books and tapes on the subject. You are encouraged to learn all that you can, and to share your ideas and experiences with others.

Review Questions

1. What is leadership?
2. What is coaching?
3. What is counseling?
4. The text suggests that the performance evaluation system is a continuous process. Explain.
5. Describe the disciplinary process in your department.
6. What are some general rules for effective punishment?
7. How should company officers deal with conflict?
8. How should a company officer respond to complaints and grievances?
9. How should a company officer introduce change?
10. How can company officers become effective leaders?

Additional Reading

Bennis, Warren, *On Becoming A Leader* (Reading, MA: Addison-Wesley, 1994).

Blanchard, Kenneth, and Spencer Johnson, *The One Minute Manager* (New York, Berkley Books, 1982).

Matzer, John, Jr., editor, *Advanced Supervisory Practices* (Washington, DC: International City Management Association, 1992).

Peters, Tom, and Nancy Austin, *A Passion for Excellence* (New York, Random House, 1995).

The National Fire Academy in Emmitsburg, Maryland, offers an excellent three-part leadership development training program for company officers. The three parts deal with personal success, supervisory success, and company success. Each of these three courses is intended for a two-day delivery format. Contact your training officer or state fire service training director for more information.

Chapter

8

The Company Officer's Role in Personnel Safety

Objectives

Upon completion of this chapter, you should be able to describe:

- The need for an ongoing concern for safety in the fire service.
- The common causes of accidents and injuries in the fire service.
- The signs and symptoms of stress among personnel.
- How to develop and implement a safety program at the station level.

INTRODUCTION

A recent article in a fire service magazine stated that the leading causes of accidents among fire service personnel are managers who don't manage, supervisors who don't supervise, and firefighters who do dumb things.[1] Safety is an organization value. Many say, "I know that is good for me but I don't need it." We also hear, "I don't have accidents, I can take care of myself." Some of the problems include apathy, attitude, and tradition.

Nearly one hundred firefighters are killed and another 100,000 are injured in the performance of their duties during a typical year in the United States. Approximately one in ten firefighters will be injured this year. In the past some firefighters boasted of these numbers, noting with pride the dangers associated with this profession. While the activities performed by the fire service are inherently dangerous and noble, a majority of the accidents that lead to these statistics are preventable.

In addition to the deaths and injuries reported each year, a growing number of firefighters are being exposed to situations that lead to occupational illnesses and conditions that have debilitating and sometimes fatal consequences. Firefighters have long had a high rate of heart and respiratory disease and cancer. Firefighters providing emergency medical care may be exposed to a wide range of hazards.

What has all of this got to do with company officers? Company officers should be intimately involved with the department's health and safety program. Most injuries occur to firefighters. Most firefighters have a company officer as their supervisor. Most fire departments, several NFPA standards, and the courts hold the supervisor responsible for the firefighter's safety.

THE VALUE OF SAFETY

One cannot put a value on safety, but you sure can put a price on the consequences. Regardless of whether the firefighter is a career employee or a volunteer member, injuries cost money. Anyone who has had recent contact with a health care system knows the cost of good medical care. Someone is going to have to pay these costs, and ultimately, it is the citizens of the community. For career departments dealing with tight budgets and minimum staffing issues, an injured employee brings an added cost in terms of providing a replacement. While the injured firefighter is in a "no duty" or "light duty" status, another firefighter is working for him, probably at overtime rates. Thus, the injured $20-an-hour firefighter is actually costing the community $50 an hour. Additional costs to the department include medical insurance, workers' compensation insurance, and repairs for damaged equipment.

There are additional impacts—indirect costs—that cannot be measured in dollars. For the injured firefighter, there is the pain, suffering, possible permanent injury, and even the embarrassment that comes from the injury. The firefighter's working colleagues also suffer, not only in working additional shifts for the

■ **Note**

One hundred firefighters are killed and another 100,000 are injured in the performance of their duties during a typical year in the United States.

■ **Note**

Regardless of whether the firefighter is a career employee or a volunteer member, injuries cost money.

BACKGROUND OF SAFETY ACTIVITY IN THE WORKPLACE

Concerns for workplace safety and health issues are not new or unique to the fire service. In the United States, awareness of workplace safety and health issues date back to the 1600s. Textbooks on industrial safety usually include a history of workplace safety concerns dating back to the start of recorded history.

As technology advanced, so did the hazards. The demand for steam power increased with the arrival of the industrial age. With the increased need for power there was an increased demand for coal. Most of the coal had to be mined, which was and still is dangerous work. Long hours, poor ventilation, and extremely demanding work conditions led to black lung disease and a host of other problems. Miners were essentially the slaves of the mine owners. Efforts to bring some balance to the mining industry were started in the 1860s. It took 60 years and the full efforts of both the United Mine Workers of America and the U. S. Congress to increase the use of safety appliances in the coal mines.

The workplace changed as the United States became more industrialized and technology improved. Instead of a company focusing on production at any cost, worker safety was also considered. In many instances, a process change brought on by a concern for safety also brought increased productivity. For example, during the early years of the industrial revolution, gears and belts were unguarded and workers were frequently required to place themselves in dangerous situations. When a worker was injured, the employee was replaced with little or no compensation. The growth of organized labor and the development of laws dealing with workplace conditions started a process that continues today.

One industry that has made remarkable progress in the area of employee safety in recent times is the railroad industry. America's railroads, possibly prodded by the government, their labor unions, and the cost of workplace accidents, have made a dramatic change in their accident statistics over recent years. These improvements have been accomplished by eliminating unsafe equipment and procedures, having and enforcing policies that provide for safety, and having highly viable recognition programs at all levels that reinforce both the company's and the industry's commitment to worker health and safety.

We can draw several parallels between the railroad industry and the fire service. Both are dangerous professions and safety and health issues are a concern for both industries. During the past 25 years we have seen increased awareness for safety in both industries. Let us look at ourselves. Are you as safe as you can be? Are you looking for and sharing ways to improve health and safety for your firefighters? Is your fire department working on programs to improve health and safety? Do you have programs that reward safety?

■ **Note**

Safety comes from eliminating unsafe equipment and procedures, having and enforcing policies that provide for safety, and having viable recognition programs at all levels that reinforce both the company's and the industry's commitment to worker health and safety.

injured employee, but in the emotional costs associated with their pain in having a hurt friend. And finally, the injured firefighter's family pays a cost for its anxiety, support, and concerns associated with the member's injury.

All of these concerns resulted in the adoption of NFPA 1500, *Standard on Fire Department Occupational Safety and Health Program*. We discuss NFPA 1500 later in this chapter.

Safety Activities in the Fire Service

By the nature of the occupation, emergency responders are at risk from accidents that can result in illness, injury, and death. Incident safety is a concern for all emergency response personnel. Firefighters are taught to deal with these problems. Why then, do we have the injury statistics we do?

Firefighters tend to be aggressive, action-oriented people, especially around other firefighters. At times this behavior is desirable, but overaggressiveness can lead to a disregard for firefighter safety. The company officer must learn to channel this enthusiasm and energy into productive activity. Although it is sometimes difficult to change attitudes about safety, some firefighters will respond to a discussion that focuses on their importance as a member of a team. In this context, every firefighter is dependent upon every other firefighter for success and survival. If one member of the team is injured, everyone is at risk. This logic can also be extended to how we serve the public. When we are not able to perform, we are not able to help our customers. For most firefighters, serving others is our mission.

There are other actions over which you have compete control. Personal fitness and good health are within the realm of everyone's personal control. Yet many injuries and even some of the fatalities could be avoided with better physical conditioning. Firefighting *is* a hazardous occupation. The action is fast paced and hot and there are often problems with the structure itself. Many ask, "What reasonable person would go *into* a burning building while everyone else is running *out*?" Going into burning buildings is no longer what the fire service is all about. Firefighters are expected to understand how fire grows and to assess the fire situation. The conditions may suggest that *not* entering the building is the right action. Interior fire fighting involves advancing hose lines to the seat of the fire. Nearly half of those who are killed during structural fire fighting operations were advancing hose lines at the time of their injury[2] (see Figure 8-1). This task is typically assigned to the newest firefighters; you cannot expect those in their first year on the job to be aware of all the dangers involved.

Some accidents are considered surprises. A collapse might be an example, however, even in these circumstances, a better understanding of fire behavior and building construction should have suggested a different course of action. During a recent fire in a commercial building, the truss roof collapsed trapping two firefighters who had ventured inside this well-involved structure. The building had been there for years. One has to wonder if the department had been there

■ **Note**
Every firefighter is dependent upon every other firefighter for success and survival.

■ **Note**
Personal fitness and good health are within the realm of everyone's personal control.

■ **Note**
Nearly half of those killed during structural firefighting operations were advancing hose lines at the time of their injury.

Figure 8-1
Firefighter fatalities by type of duty, 1992–1996. From the annual reports on firefighter fatalities and firefighter injuries, published annually in the NFPA Journal.[3] Used with permission, National Fire Protection Association, Quincy, MA 02269.

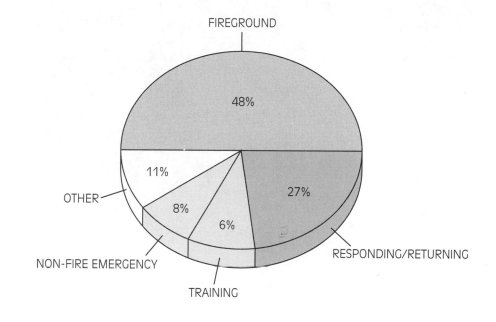

to survey and preplan, or if anyone told the firefighters about the dangers of a truss roof. The collapse of a truss roof during a fire should not come as a surprise.

Many fatalities occur as a result of physiological and psychological stress (see Figures 8-2 and 8-3), manifested as a heart attack or stroke, rather than some traumatic event. The work *can* be very demanding at times and is often undertaken in unfamiliar settings with uncertain outcomes. These facts suggest that proper rest, good physical condition, proper health care, and proper lifestyle are just as important as protective clothing and an effective incident command system.

Many of these factors—attitude, conditioning, health care, and lifestyle—are individual concerns. Supervisors should set a good positive example, encourage fitness and good health practices, and even provide on-duty opportunities for improving every employee's fitness and health.

Knowledge of the job is also a matter of personal concern and initiative; it all comes down to the firefighter's attitude and awareness. In many cases the firefighter may need motivation and some help from the supervisor, and for most firefighters, the supervisor is the company officer.

■ Note
Proper rest, good physical condition, proper health care, and proper life style are just as important as protective clothing and an effective incident command system.

ANNUAL NFPA REPORTS ON FIREFIGHTER SAFETY

Each year, the NFPA publishes two reports in firefighter safety. One of the reports focuses on firefighter injuries; the other on firefighter fatalities. Both reports are published in the *NFPA Journal*. Although these topics are not pleasant, these reports should be reviewed by all fire service professionals.

Figure 8-2

Firefighter fatalities by nature of injury, 1992–1996. From the annual reports on firefighter fatalities and firefighter injuries, published annually in the NFPA Journal. *Used with permission, National Fire Protection Association, Quincy, MA 02269.*

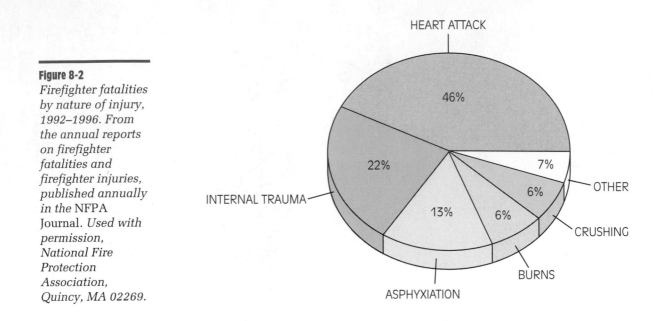

Figure 8-3

Firefighter fatalities by cause of injury, 1992–1996. From the annual reports on firefighter fatalities and firefighter injuries, published annually in the NFPA Journal. *The years of data available at the time of publication. Printed with permission, National Fire Protection Association, Quincy, MA 02269.*

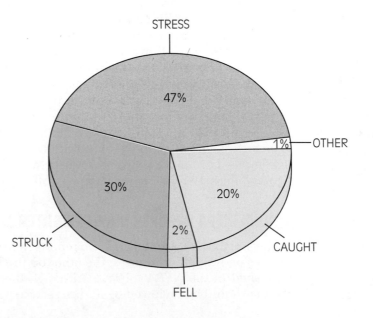

Safety professionals look at past statistics for several reasons. First, they provide considerable information about what causes the accidents that lead to firefighter deaths and injuries. The published safety statistics are indications of the activities that are causing the greatest problems for the fire service at the national level. However, these statistics can indicate problems that may not yet be apparent in a particular department. In such cases, these statistics should be a wake-up call and suggest changes in policy or the need for training.

Departments that do not enforce their policies (or do not have any policies) regarding the use of personal protective equipment, or that do not have and routinely use a good incident command system at the scene of every event cannot expect positive results from members regarding occupational safety. How do your department's accident statistics compare with the published data?

Understanding the Nature of the Injuries

The largest percentage of firefighter injuries and deaths occur at the scene of fires. However, statistics show that every aspect of the firefighter's work routine has the potential for injury. Each year, a significant number of firefighters are killed and injured while responding to and returning from calls, as well as at the scene of emergencies (see Figures 8-1 and 8-4). With departments providing a greater variety of service to their communities, firefighters are also being injured at the scene of nonfire emergencies. Many injuries occur at fire stations. Injuries and even a few deaths occur during training activities.

Figure 8-4
Firefighter injuries by type of duty, 1992–1996. From the annual reports on firefighter fatalities and firefighter injuries, published annually in the NFPA Journal. *Used with permission, National Fire Protection Association, Quincy, MA 02269.*

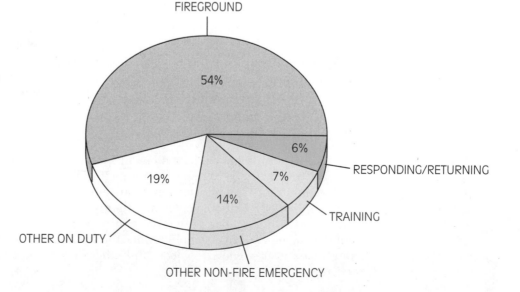

Figure 8-5
Firefighter injuries by the nature of the injury, 1992–1996. From the annual reports on firefighter fatalities and firefighter injuries, published annually in the NFPA Journal. Used with permission, National Fire Protection Association, Quincy, MA 02269.

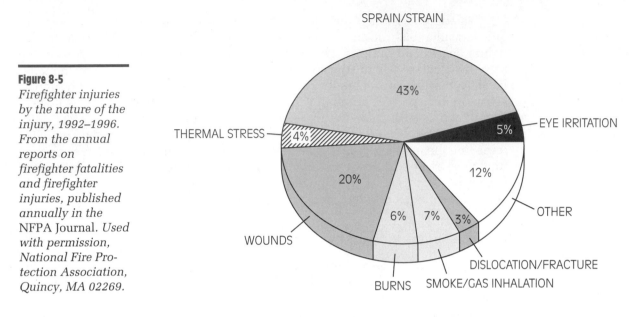

■ **Note**

The most frequent injuries are sprains and strains.

The most frequent injuries are sprains and strains (see Figure 8-5). These "invisible injuries" usually are the result of improper lifting. A second large category of injuries is wounds. Wounds include all kinds of skin injuries from cuts to the hands to stepping and falling on sharp objects. Given the nature of these injuries, it would seem logical that most of the injuries occur at the scene of fire-related emergencies.

Check your own department's safety records. The statistics should suggest opportunities for corrective action to improve safety performance. If 50% of your department's injuries are back injuries and are occurring as the result of improper lifting, it makes sense to focus corrective actions in this area. Selecting the biggest target usually provides the greatest opportunity for improvement.

Reducing These Injuries

Sprains and strains can be reduced by proper physical conditioning and by using proper lifting methods. When a department is experiencing a large percentage of sprain-and-strain-type injury from a particular activity, some revision in procedure is indicated. Reducing cuts and other wound-type injuries may require providing and using proper protective equipment. Effective supervision and fire scene management can also help reduce the firefighter's injuries from falling, and from the exposure to sharp objects and other dangerous conditions that exist on the fireground.

FIRE DEPARTMENT OCCUPATIONAL SAFETY AND HEALTH PROGRAM

In 1987, the membership of NFPA adopted a new and somewhat controversial document entitled NFPA 1500, *Standard on Fire Department Occupational Safety and Health Program*. Although NFPA 1500 introduced many new concepts to the fire service, much of its content is taken from well-accepted practices of private industry. NFPA 1500 is the start of a major undertaking; it provides a framework for a series of new standards that would help fire departments plan, implement, and manage effective health and safety programs.

■ **Note**

NFPA 1500 was designed to help fire departments reduce the frequency and severity of accidents and injuries to their members.

NFPA 1500 has had a significant impact on the fire service. NFPA 1500 was designed to help fire departments reduce the frequency and severity of accidents and injuries to their members. It has increased the awareness level about safety in every aspect of the job. Equipment has changed, procedures have changed, and new incident management systems and personnel accountability systems have been added. But no law or standard can make you safe. In spite of all of these advances, personal safety is still a matter of one's own attitude.

NFPA 1500 contains the "minimum requirements" for a fire department health and safety program. The term "fire department" includes all organizations that provide fire suppression functions, including private, government, military, industrial, and public sector organizations. As used in the standard, the term "member" applies to volunteers, whether paid or unpaid, and paid personnel whether full time or part time of any such department.

NFPA 1500 consists of eleven chapters and runs about fifty pages. Company officers should be familiar with its contents. The following is a brief summary of its contents with some commentary.

Chapter 1. Administration

The first chapter of NFPA 1500 defines the scope, purpose, and application of the standard and defines the terms used in the following chapters.

Chapter 2. Organization

Chapter 2 of NFPA 1500 explains risk management, provides information for starting a health and safety program, and defines the role of the various officials. It requires a written health and safety policy and establishes goals to help reduce accidents and injuries. This chapter also discusses the concept of risk and risk management, the role of the department's health and safety officer, and the value of accident investigations.

health and safety officer
a person assigned as the manager of the department's health and safety program

The Health and Safety Officer   According to NFPA 1500, every fire department should have an individual assigned to the duties of a health and safety officer. The **health and safety officer (HSO)** must be an officer from within the agency and should be formally appointed by the fire chief or the head of the agency. The health and safety officer must have access to the chief and other senior officers in the department. As such, the HSO also needs the personal skills to work with people and build strong collaborative relationships.

The health and safety officer's role varies, depending upon the direction of the chief and the support one gets from other members of the staff. The HSO should have a thorough knowledge of activities undertaken by the department and the hazards associated with these activities. The HSO should also be familiar with the department's policies as well as the laws, standards, and regulations that affect firefighter safety.

The HSO's most important job is the implementation and management of the department's safety program. In some corporations, this person is frequently called the organization's "risk manager." Although the ultimate responsibility for the safety of the department's personnel lies with the fire chief, the chief can delegate some of the daily management activity of the health and safety program to the HSO.

While the HSO's normal activity is administrative in nature, the HSO should also be available to assist at the scene of an emergency as the incident safety officer (ISO). Such events allow the HSO to see firsthand the activities of the fire department's personnel at work. At times like these, training and procedural deficiencies become evident. The incident commander may not have time to notice or record these problems, but the ISO should. Corrective action can be implemented later.

Investigating the Cause of Accidents   One of the HSO's duties is to investigate accidents. Accidents should be investigated whenever someone is injured or when fire department equipment or property is damaged. Accident investigations tend to be clouded with a negative gloom, especially when a department uses the investigation to find fault with an individual rather than with the environment in which the accident occurred. As a result, many people get defensive when the department starts asking questions about their performance. Accident investigations play an important part in a department's overall safety program. Accident investigations should focus on finding ways to improve workplace safety and become the basis for developing new procedures and training programs to help prevent similar circumstances from happening again. In many cases the accident investigation and its report are required by law.

Health and Safety Committee   The health and safety officer is also a member of the department's safety committee. An important provision for improving the health and safety of members in any organization is the creation of a safety committee. To be effective, the committee should include members from management, member organizations, and the safety officer.

Safety committees have several functions: They recommend policy, respond to reports of unsafe conditions or practices, provide information for the organi-

■ **Note**

The HSO's most important job is the implementation and management of the department's safety program.

■ **Note**

Accident investigations should focus on finding new ways to improve workplace safety and become the basis for developing new procedures and training programs to help prevent similar circumstances from happening again.

zation and its members, and provide research and recommendations in matters pertaining to health and safety issues.

DUTIES FOR THE HSO

- Develop standard operating procedures (SOPs) for activities that present risks to the organization. These might include procedures to deal with both routine and unusual events.
- Develop and provide safety training programs.
- Develop and manage injury and exposure reporting systems.
- Review draft directives to ensure that safety considerations are adequately addressed.
- Develop and manage a viable fitness program with periodic testing for all personnel in the organization.
- Review regular and special reports that deal with firefighter health and safety issues.
- Research, test, and evaluate planned purchases of such items as apparatus and protective clothing and equipment.
- Keep up to date on new laws, regulations, and standards that affect firefighter safety and health issues and keep appropriate members of the department informed of the impact of these the documents.
- Investigate accidents resulting in injury or property damage.
- Respond to working incidents, including fires, hazardous material incidents, and technical rescue activities where firefighter safety is at risk, where they shall identify and mitigate such risks.
- Attend critiques of such incidents.
- Attend safety committee meetings.

Chapter 3. Training and Education

Chapter 3 of NFPA 1500 provides direction for training programs at all levels. Firefighter safety is closely related to good firefighter training. When a new recruit enters the fire service, he or she enters a new culture that includes training on a variety of basic procedures, many of which will remain with the firefighter for the rest of his or her life. That training must include information about firefighter safety. While safety should be a topic of instruction itself, it is also important to incorporate safety into all training activities. For example, when teaching the proper use of power tools, safety considerations must always be included.

If firefighters are trained properly, they will know the correct actions to take during emergency activities. If the training is sufficient, they will have acquired good, safe working habits, and hopefully forgotten any "bad habits" that they may have brought with them before entering the fire service. During times of stress, and sometimes when no one else is looking, firefighters may have the tendency to forget some of the good habits they learned during their recruit training. Many accidents occur during these moments. Safety practices and good habits learned during recruit training must be reinforced throughout the firefighter's career. The company officer has a leading role in this activity, both by setting a positive personal example and by reinforcing safety at all times. Correcting unsafe acts during training, and even during actual events, reduces accidents among firefighters. The approach should be positive and constructive.

■ **Note**
Correcting unsafe acts during training, and even during actual events, reduces accidents among firefighters.

> ### UNNECESSARY DANGERS CREATED DURING TRAINING
>
> We continue to see evidence of training efforts that, although well intended, provide dangerous situations, especially for the newest members of our department. Some departments look upon live fire training for their recruits as an initiation ceremony or rite of passage in which the instructors build fires that will cause at least a few of the new members to retreat. (Instructors with melted visors should be considered suspect.) Every year we see reports of injuries and occasionally even deaths resulting from such reprehensible conduct.

Where these training activities are conducted in acquired structures, extreme care must be given to the condition of the building. These structures are typically old and burn readily. Where these training activities are conducted in training facilities, care must be given to the type and size of fire that is built. Even the best live fire training facility will fail when repeatedly subjected to intense fires and firefighting activity. NFPA 1403, *Standard on Live Fire Training Evolutions in Structures,* provides directions for safe training in either environment, and must be understood and followed if training is to be conducted safely. In addition, it is important that the training be conducted in accordance with a plan and under the supervision of persons who have the experience needed to conduct such training activity.

■ **Note**
The training should be conducted in accordance with a plan and under the supervision of persons who have the experience needed to conduct such training activity.

Chapter 4. Vehicles and Equipment

Chapter 4 of NFPA 1500 provides directions for the safe design and operation of fire department vehicles, tools, and equipment. Probably the most visible impact of NFPA 1500 is apparent in the design of today's fire apparatus.

VEHICLE SAFETY IS IMPORTANT, TOO

The second most frequent situation in which firefighters are killed is while responding to and returning from incidents. Some of these are injuries that occur while operating emergency vehicles and some occur while volunteers are operating their private vehicles en route to the station or emergency scene. In either case they represent needless waste that does nothing to extinguish the fire or rescue the victim. In many cases, there is an accident involving injuries and property damage to civilians as well as the firefighters. This activity is inconsistent for an organization that has a mission of saving lives and protecting property.

While you may think of accidents involving apparatus and volunteers' privately owned vehicles racing to the scene as the source of these injuries, many of these injuries occur as the result of less spectacular acts. Firefighters sometimes slip and fall while *running* to board apparatus. Watching the firefighters' response at a well-disciplined fire station is almost a nonevent; the firefighters *walk* rapidly to their assigned apparatus, deliberately put on their gear, and mount up with little if any conversation. All of this is easily done in less than a minute. As a result, the crew leaves the station without gasping for air or pounding hearts. Their trip to the scene, and their performance at the scene, are certain to be safer.

While the driver of the emergency vehicle is responsible for the safe operation of the vehicle, the company officer is responsible for the safety of all of the personnel in the company and the citizens with whom they come in contact. So in the final analysis you, as the officer, are responsible for the safe operation of the vehicle.

Before the apparatus moves, all firefighters should be seated and their seat belts properly secured. This is usually routinely accomplished when the company is departing the station but sometimes forgotten when the response starts from another site. Take that extra moment to make sure everyone is present and they are all properly secured. Once the response starts, you will be busy with traffic, both radio and vehicular.

Surprisingly, accidents also occur when crews are returning to quarters. Some of this is due to fatigue and the letdown that comes after the excitement has passed. You must keep your guard up for such events. Once the apparatus is back in quarters all hands should work together to help restore the vehicle to a "ready" status for the next call.

We must remember that issues involving apparatus design and use were on everybody's mind in the mid-1980s when NFPA 1500 was developed and adopted. Chapter 4 also addresses the role of the driver as well as the design of the vehicle. Provisions of NFPA 1500 indicate that the driver is responsible for the safe operation of the vehicle; for seeing that all firefighters are seated and secure before moving the vehicle; and for obeying all traffic laws, rules of the road, and all department policies regarding the use of emergency vehicles. The standard also indicates that the officer is responsible for the overall safety of assigned personnel and equipment, including the actions of the driver (see Figure 8-6).

NFPA 1500 provides some very specific policies regarding emergency response, namely that the vehicle will be brought to a complete stop when:

- Approaching a red traffic light or stop sign
- At blind intersections, or when intersection hazards are present, or when the driver cannot account for all lanes of traffic
- When meeting a stopped school bus, at all unguarded railroad crossings, or when directed by a police officer

These policies are the result of reports of serious accidents involving emergency apparatus while responding. For those who say, "We do not need to follow those rules because we have not adopted NFPA 1500," you might start thinking about the defense you will use after an accident in which you or one of your personnel was involved, and the attorney for the state or the injured party asks about your awareness of a nationally recognized standard on firefighter safety and health.

Figure 8-6 *Although vehicle accidents occur, good safety programs can have a significant impact on the outcome. In this case, all of the firefighters were wearing seatbelts. All walked away from the event. Photo courtesy of the Forth Worth Fire Department.*

■ Note

With the arrival of NFPA 1500, the standards, for the first time, addressed human behavior. Other NFPA standards provide specifications regarding protective clothing, apparatus, and so on. In NFPA 1500 we see a requirement for departments to issue this equipment and for firefighters to use it.

Chapter 5. Protective Clothing and Equipment

Chapter 5 of NFPA 1500 sets requirements for fire departments regarding providing and requiring the wearing of appropriate personal protective equipment (PPE) (see Figure 8-7). Although the standard deals with some aspects of the design of protective clothing, it also states that members shall be provided with protective helmets, hoods, coats, trousers, gloves, and shoes to make a complete protective ensemble for the firefighter, and that departments shall establish policies requiring members to wear their protective clothing when exposed to the dangers of the job.

NFPA 1500 declares that self-contained breathing apparatus (SCBA) must be worn (and used) when the atmosphere is hazardous, or suspected of being hazardous, or may rapidly become hazardous. To ensure a good seal, the standard prohibits beards, spectacles, and other objects that could interfere with the seal

Figure 8-7 *A well-dressed firefighter ready for business. To get the intended benefit from PPE, one must use it properly.*

of SCBA against the face. The standard also provides for providing and using a personal alert safety system (PASS) device.

PROTECTIVE EQUIPMENT ONLY WORKS WHEN YOU USE IT

Firefighters are expected to perform superhuman feats. Instead of wearing a smart looking Spandex outfit like Superman wears, firefighters are encased in a suit of protective clothing that, despite considerable technological advances, is considered by some to be about as comfortable as medieval body armor. That discomfort is a small price to pay for the protection from the fire and other sources of injuries offered by today's modern PPE. But it only works if it is worn. Having an SCBA on your back is not good enough; you have to be breathing out of it to get the intended benefit!

And wearing a PASS device is not going help others find you unless it is turned on.

When firefighters repeatedly fail to properly use their SCBA or PASS device and the company officer does nothing to change the habit, who is responsible for injuries to the firefighters when they are caught in a situation where they lose their life because the SCBA was not used or the PASS device not activated?

■ **Note**
Wearing a PASS device is not going to help others find you unless it is turned on.

NFPA 1500 also provides that the equipment shall be cleaned at least once every six months. By design, turnout gear protects you from a host of hazards. In the process of providing you with that protection, the protective clothing becomes contaminated with all the products encountered during the emergencies you encounter, including carbon monoxide, hydrogen cyanide, asbestos, bodily fluids, gasoline, diesel fuel, and other contaminants. You then expose yourselves, your colleagues, other citizens, and even your family to all of that material. NFPA 1581, *Fire Department Infection Control Program,* provides additional important information.

Chapter 6. Emergency Operations

Chapter 6 of NFPA 1500 provides that emergency operations and other situations, including training, shall be conducted with appropriate safety considerations and requires the use of incident management and personnel accountability systems. Most injuries occur while firefighters are at the scene of fires and other emergencies. Here the company officer can have a great impact on the accident rate in the department. The company officer must set a positive personal example in conduct, following the department's policies, wearing personal protective clothing, supporting the incident command system, and keeping cool when the heat is on. Firefighters will see and remember a lot more from your example than they will from any lecture they may have heard in recruit school. Once the pattern is set, all members of the company must follow the officer's example.

Firefighters should be continually reviewing the latest information regarding fire behavior, building construction, fireground tactics, the proper use of tools, and of course, the proper use of the incident command system (see Figure 8-8). Many fire departments get few opportunities to practice these activities in real fire situations, and as a result, some companies become weak on basic skills. Company officers should recognize this fact and use every opportunity for training and safely practicing those basic skills.

Recognizing that emergency activities present the greatest threat to the firefighter, NFPA 1500 provides policies for managing such events. These policies indicate that an incident management system must be used and the incident commander is responsible for the overall management of that system.

Personnel accountability is also addressed in NFPA 1500. **Personnel accountability** during emergency operations has become an important topic for the fire service in recent years, and much has been done to enhance firefighter safety though better tracking of personnel during emergency events. To enhance accountability and personnel safety, NFPA requires that firefighters work in teams of two or more. During the initial phases of an event, such as a structural fire, where a team of two may be required to enter the structure, a backup team of at least two individuals must be outside the structure, immediately ready to assist the members who are inside the hazardous area. This is the "two in, two

personnel accountability
the tracking of personnel as to location and activity during an emergency event

■ Note
The requirements for two in, two out and for Rapid Intervantion Teams enhance firefighter safety by dedicating some of the on-scene resources for the rescue of firefighters who may need assistance.

Figure 8-8 *The incident commander is responsible for managing the overall incident and for the safety of the members at the scene.*

out" policy. These personnel must be in constant contact with one another by radio or other means.

The emphasis on accountability and safety during emergency operations has led to the requirement for **rapid intervention teams (RIT)**. NFPA 1500 indicates that at least one team, consisting of at least two individuals, shall be kept immediately available for purposes of rescuing firefighters if the need arises.

While accounting for and ensuring the safety of firefighters during emergency operations are vital for the safety of our personnel, caring for firefighters also includes rotating crews out of action, offering them food and liquids, rendering medical examinations and care, and furnishing shelter from extreme weather conditions (see Figure 8-9). These activities are referred to as **rehabilitation** or, simply, rehab.

Chapter 7. Facility Safety

Chapter 7 of NFPA 1500 requires that all fire department facilities must meet all applicable health, safety, building, and fire code requirements and that all facilities shall be inspected at least once a year to determine their compliance.

When you look at the injury statistics, you will notice that "other on duty" is the second largest category after "fireground" as the place where firefighters are injured. Even in the busiest companies firefighters spend most of their on-duty time at the fire station (see Figures 8.10 and 8.11). This area should be an easy target for improving occupational safety and health.

rapid intervention team (RIT)
a team or company of emergency personnel kept immediately available for the potential rescue of other emergency responders

rehabilitation
as applies to firefighting personnel, an opportunity to take a short break from firefighting duties to rest, cool off, and replenish liquids

Figure 8-9 *Fire departments should develop a systematic approach for providing rehabilitation for personnel at the scene. Photo courtesy of the Fairfax County Fire and Rescue Department.*

IMPROVING SAFETY FOR OUR FIREFIGHTERS

The current edition of NFPA 1500 indicates that a minimum of four individuals is required to start interior operations in a hazardous environment, which means that first-arriving units must assemble at least four members at the scene before entering a hazardous area. Although this policy may appear to limit the practices of some fire departments, it provides a significantly safer working environment for all members at all situations. The policy was established after careful consideration and considerable debate. In the final analysis, the purpose is to enhance firefighter safety.

A hazardous environment is defined here as an area where there is immediate danger to life and health (IDLH). It includes areas inside of a structure where SCBA is required for protection from smoke, the by-products of combustion, particulate matter given off by any materials, and oxygen-deficient atmospheres.

During the initial stages of the event, two individuals, working as a team, may enter the hazardous area. The other two individuals must remain outside, ready for rescue assistance. The outside team must remain in direct contact and have constant awareness of the activities inside. This includes the number, identity, location, and activity of those inside, including their time of entry. The standby team must have all of their PPE and the tools needed for rescue activities.

Once additional personnel arrive on scene, the incident is no longer considered in its initial stage; at this point at least one RIT is required if members are still operating in the hazardous environment.

The RIT shall consist of at least two members who shall be immediately available to rescue one or more firefighters if the need arises. Such personnel shall have complete PPE and the tools they will likely need to facilitate a rescue. Members should be familiar with the building and the ongoing operation.

Ideally, the members of the RIT are members of the same company and are accustomed to working together. One of the members should be an officer. All of the RIT members should have radios and they should remain at or near the command post until needed, or the incident is deemed stable.

■ **Note**
After the fireground, other-on-duty infuries rank second. Since personnel spend a significant part of their duty time in the station, the station environment is an opportunity for improving firefighter safety.

Most fire stations are reasonably safe and healthful. In many cities however, older fire stations have suffered from years of neglect. Equipment does not work properly, floors and steps are smooth from years of wear, and many stations still have the traditional brass pole. Stations should be designed and maintained so that tripping hazards, areas that are poorly lit, and problems resulting from poor ventilation are eliminated.

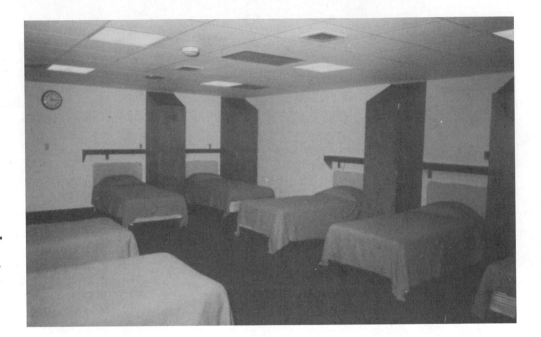

Figure 8-10 *Modern fire stations provide clean and comfortable accommodations for assigned personnel.*

Figure 8-11 *A modern fire station. The glass windows provide warmth and lighting inside the station during the daytime, and attractive artificial lighting provides a safe environment at night. While attractive, these features all indicate that resources are ready to respond to the community's needs. Both the fire service and the community can take pride in such a facility.*

HARMFUL EXHAUST EMISSIONS

We have known about the harmful emissions of our apparatus for years, yet in many fire stations firefighters are expected to eat and breathe the stuff that comes out of the vehicles' exhaust pipes. Some firefighters, especially younger ones, will complain. Older firefighters, and that might include the company officer, will say, "It has always been that way." But it does not always have to be that way. Rub your hands over the walls and across the tops of the door frames at your fire station. What you find there is in the air and in your lungs! In this case the fire department has a responsibility to help by providing the technical assistance and the financial resources to correct this type of problem.

Since firefighters "live" in this environment, they tend to become tolerant and take many of these hazards for granted. Firefighters spend a lot of time in the station, and it is their "home" if you will, while working. The combination of increased exposure and decreased vigilance provides the environment in which accidents are likely to occur. Although everyone has a responsibility for helping keep their fire station safe, you, as the company officer are clearly the person who will have the greatest impact. If you tolerate a sloppy and unsafe environment, few others will take the initiative to correct the problems. On the other hand, if you set a good positive example and take an active interest in keeping the fire station safe as well as presentable, others will be more likely to join in the effort.

■ Note
If you set a good positive example and take an active interest in keeping the fire station safe as well as presentable, others will be more likely to join in the effort.

TWO OPPORTUNITIES FOR THE COMPANY OFFICER

During a rainstorm and because someone had removed the door mat to sweep, personnel entering the station tracked the floor with their wet shoes. After several hours of this, a member, ironically returning from a doctor's appointment, entered, slipped, fell, and was seriously injured.

In another station, the lighting in the apparatus area was either on or off. Turning all the lights on at night blinded those coming from the darkened quarters, and left drivers momentarily blind as they entered traffic. Leaving the lights off presented dangers to firefighters moving rapidly to their apparatus.

Both cases had easy solutions, but in each case, someone had to take the initiative. In the first case, the problem could be easily fixed. In the second, several additional light fixtures may be needed and that may require some help. In both cases someone, likely the company officer, should take the initiative. Once the idea catches on, others will offer ideas and take action to make the station a safer workplace.

■ **Note**

Company officers, as part of their supervisory duties, should be taking an active role in the station's safety management activities to be sure that such accidents are eliminated.

Company officers should tour the station every day and inspect it once a month. Look for unsafe acts as well as unsafe conditions. Unsafe acts should be corrected immediately in a constructive manner and many unsafe conditions can be fixed right away. Other problems may take longer.

The fire department should also have a program of regular station visits conducted by the fire chief or other senior officers. The department's health and safety officer should also be involved, helping locate and document (see Figure 8-12) problems, and helping find solutions to correct them. The safety officer is the department's "expert" on station safety. Problems that are noted at one station are often present at others. Conducting these safety inspections becomes far more efficient and effective as one becomes experienced in doing so.

Fire departments should provide appropriate training to enhance station safety. Fire departments and company officers should work together to obtain, maintain, and provide practical procedures regarding the safe use of all equipment, including the physical fitness equipment in the station. Fire departments should also provide positive motivation to company officers and firefighters for fire station safety by recognizing the safest companies and stations.

Firefighters injured at fire stations are just as hurt as those who were injured at the fireground. These injuries result in the same lost time and medical costs as the injuries that occur during operational activities. Fire station injuries are embarrassing and are almost always avoidable. Company officers, as part of their management duties, should take an active role in improving the safety conditions in the station to be sure that such accidents are eliminated.

Chapter 8. Medical and Physical

Chapter 8 of NFPA 1500 requires that all members shall be given a physical examination before they start their recruit training and periodically thereafter; that departments shall establish and support programs that enable members to develop and maintain appropriate levels of physical fitness needed for the job; and that they shall establish physical performance standards for candidates and members who engage in emergency operations.

Firefighting and many of the other tasks that firefighters perform are physically demanding. These activities require both good physical condition and strength. In many ways the work is similar to that performed by professional athletes. Good professional athletes work out all year to maintain their excellent physical condition. They also have the advantage of having an opportunity to warm up before their workouts and games. Here we see a major distinction between the athlete and the firefighter: Firefighters are frequently called upon to perform at their best at the very start of an event, with no opportunity for mental or physical warmup. We wonder why strain and sprain injuries represent 40% of all injuries (see Figure 8-13), and why heart attacks and strokes represent nearly 50% of all deaths in the fire service when firefighters do not have an opportunity to warm up.

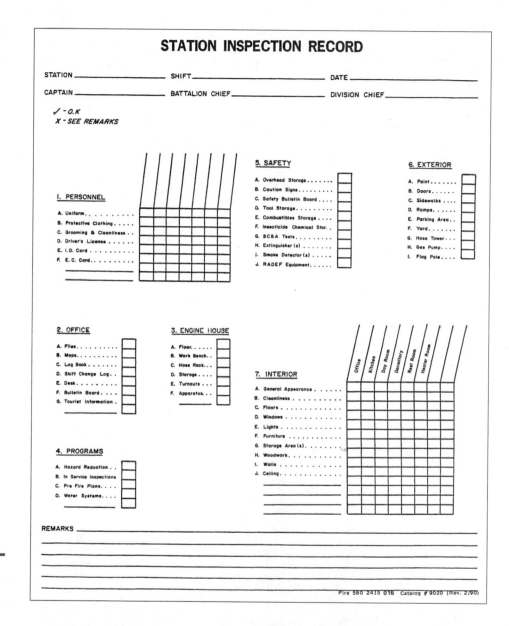

STATION INSPECTION RECORD

STATION _____ SHIFT _____ DATE _____

CAPTAIN _____ BATTALION CHIEF _____ DIVISION CHIEF _____

✓ - O.K
X - SEE REMARKS

I. PERSONNEL

A. Uniform . . . .
B. Protective Clothing . . . . .
C. Grooming & Cleanliness . .
D. Driver's License . . . . . .
E. I.D. Card . . . .
F. E.C. Card . . . .

5. SAFETY

A. Overhead Storage . . . . . .
B. Caution Signs . . . .
C. Safety Bulletin Board . . . .
D. Tool Storage . . . .
E. Combustibles Storage . . . .
F. Insecticide Chemical Stor. .
G. SCBA Tests . . . .
H. Extinguisher (s) . . . . . .
I. Smoke Detector (s) . . . . .
J. RADEF Equipment . . . . . .

6. EXTERIOR

A. Paint . . . .
B. Doors . . . . . .
C. Sidewalks . . . .
D. Ramps . . . . . .
E. Parking Area . .
F. Yard . . . .
G. Hose Tower . . .
H. Gas Pump . . . .
I. Flag Pole . . . .

2. OFFICE

A. Files . . . .
B. Maps . . . .
C. Log Book . . . .
D. Shift Change Log . .
E. Desk . . . .
F. Bulletin Board . . . .
G. Tourist Information .

3. ENGINE HOUSE

A. Floor . . . . . .
B. Work Bench . .
C. Hose Rack . . .
D. Storage . . . .
E. Turnouts . . .
F. Apparatus . . .

7. INTERIOR

	Office	Kitchen	Day Room	Dormitory	Rest Room	Heater Room
A. General Appearance						
B. Cleanliness						
C. Floors						
D. Windows						
E. Lights						
F. Furniture						
G. Storage Area (s)						
H. Woodwork						
I. Walls						
J. Ceiling						

4. PROGRAMS

A. Hazard Reduction . .
B. In Service Inspections
C. Pre Fire Plans . . . .
D. Water Systems . . . .

REMARKS _____

_____

_____

_____

Fire 580 2415 016 Catalog # 9020 (Rev. 2/90)

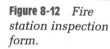

Figure 8-12   *Fire station inspection form.*

Having an opportunity to warm up would be beneficial, but other things can be done to reduce the risks firefighters face. All kinds of fitness programs are available, and while some are better than others, any program is likely to be better than none. Good physical condition (see Figure 8-14) does not eliminate the risk of injury but it certainly reduces it, and good personal fitness usually reduces the recovery time when injuries do occur. Good body strength and flexibility reduce the likelihood for most of the types of accidents that lead to strain and

Figure 8-13
Fireground injuries by cause, 1992–1996. From the annual reports on firefighter fatalities and firefighter injuries, published annually in the NFPA Journal. Used with permission, National Fire Protection Association, Quincy, MA 02269.

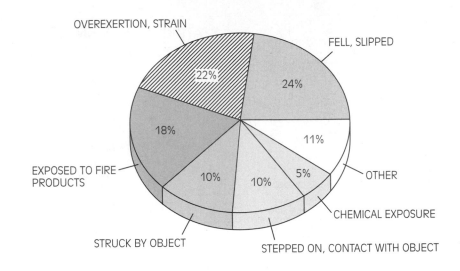

OVEREXERTION, STRAIN

22%

FELL, SLIPPED

24%

11%

18%

EXPOSED TO FIRE PRODUCTS

10%

10%

5%

OTHER

CHEMICAL EXPOSURE

STRUCK BY OBJECT

STEPPED ON, CONTACT WITH OBJECT

Figure 8-14 *Good physical conditioning is a part of good health.*

sprain injuries. Good aerobic conditioning, along with body strength, reduce the effort required for any task.

NFPA 1500 indicates that all fire departments shall have a fitness program "to enable members to develop and maintain an appropriate level of fitness to safely perform their assigned functions, and . . . shall require the structured participation of all members in the physical fitness program." Notice that the emphasis is on safety—"to enable members . . . to safely perform their assigned functions." This suggests a program that benefits the member. Many departments are showing the benefits of fitness to recruits when they join the department and are providing opportunities for them to develop to peak fitness and maintain that fitness until they retire.

Fitness testing should offer events that represent the tasks that firefighters are expected to perform on the job. Tests that combine such tasks as lifting objects, carrying something that represents an unconscious adult victim (see Figure 8-15), dragging a hose, and simulating the motions of cutting a roof with an axe, are all fair and legal. Every fire department should have this type of testing in place for all hands.

Along with a good fitness program, good nutrition can have a profound impact on how you feel and how well you perform. The firefighter's lifestyle has

Figure 8-15
Participating in a firefighter combat challenge can be one way of keeping fit.

Figure 8-16 *Good nutrition will improve the way you feel, the way you look to others, and the way you perform.*

changed from one of sitting around drinking coffee and smoking cigarettes to a routine that allows little time for either. Many firefighters are also turning away from the traditional heavy meals that added inches around their waists (see Figure 8-16).

Company training programs should include information and activities to improve the firefighters' health and safety through exercise and proper diet. As a company officer, you can do a lot to improve your firefighters' health just by setting a positive personal example for your subordinates to follow.

■ **Note**

As a company officer, you can have a significant impact on the health and safety of your personnel by just setting a positive example for them to follow.

Chapter 9. Member Assistance and Wellness Program

Chapter 9 states that departments shall provide programs to help members and their families deal with problems associated with substance abuse, stress, and personal problems that might affect a member's on-the-job performance. Members should have access to member-assistance programs that provide the training and counseling needed to deal with the physical and mental problems associated with the job.

Chapter 10. Critical Incident Stress Program

Chapter 10 of NFPA 1500 deals with critical incident stress (CIS), and states that departments shall have a program for dealing with CIS.

We are seeing more and more about this topic in the trade journals. Only recently has the fire service recognized that mental condition, as well as physical condition, is a part of the firefighter's overall well-being. Many firefighters, especially those providing EMS care, tend to burnout from the routine aspects of the job. This gradual burnout is just as real and just as debilitating as the stress that comes from "the big one."

We usually recognize the stress that comes from large-scale events—the event that results in prolonged operations and multiple loss of life. For example, airplane crashes are probably near the top of the list, but everyday exposure to lesser stuff can have the same long-term impact. More typical situations may include fatalities of all kinds, but especially those involving children or fire depart-

---

FIVE GENERALLY ACCEPTED TOOLS FOR DEALING WITH CIS

Training. Fire departments have an obligation to prepare their personnel for what they will likely encounter. This does not need to be a course in the macabre, but rather a realistic presentation of what firefighters will see.

Scene management. At the scene of emergencies, company officers can reduce the impact of surprise of gruesome scenes by telling firefighters what they can expect to see, covering bodies, controlling entry, and otherwise reducing the unexpected and unwelcome sight of human carnage.

Peer Support. Peer support comes from other members of the department. Following a critical incident, and sometimes even after several minor events, firefighters might be getting "near the edge." This is a critical point, one that can turn either way. The company officer should recognize the signs and take quick action to provide peer support for endangered firefighters.

Debriefing. As used here, a "debriefing" is frequently conducted *after* an incident where there is potential for a stressful impact on the members. The debriefing team may consist of fellow firefighters, trained mental health professionals, or others qualified in dealing with such situations. All members exposed to the stressful situation should go through this process.

Counseling. Counseling involves a mental health professional. Counseling may be needed for an especially stressful event or for the compounding effects in a series of events.

As a supervisor, you should know that a firefighter who is having problems at home is less likely to be immune to stress in the workplace. Since you are responsible for the members' overall performance, health, and safety, you should be interested in helping members resolve their problems, regardless of the source.

■ **Note**
Critical incident stress cannot be avoided, but its effects can be lessened by the support system in place.

ment members, prolonged unsuccessful rescue efforts, and others. What are you and your fire department doing to protect members from the effects of these events, and to enhance the psychological or physical well-being of your employees?

Critical incident stress cannot be avoided, but its effects can be lessened by the support system in place. Once again, the company officer has a pivotal role in helping firefighters cope with stress. As supervisors, the company officers must know and understand the mental makeup of their firefighters. They must be sensitive to the firefighters' needs and supportive of programs that will help the firefighters avoid or overcome mental stress. Discussion about the event, the emotions that follow, and assurance that these feeling are normal helps reduce more severe and long-term stress.

Chapter 11. Referenced Publications

This section, as well as three appendixes, provide additional information to help fire department and individuals to understand and implement NFPA 1500.

TEN SUGGESTIONS TO HELP COMPANY OFFICERS IMPROVE THEIR FIREFIGHTERS' SAFETY

1. Know the job. Company officers should constantly be working on improving their knowledge and skills.
2. Be observant of the actions of others.
3. Be a team player. Build team spirit and interdependence among all of the players.
4. Set a good personal example. Don't just talk the talk; walk the walk, too.
5. Be obviously concerned about safety at all times.
6. Provide constructive feedback to firefighters.
7. Communicate effectively in all your dealings, but especially in the area of safety.
8. Be assertive when necessary! Sometimes it is the only way you can get compliance. This is not a time to be timid about your feelings.
9. When all else fails, feel free to take appropriate disciplinary action.
10. Be truly positive about safety and health issues.

Summary

While individual firefighters have a responsibility for their own well-being, company officers also have a moral and legal responsibility for the safety and survivability of their firefighters. If an accident occurs, company officers must explain their action, or lack thereof.

NFPA 1500 compressively addresses a fire department health and safety program. As fire departments start to implement some or all of the procedures outlined in the standard, they will see improvements in the health and safety of their members. Many departments have already undertaken these steps and the results are already apparent: There has been a significant reduction in the number of injuries and deaths in the fire service.

Providing a safe and healthful workplace should be the goal of every employer. The results of such an approach usually mean fewer accidents, higher morale, more productivity, and reduced liability resulting from accidents. This proactive approach can result in reduced risk to the members of the fire service as well as a reduced risk to their departments. Many fire departments tend to be reactive, rather than proactive, when it comes to firefighter safety. This position is unfortunate for all.

Regardless of the fire department's organizational attitude about safety, company officers can have a significant impact on their own and their subordinates' health and safety. Many of the requirements of NFPA 1500 can be introduced at the company level. As with many of the topics discussed in this book, setting the tone and enforcing the department's policies regarding safety is one of the company officer's many responsibilities. During emergency response, whether the call be a single-unit response for a dumpster, or a three-alarm event, the company officer is responsible for the safety of the entire crew. At the scene of an emergency as well as in the fire station, firefighter safety is ultimately the company officer's responsibility.

SOMETHING TO REMEMBER

- It takes a minute to write a safe rule.
- It takes one hour to hold a safety meeting.
- It takes one week to plan a safety program.
- It takes one month to put the plan into full operation.
- It takes one year to win the chief's safety award, and
- It takes just a second to destroy it all with an accident!

  Hey, let's be careful out there!

Review Questions

1. About how many firefighter injuries and fatalities occur each year?

2. Where do most firefighter injuries and fatalities occur?

3. List the most common causes of firefighter fatalities.

4. List the most common causes of firefighter injuries.

5. Under what conditions do these injuries occur?

6. What are some of the conditions that bring stress to firefighters? What are the signs and symptoms of stress?

7. What actions can company officers take to reduce injuries at the fire station?

8. What actions can company officers take to reduce injuries while responding to emergencies?

9. What actions can company officers take to reduce injuries at the scene of emergencies?

10. What is your department's commitment to firefighter safety and health?

Additional Reading

Firefighter Fatalities in the United States in 1995 (Emmitsburg, MD: United States Fire Administration, 1996).

Kipp, Jonathan D., and Murrey E. Loflin, *Emergency Incident Risk Management: A Safety and Health Perspective* (New York: Van Nostrand Reinhold, 1996).

Peterson, William, "Fire Department Occupational Safety and Health," *Fire Protection Handbook, Eighteenth Edition,* (Quincy, MA: National Fire Protection Association)

The following NFPA standards pertaining to firefighter safety are available from the National Fire Protection Association, Quincy, Massachusetts:

NFPA 1403 *Live Fire Training Evolutions in Structures,* Latest Edition.

NFPA 1451 *Fire Service Vehicle Operations Training Program,* Latest Edition.

NFPA 1500 *Fire Department Health and Safety Program,* Latest Edition.

NFPA 1521 *Fire Department Safety Officer,* Latest Edition.

NFPA 1561 *Fire Department Incident Management System,* Latest Edition.

NFPA 1581 *Fire Department Infection Control Program,* Latest Edition.

The National Fire Academy has two excellent two-day hand-off training programs that can help any fire department start and improve its safety program. These are entitled *Health and Safety Officer* and *Incident Safety Officer.* Contact your training officer or your state fire training director for additional information.

Notes

1. Frank C. Schaper and Gregg Gerner, "Why Firefighters Continue to Get Injured and Killed," *Firefighter's News,* March, 1996, pp. 1 and 2.

2. *Firefighter Fatalities in the United States in 1995,* (Emmitsburg, MD: United States Fire Administration, 1996) p. 25.

3. For Figures 8-1, 8-2, 8-3, 8-4, 8-5, and 8-13 the information is the result of averaging the most recent five years of data available at the time. (Used with permission of National Fire Protection Association, Quincy, MA 02269.)

The Company Officer's Role in Fire Prevention

Objectives

Upon completion of this chapter, you should be able to describe:

- The common causes of fire and fire growth.
- The requirements for fire prevention and building codes.
- The fire prevention ordinances applicable to your jurisdiction.
- How to deal with inquiries from the public on fire prevention issues.

INTRODUCTION

Most fire service personnel joined their fire department to fight fires; few members of the fire service joined their organization to prevent fires. In most fire departments, fire prevention activities take second place to the organization's fire suppression activities. However, there are at least two ways to fight fires: One is difficult, sometimes dangerous, and usually results in property loss; the other, called fire prevention, is easier, safer, and less costly to the community. Fire prevention has an important place in the fire service.

As you will see from the information in this chapter, firefighters are expected to be enthusiastic about fire prevention without having a clear understanding of what fire prevention is all about, or an understanding of the fire problem in their community. The information in this chapter is intended to address both of these problems.

If we expect firefighters to be enthusiastic about fire prevention, we have to help them understand the community's fire problems, help them understand how to reduce these problems, and help them learn to effectively communicate this information to the public we serve. Once again, the company officer should take the lead, set a good personal example, and give enthusiastic support to the department's fire prevention efforts.

UNDERSTANDING THE NATION'S FIRE PROBLEM

America Burning, published in 1973, was a report of a study by the National Commission on Fire Prevention and Control. The report noted that fire prevention activities were frequently assigned a low priority in fire departments, compared with the fire department's other activities.[1] This condition often leads to a poor image of the department's prevention organization, regardless of its accomplishments. Understandably, members of the fire service did not rush to join in their department's fire prevention activities. In addition, those working in suppression assignments often had better working hours, more overtime opportunities, and better promotion opportunities. Few were willing to give up these benefits to work in prevention.

Highlights of the Problems Identified by the Commission[2]

- A need for more emphasis on fire prevention
- A need for better training for the fire service
- A need for more fire safety education for the public

To some extent, the problems identified by the commission are still true. In some fire service organizations, fire prevention is treated like a poor relative. In some volunteer fire departments, and quite understandably because of a lack of time, fire prevention activities get too little attention. But just because it has always been this way does not mean that it should continue this way. Fire suppression, while sometimes fun, suggests that there has been a failure somewhere.

PREVENTION DIVISION ATTRACTS THE BEST OFFICERS

A progressive New England fire department serving a city of 35,000 provides a unique program for getting their best officers involved with prevention activities. All newly promoted lieutenants are required to successfully serve at least a year in the prevention division. Although they give up some of the benefits of shift work, the promotion and other perks attract the best people. As a result, fire prevention activities benefit from the talents of the "rising stars" of the organization. Although many of these officers return to suppression activities when their fire prevention assignment is complete, they take with them a continuing interest in fire prevention in their community.

These officers have an opportunity to meet members of the business community while the officers are conducting inspections and other fire department business. Those relationships continue when the officers return to suppression. As a result, there is a strong personal relationship between the members of the department and the citizens of the community. It is a win–win situation.

---

fire suppression
action taken to control and extinguish a fire

fire prevention
action taken to prevent a fire from occurring, or if one does occur, to minimize the loss

Fire protection consists of two types of activities, namely fire suppression and fire prevention. **Fire suppression** is a reaction—resources are mobilized following an event. **Fire prevention** is proactive; it involves the activities that help keep fires from occurring, and if they do occur, helps reduce the size of the loss.

The subject of fire prevention received a lot of attention in *America Burning*. The report recognized that there were limits to the capabilities of suppression personnel to control fires and noted that where communities rely on suppression efforts for fire protection, they may be assuming a higher risk. The report also noted that heavy emphasis on suppression forces was not effective; that is, increased resources for fire suppression would not likely bring a corresponding decrease in fire loss.

America Burning included ninety recommendations that could help reduce fire loss in America. The report indicated that a reduction of deaths, injuries, and property loss by as much as 50% was possible within a generation.[3]

In 1987, another committee was assembled to examine the progress that had been accomplished since the first report was published. While a generation of time had not yet passed, the new committee noted that much progress had been made in the last 15 years, but there still remained much to do. The second committee also published a report, *America Burning Revisited*. The latter report noted that many of the nation's fire departments had increased their fire prevention efforts and as a result of these efforts, there appeared to be a decline in both the number of fires and the number of fire deaths. However, the report concluded that if the nation is to continue the downward trend in fire losses, several critical issues must be resolved. These issues are:

- a lack of knowledge and recognition of the fire problem, and the value of fire prevention, and
- a lack of support for a standardized reporting system to document needs, problems, and program results.[4]

More than 10 years has passed since that report, and considerable progress has been made on many fronts, but these problems still exist. As a result the United States has one of the highest fire loss rates in terms of both fatalities and dollars in the industrialized world.

■ **Note**
The United States has one of the highest fire loss rates in terms of both fatalities and dollar loss in the industrialized world.

America Burning Revisited identified several problem areas, including:

1. A lack of funds and personnel to establish community relations services that allow fire service personnel to conduct prevention activities.
2. A lack of productive use of firefighters' time for fire prevention.
3. A lack of fire safety education programs in schools.
4. A lack of widespread public understanding of the hazards of fire and its cost impact on the community.
5. A lack of funding for studying the impact of fire in a community, in other words, a cost–benefit analysis that would support fire prevention activities.
6. A lack of commitment to fire prevention programs during times of budget reductions.
7. A lack of recognition within the fire service and the public of the scope of the fire problem or the value of fire prevention.
8. A lack of widespread support for automatic residential sprinklers.

WHAT IS THE FIRE PROBLEM?

Let us briefly examine the nation's fire problem. To do that we have to look at some fire loss statistics. The national statistics may not reflect precisely the fire loss data for your community, but chances are that they are pretty close.[5] The data reported here comes from two sources: the National Fire Protection Association (NFPA) and the U. S. Fire Administration (USFA). NFPA collects data from more than 3,000 fire departments. Although this represents only 10% of all departments, the data represents most major metropolitan areas and represents fire departments serving nearly 90% of the population. The sample data is projected to represent the entire country. On the other hand, the USFA collects data from nearly 25,000 departments in forty states through the National Fire Incident Reporting System (NFIRS). Although neither system is exact, both systems provide huge quantities of very useful information. Both sources of data were used to gather the information that follows.

Recording Fire Loss

Fire loss data is generally recorded in four ways:

- from the number of fires attended
- from the dollar loss from these fires
- from the number of fire-related civilian deaths
- from the number of fire-related civilian injuries

■ **Note**

In the United States each year, fire departments respond to approximately 2,200,000 fires.

Number of Fires In the United States each year, fire departments respond to approximately 2,200,000 fires.[6] These fires range from very many small grass fires where there is little, if any, property damage, to a very few multimillion dollar losses. Collectively, these fires occur in structures, vehicles, and the outside environment where neither structures nor vehicles are involved (see Figure 9-1). "Outside" fires can include materials of considerable value, such as outside storage, timber, and crops, as well as materials with little or no value such as brush, grass, and rubbish. The largest percentage of fires occurs in the outside environment.

Figure 9-1 *Fires by type of property. Source: Fire in the United States, 1985 to 1994, (Emmitsburg, MD: United States Fire Administration, 1997). Data shown is for the calendar year 1994. The distribution remains relatively stable from year to year.*

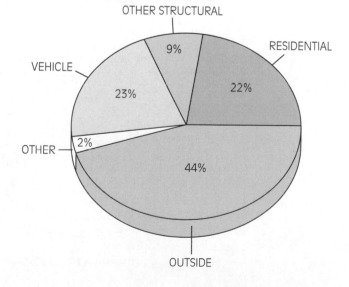

■ **Note**
Of far greater concern to most fire departments is the number of structural fires, particularly those involving residences.

■ **Note**
In terms of dollar loss, the annual national fire loss is estimated at about $9 billion.

■ **Note**
The annual loss from fires surpasses the loss from all natural disasters combined, including hurricanes, floods, and earthquakes.

In recent years we have become more aware of outside fires as a result of large wildland fires. But like the fires in other categories, these few large-loss wildland fires, as catastrophic as they are, do not represent the typical situation. Many of the reported outside fires are very minor in nature and are the result of a casually discarded cigarette, a careless camper, or the results of children playing with matches or fireworks that ignite the surrounding vegetation.

Of far greater concern to most fire departments is the number of structural fires, particularly those involving residences. As can be seen, structure fires also represent a large part of the total number of fires. We will soon see the important role that residential fires play in every category of fire loss data.

Dollar Loss In terms of dollar loss, the annual national fire loss is estimated at about $9 billion.[7] Again, that value represents a few very large-loss fires and many smaller ones. Typically the reported dollar loss represents the worth of the property in terms of the insurance loss. It does not include the cost of fire insurance or the cost of the fire suppression effort, nor does it deal with the many other losses that follow a fire, such as medical costs, loss of jobs, loss of taxes normally paid to the local community as well as to state and federal governments by the company and by the employees, and so on. Some reports suggest that the real total loss from fires, including these factors, is five to ten times the *reported* loss.[8]

Looking at that $9 billion loss, we can see from Figure 9-2 that a majority (76%) of the dollar loss occurs in structure fires and a majority of all structural fire loss occurs in residential occupancies. The average loss is about $12,000. To help put this $9 billion figure in perspective, the annual loss from fires surpasses the loss from all natural disasters combined, including hurricanes, floods, and earthquakes.[9]

Each year, the NFPA publishes a report on large-loss fires. By the NFPA's

Figure 9-2 *Value lost by type of property. Source:* Fire in the United States, 1985 to 1994, *(Emmitsburg, MD: United States Fire Administration, 1997). Data shown is for the calendar year 1994. The distribution remains relatively stable from year to year.*

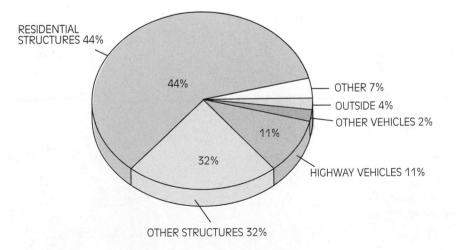

RESIDENTIAL STRUCTURES 44%

44%

OTHER 7%

OUTSIDE 4%

OTHER VEHICLES 2%

11%

HIGHWAY VEHICLES 11%

32%

OTHER STRUCTURES 32%

Table 9-1   *Recent large-loss fires in the United States.*

Event	Location	Date	Dollar Loss
Fire storm	Oakland, Ca.	October 1991	1, 500,000,000
Chemical plant	Pasadena, Tx.	October 1989	750,000,000
Riot	Los Angeles, Ca	April 1992	567,000,000
Wildland fire	Laguna, Ca.	October 1993	528,000,000

■ Note

In recent years, approximately 4,600 people died as a result of fire. A majority (71%) of these fire deaths occurred in the residential environment.

definition, a large-loss fire is an event with $5 million or more in property damage (as shown in Table 9-1). Approximately fifty to fifty-five such fires occur each year. These "big ones" range from fires in large industrial complexes to large wildland losses (see Figure 9-3). While these events represent less than 1% of all reported fires, the total financial impact of the loss in property from these few incidents typically runs from 10 to 20% of the total annual loss from all fires.[10]

Civilian Fatalities   Possibly the one fire statistic that does get some attention is the number of fire fatalities. In recent years, approximately 4,600 people died in the

Figure 9-3   *The Oakland, California, fire in the fall of 1991 is one of the largest fire losses in American history. This picture was taken one year after the fire destroyed an entire community. Photo courtesy of Kim Smoke.*

Figure 9-4 *Distribution of fire-related fatalities. Source:* Fire in the United States, 1985 to 1994, *(Emmitsburg, MD: United States Fire Administration, 1997). Data shown is for the calendar year 1994. The distribution remains relatively stable from year to year.*

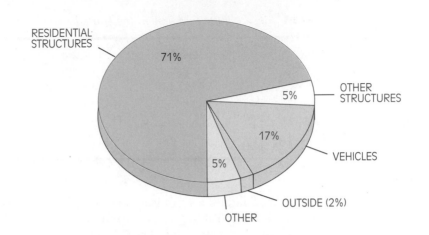

United States as a result of fire. Again, a majority (71%) of these fire deaths occurred in the residential environment (see Figure 9-4).

During the twentieth century the number of fire fatalities in this country has continued to decline. Over 10,000 Americans died each year from fires in the early 1900s. Today, it is less than half that number. Meanwhile the population of this country has grown nearly fourfold. Although we have made significant progress, the United States still has one of the highest per capita fire death rates in the world.

The very young and the very old are often the victims of fire. Statistically these sectors of our population have a higher risk and represent about a third of all fire statistics. Children under 5 years of age have twice the fire death rate of adults; survival improves significantly after a child turns six. Senior citizens are also a serious risk, because those past 70 years of age have a fire-related death rate that is 1.5 times the national average. For those past 80 years, the rate goes up even more.[11]

Looking at the population in another dimension, twice as many men as women die in fires.[12] In general, this statement applies to all age groups. This part of the fire problem cuts across ethnic and racial lines. The statistics also reveal that those who live in the inner city and those who live in the rural areas tend to have twice the risk of fire as those in the suburbs.[13] Since these two very diverse areas represent the localities where many of our poorest citizens live, we also see these communities as the scene of many fires. We examine the cause of these fires in the next section.

■ Note

The very young and the very old are often the victims of fire.

Civilian Injuries In contrast to the data on fatalities, fire-related injuries tend to peak while people are in their 20s. Young adults have a fire injury rate that is 1.5 times the average for all age groups. Approximately 29,000 fire-related injuries are reported each year (see Figure 9-5). Most fire prevention specialists agree that this

Figure 9-5 *Distribution of fire-related injuries. Source:* Fire in the United States, *1985 to 1994, (Emmitsburg, MD: United States Fire Administration, 1997). Data shown is for the calendar year 1994. The distribution remains relatively stable from year to year.*

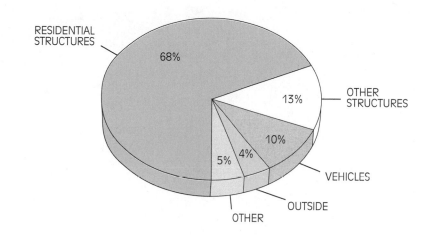

RESIDENTIAL STRUCTURES 68%

OTHER STRUCTURES 13%

VEHICLES 10%

OUTSIDE 4%

OTHER 5%

number is low in comparison to the actual number of injuries suffered as a result of fire. Some estimates are that the number of injuries is ten times that number, or 290,000.[14]

SOCIOECONOMIC FACTORS AND THE INCIDENCE OF FIRE

A recent U.S. Fire Administration report indicates that socioeconomic characteristics in the lower levels of income correlate to an increased risk of fire. Three factors appear significant in contributing to these conditions: parental presence, or the number of children under the age of 18 living with two parents; undereducation, or the percentage of adults with less than eight years of education; and poverty.[15]

Why the variation in these statistics? As in the previous categories, a majority (more than 80%) of the fire-related injuries that are reported occur in structural fires; more than 60% of all fire-related injures occur in residential fires. Many of these minor burn injuries are treated at home. Those burn injuries that require professional medical care may be treated in a doctor's office or in a hospital emergency room. In either case, unless the patient was treated or transported by the fire department, it is unlikely that their fire-related injury is reported or included in these statistics.

At this point, we should be able to see that the number of fires, the number of civilian deaths, and the number of civilian injuries all point to one occupancy class—residential (see Figure 9-6). Most of these fires occur in one- and two-family residences, where most Americans live. On a per capita basis, apartment fires also account for a large share in the total number of fires. Mobile homes have a fire fatality rate that is more than twice that for other types of residences.[16]

■ **Note**

The number of fires, the number of civilian deaths, and the number of civilian injuries all point to one occupancy class—residential.

Figure 9-6 *Even with today's modern fire protection, total losses like this still occur, especially in the rural environment.*

■ **Note**

More than two-thirds of the firefighter injuries occur in residential fires.

Indeed, residential fires are the major problem for the fire service More than two-thirds of the firefighter injuries occur in residential fires.[17] Residential fires also present a major opportunity for fire prevention for the fire service. In most communities, the fire department exercises little control over the activities that take place within the residential setting. As citizens, we tend to defend our actions in our own homes with the statement, "It's mine, I'll do what I want!" Only persistent efforts by the fire service to encourage fire safety practices will have an impact.

Causes of Fires

■ **Note**

The leading causes of fires include arson, children playing, cooking, electrical problems, heating, and the careless use of and discarding of smoking materials.

The leading causes of fires include arson, children playing, cooking, electrical problems, heating, and the careless use of and discarding of smoking materials.[18] Careless smoking and the careless discarding of smoking materials is the leading cause of fire-related fatalities in the residential environment. More than 20% of all fire fatalities in residences are the result of smoking.[19]

Heating is the second leading cause of residential fires and significantly contributes to both fatalities and property damage. This classification includes those fires where the equipment involved in the ignition sequence is a central heating source or fireplace, or where there is a portable heating device. Few of these fires are associated with central heating units; many of these fires are associated with various types of portable heating devices. Inappropriate use of the heating devices is often the cause of the fire. It appears that fuel spills are a major contributor of fires in the case of kerosene heaters. Mechanical failure or malfunction is the likely cause of the ignition sequence in portable electric heating equipment.

One of the less obvious causes of fire is a failure of the electrical distribution system. Electrical fires frequently occur during the early evening hours when meals are being prepared, and when there is a significant demand on the electrical system to either provide heat or cooling during extreme weather conditions, to heat water for bathing, or to prepare meals. Electrical fires contribute to all categories of fire loss, with some 40,000 fires, 250 deaths, 1,200 injuries, and $454 million in property loss. The latter statistic is sufficient to bring residential electrical fires into the top three categories in terms of dollar loss. Although electricity is the source of many fires, electricity is also frequently blamed for fires that really were the result of other causes.

Kitchens are the most common site of fires in the home (see Figure 9-7). Most of these fires are the result of cooking. Cooking remains the leading cause of fires in residences and cooking is by far the leading cause of fire-related injuries in residences.[20] Many of these fires are the result of unattended cooking. The second leading cause of injuries in residential occupancies is children playing with matches and other fire-related activities.

arson
a legal term denoting deliberate and unlawful burning of property

So far, we have discussed only accidental fires. Another major class of fire is generally called "incendiary" or "suspicious" or simply "arson." **Arson** is a legal term that describes the criminal process of deliberately setting a fire. Arson

Figure 9-7 *Kitchen fire. Kitchens are a common site for fires in the home. Photo courtesy of the Fairfax County Fire and Rescue Department.*

Table 9-2 *Leading causes of residential fires.*[21]

Rank	Civilian Deaths	Civilian Injuries	Dollar Loss	Number of Fires
1	Smoking	Cooking	Arson	Cooking
2	Arson	Children	Heating	Heating
3	Heating	Arson	Electrical	Arson

■ **Note**

Arson represents a significant portion of the United State's fire problem.

represents a significant portion of this country's fire problem. Arson is generally thought of as a fire involving a business or commercial occupancy, but there is considerable arson activity in residences as well. The impact of arson activity in the residential area is tremendous (see Table 9-2).

By now you should be able to see that the focus of our concerns should be preventing fires in residential occupancies, as this is where the greatest loss occurs.

THE U.S. HAS ONE OF THE HIGHEST FIRE LOSS RATES IN THE WORLD

Despite the impressive gains made in recent years, the United States still has one of the highest per capita fire death rates among the world's fourteen leading industrialized nations. In fact the most recent study, conducted in 1992, indicated that the United States is still 30 to 50% higher than its peer nations, those countries that analysts say are most like the United States. Many feel that there is little reason for the United States, which possesses a wealth of advanced fire suppression technology and fire service delivery systems, to lag so far behind other nations in terms of fire safety. However, we should note that most of this advanced fire technology is installed in public places. The fire problem is in the home.[22]

SOLUTIONS TO THE NATION'S FIRE PROBLEM

Solutions to the nation's fire problem rest with ongoing fire prevention activities. Fire prevention includes three essential elements, sometimes referred to as the three E's of fire prevention:

- Education
- Engineering
- Enforcement

life safety education
the first priority in emergency operations, life safety addresses the safety of occupants and emergency responders

■ **Note**

The lack of a clear, consistent message affects not only your effectiveness in getting the life safety message out to the public, but also the way you are perceived by the public.

■ **Note**

Four of the most common causes of fire are the result of *human behavior.*

Education

Education, in this context, involves efforts to get the fire safety message to the public. We often call this **life-safety** or **fire-safety education**. Unfortunately, not many firefighters or fire officers really understand the community's fire problem. Therefore, they may not be in a position to spot problems or answer questions of their citizens. There is good reason for their confusion. We frequently hear statistics about fire safety issues, but because of the various information we hear, and because of the inconsistency of that information, we tend not to understand or remember the important information. The lack of a clear, consistent message affects not only your effectiveness in getting the life safety message out to the public, but also the way you are perceived by the public (see Figure 9-8). In many fire departments, there is no overall strategy for dealing with fire prevention issues.

Four of the most common causes of fire are the result of *human behavior.* These include careless smoking, the inappropriate use of heating equipment, unattended cooking, and children playing. All of these human behavior problems present good opportunities for fire safety public education programs. Fire departments should inform citizens of the risks associated with these activities and thus reduce these losses. But experience has shown that changing human behavior is a difficult task.

U.S. Fire Administration and Fire Safety Information for Kids

Each year children set more than 10,000 fires. Children make up between 20 and 25% of all those killed in fires. Much of this problem can be attributed to a lack of education, guidance, and supervision.

Figure 9-8 *Public education programs can be an effective way of talking to the community about fires resulting from human behavior.*

The U.S. Fire Administration has recently unveiled a new fire safety web site for children. Entitled KidsPage, it provides child-friendly graphics, games, and other activities. There is also a section for parents and teachers that explains how to use the new webpage, discussion points for talking about fire safety. Click on the KidsPage at hppt://www.usfa.fema.gov/kids.

School children are routinely taught fire prevention and how to react when fire does occur. Although a whole generation has now been exposed to the benefits of these very worthwhile programs, we still have a fire problem in this country. Again, to cite *America Burning*, "Americans are careless about fire as a personal threat. It takes a careless or unwise action of a human being in most cases to begin a distructive fire, and . . . though (they) are aroused to issues of safety in consumer products, fire safety is not one of their prime concerns."[23]

USING TECHNOLOGY TO TEACH FIRE SAFETY

At least one fire department is using the Internet to provide safety information for the community. Entitled *The Blazer,* the newsletter offers information on everything from smoke detectors to business inspections. In addition to information for adults, the newsletter provides information and activities for children. Click on *The Blazer* at http://www.bvfrd.org.

Engineering

The second element of fire prevention is engineering. To best understand the role of engineering in fire prevention, we should look at fire prevention in a systematic way. Today, fire protection professionals recognize that there are some key components in fire-safety and life-safety system design. With regard to fire, these components are generally recognized as:

- Prevent ignition
- Control fire growth
- Provide detection and alarm
- Provide automatic suppression
- Confine the fire and its products
- Provide for the safe evacuation of the occupants

■ Note
If we can keep fuel, heat, and oxygen from meeting under adverse circumstances, we can prevent fire from occurring.

Prevent Ignition In the very early days of your fire department training program there was a discussion on fire behavior. Probably that discussion included the role of each of the three sides of the fire triangle (see Figure 9-9) in the ignition process. If we can keep fuel, heat, and oxygen from meeting under adverse circumstances, we can prevent fire from occurring.[24] When these three components finally meet,

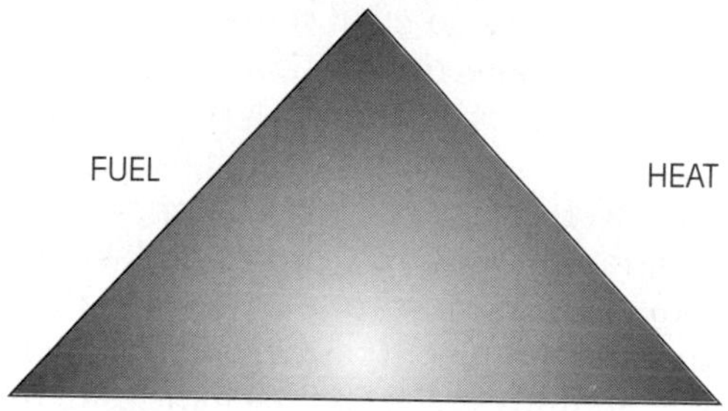

Figure 9-9 *The fire triangle helps explain the ignition and extinguishment of fires. It also explains fire prevention. Fire prevention regulations attempt to prevent heat and fuel from coming together. When these come together fire is always a threat.*

FIRE TRIANGLE

FUEL

HEAT

AIR

there has frequently been a human act or omission that set the stage for the fire. You cannot do much about the oxygen that surrounds us; you can attempt to control fuel, the heat, and the human behavior.

Human behavior has already been discussed. You can provide education programs that will help humans keep from making mistakes that lead to fire. Through engineering efforts, you can focus on controlling fuel and heat.

Fuel can be controlled by what we allow into our surroundings. We can exclude things that are highly flammable by product regulation. We can control the use of other products through laws and programs that permit dangerous activities only under controlled conditions. We can control heat by regulation and product design. In many cases, the regulatory systems have already made decisions regarding these safety issues before the product reaches the marketplace.

Control Fire Growth Having failed to prevent the fire, you should next focus on controlling the fire. The goal should be to keep the fire from extending beyond the room of origin. Statistics clearly show that when the fire can be contained in the room in which it started, or the "room of origin," both human and dollar loss are significantly reduced. We are all familiar with strategies induced by building and fire codes that help accomplish this goal by limiting the use of flammable materials in room contents and furnishings, and by using fire resistive barriers to surround the room.

Provide Detection and Alarm Systems Along with controlling the fire, it is important to provide timely notification of its presence. Automatic fire detection and alarm equipment provides this service. Humans are capable of recognizing a fire when they are present, alert, and awake. However, when they fail to meet one of these

■ **Note**
Statistics clearly show that when the fire can be contained in the room in which it started, or the "room of origin," both human and dollar loss are significantly reduced.

three requirements, their ability to recognize a fire and act appropriately is greatly reduced. Thus, if they are asleep, or intoxicated, or under strong medication, they may not be able to recognize the threat or react to it. Obviously, when no one is present, we also need some form of automatic fire detection.

Fire detection equipment has been around for years and is required in nearly all occupancies except private residences. Starting in the 1970s, residential smoke detectors became readily available. In most jurisdictions, they are now required by law. Smoke detectors do not prevent or control fires, but they do provide notification of the fire's presence. This notification provides opportunity for safe exit and quicker fire suppression efforts. Smoke detectors are credited with cutting the number of residential fire fatalities approximately in half and the reduction in the number of fire fatalities in this country can be directly correlated with the recent increase in smoke detector use.

In spite of these accomplishments, we still have problems. Although the benefits of smoke detectors are generally well known, the number of homes without an operating smoke detector should be a concern to every fire professional, for a majority of residential fire fatalities occur in homes without operating smoke detectors.[25] There are many reasons for these problems. Some citizens claim that they cannot afford a smoke detector. Where smoke detectors have been installed, the occupant's failure to replace a worn-out battery, or the removal of the battery for other uses or because of false alarms, lead the occupants to a false sense of security. Again, here is an opportunity for continuing involvement by the local fire department.

Provide Automatic Fire Suppression Having failed to prevent the fire from occurring, and having provided notification, we next need to provide for timely automatic fire suppression. Automatic sprinklers have been around for more than a century. Sprinklers were initially installed to protect property rather than human life. Some modern sprinkler systems, such as those designed for use in residential occupancies, are primarily to enhance life safety. Modern automatic fire sprinkler systems can be installed to provide protection in every occupancy class.

Fire extinguishers are another effective tool for reducing fire loss. Although required in many occupancy classes, they remain only in the "recommended" status in homes. Fire departments should encourage homeowners to purchase fire extinguishers and provide homeowners with information regarding the proper selection and use of fire extinguishers.

Confine the Fire and Its Products As the size of the fire progresses, designing safety features to cope with the problem becomes more complex. Clearly, here, if not before, we see the value of a systems approach to fire safe engineering practices. One has to look at the design as well as the use of the building. The designer can build a safe building. But when a fire door is left open, or when a wall or ceiling is penetrated, fire will find a path for extension. Again, fire loss data reveals that as the fire grows beyond the room of origin, and even beyond the floor of origin, life and property loss increase dramatically. It is important to recognize that these efforts must not only control the fire, but also its products. Some proven strategies for fire confinement include:

- Use solid construction barriers.
- Use doors and windows that help preserve the integrity of the room, area, and floor where fire starts.
- Use the airhandling system to remove harmful products and preclude it being used to transfer these products to other portions of the structure.
- Provide for the safe evacuation of the occupants.

Provide for Safe Evacuation Regardless of the effectiveness of any of the previous steps, our most important task is to protect human life. In this regard you must consider both the protection of the occupants and their safe evacuation from the building. Evacuation involves consideration of both building design and human behavior. The building must provide for at least two effective means of egress that are properly marked, adequate in capacity, and protected from the conditions that prompted the evacuation. The next step is to be sure that all occupants know how and when to use these exits. Many facilities hold practice fire drills just to be sure that the occupants understand the importance of their rapid and safe evacuation. How many families undertake the same drill in their homes?

Enforcement

in-service company inspections
using suppression companies to conduct fire safety inspections in selected occupancies, usually within the company's first-due assignment area

Enforcement, the third fire prevention element, is often represented by the inspection programs that motivate a lot of other fire prevention activities. Using the enforcement process generally does not mean that the other fire prevention efforts have failed, because it really takes all three—public fire safety education, engineering, and enforcement—to make an effective fire prevention program work. Inspection activities (see Figure 9-10) are the means by which fire departments can monitor, and if necessary correct, any deficiencies that may pose a hazard to the community. While some inspection activities require the resources of the department's prevention division, many fire departments are effectively using their suppression personnel to help in a prevention role.

■ Note
Fire suppression personnel can effectively inspect many occupancies to detect and even facilitate correcting common fire code violations.

Fire suppression personnel can effectively inspect many occupancies to detect and even facilitate correcting common fire code violations. These inspections are frequently referred to as **in-service company inspections**, for the company is in service and available for a call, just as much as if they were at the fire station.

Permits, Certificates, and Licenses Inspections should include places of public assembly, business and commercial locations, as well as educational, institutional, industrial, and storage facilities. In-service companies should visit all such occupancies in

Figure 9-10 *Fire suppression personnel can effectively inspect many occupancies to detect fire hazards.*

permits
fire prevention tool required where there is potential for life loss, or where there are hazardous materials or hazardous processes

certificates
a document serving as evidence of the completion of an educational or training program; term also describes a document issued to an individual or company as a fire prevention tool

licenses
formal permission from an authority to participate in an activity

their first-due area, checking the means of egress, checking for safe activities, and checking for the proper storage and use of materials. During these inspections, permits, certificates, and licenses that are in effect should be checked to see if they are current and if the requirements of these documents are being satisfied.

Permits are usually issued by the fire prevention division of the fire department. They authorize a specific activity, provide the conditions that must be satisfied for that activity to take place, and provide notice to the nearest fire company and other government agencies of the ongoing activity. **Certificates** are issued to individuals or companies who engage in activities of interest to the fire department. These may include those who install and service fire protection equipment as well as those who handle explosives. **Licenses** are issued to individuals to engage in business activity.

All three of these documents provide a means by which the fire department can monitor and enhance the safety of business activity within the community, ensure compliance with applicable laws and regulations, and provide a system of notification to government agencies of such activity. Many businesses have all three documents.

Fire service personnel should also check to verify if the installed fire protection equipment is being properly inspected, maintained, and tested (see Figure 9-11). The installed fire protection equipment may be the fire department's greatest ally during an emergency.

While all of this activity is going on, fire service personnel are getting two added benefits. First, they all get a chance to see the inside of the property first-

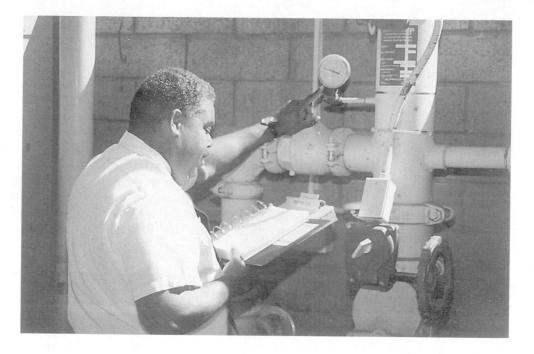

Figure 9-11 *Fire suppression personnel should always check installed fire protection equipment.*

hand. While we discuss the value of preplanning and site survey information in Chapter 12, you should never miss an opportunity to tour a facility within your first-due assignment area.

The second benefit is a little less tangible, but just as important. Fire prevention activity allows the community to see fire service personnel under much different conditions than when they are observed during emergency activities. Many citizens will see their fire department for the first time up close, will see its human side, will learn more about their firefighters, and about their mission. When many citizens see a company in the community performing a fire prevention inspection, they may realize for the first time that those firefighters really do care about their community. It all pays off: A good fire prevention inspection, an opportunity to check the prefire plan, and some good public relations, all in one visit. Many departments, and many companies within fire departments, find that this is an excellent use of their personnel.

Some fire departments have carried the benefits of their fire prevention inspection program to the residential setting. In some communities, fire prevention inspections of private residences are authorized by law. Even where such activity is not authorized by law, voluntary residential fire inspections are an effective means for reducing fires. As we have already noted, we have the greatest number of fires, the greatest number of lives lost, and the greatest number of injuries in the residential setting. The fact that most other occupancy classes have much better fire safety records is due in part to past inspection efforts.

■ Note
Fire prevention activity allows the community to see fire service personnel under much different conditions than when they are observed during emergency activities.

RURAL DEPARTMENT HELPS ITS NEIGHBORS

Challenged with updating the mapping of their response area, one combination fire department in a rural area combined its mapping activity with preplanning and a great fire safety public education program. The members divided the county into sections and developed a plan to cover every road in the county. While the units (fire trucks and ambulances) were traveling these roads, they made maps, developed preplans, and visited every home and business. During these visits they provided citizens with fire safety literature and offered to check smoke detector installations. Where necessary, the department provided replacement batteries and even provided smoke detectors for its citizens. Several large businesses in the community helped with the costs of these activities in exchange for modest recognition.

In case you are wondering what all of this has to do with the average company officer, consider that at every step of the education, engineering, and enforcement process we have discussed, there is an opportunity for every member of every company of every fire department to be working in their community on fire prevention through public fire safety education programs and through understanding and supporting the engineering systems approach to fire prevention. Most of this activity can be done by firefighters already on duty at little added cost to the department.

THE ROLE OF BUILDING AND FIRE CODES IN FIRE PREVENTION

Fire department personnel have the authority to set standards for fire safety engineering practices, to require permits for certain hazardous activities, to undertake code enforcement activities to make sure that fire-safe practices are being followed, and when necessary, take appropriate action to stop activity that might lead to fire or explosion and endanger the community. How does the fire department get such authority?

The authorities and responsibilities of fire officials are typically defined by state or local laws. These laws frequently provide for the adoption of a building code and a **fire prevention code**. These codes provide for enforcement authority, the right for fire officials to enter and inspect property, the right for them to issue violations, and the right for them to order the evacuation of unsafe buildings when justified. These are significant powers. To see how they impact on our fire prevention activities, let us briefly look at how they evolved.

Evidence of building codes can be found as far back as 1700 B.C.[26] Many of the early building codes focused on structural collapse, but as cities grew and became more congested, the concern for fire brought provisions that would reduce the likelihood of a conflagration.

fire prevention code
legal document that sets forth the requirements for life safety and property protection in the event of fire, explosion, or similar emergency

Figure 9-12   *Fire was a significant threat in colonial America.*

Fire codes go back to 300 B.C.[27] In spite of these efforts, large fires have destroyed substantial portions of most of Europe's larger cities during the last 2,000 years.

When America was being colonized, the settlers were well aware of the hazards associated with fire (see Figure 9-12). The reality of fire was a threat from the very outset. Fire struck at Jamestown in the first year of that settlement and shortly thereafter in Boston. As a result, some of the very first laws enacted in the new colonies were concerned with fire safety. These laws were intended to prevent the kind of fires that had occurred in Europe.

building code

law or regulation that establishes minimum requirements for the design and construction of buildings

WHAT IS A BUILDING CODE?

A **building code** is a legal document that sets forth requirements to protect the health, safety, and general welfare of the public as they relate to construction and occupancy of buildings and structures. The typical building code regulates the construction, alteration, and demolition of buildings, and covers structural design, exits, fire protection, lighting, sanitation facilities, and ventilation.

Many American cities developed building and fire codes and some even made an effort at code enforcement. However, many other American cities grew without the benefits of fire codes. Many professionals saw the risk of fire in

Chicago before the fire in 1871, and as a result, some insurance companies had stopped writing policies in that city. The fire that started on October 8, 1871, resulted in what would be a $2 billion fire in today's dollars. Needless to say, the insurance companies that remained took considerable losses.

Many of the early codes were developed by the insurance industry as guides for local governments in the attempt to stimulate regulations regarding the construction of buildings within their jurisdictions. The insurance industry had an interest in this process—it was an effort to help them control the risks they faced. These codes were the forerunners of the codes we see today that deal with health, safety, and public welfare. Frequently, a tragic event prompted the writing of a code to prevent a recurrence. But a code or practice developed by an industry does not have the effect of law. How then, does the fire department and its representatives have the authority to impose laws on the citizens?

The Code-Making Process

When examining the laws in the United States, we must start with the U. S. Constitution. In its original form, that remarkable document had only seven articles and would easily fit on two pages of this book. Those original articles set up the form of our federal government. Shortly after the Constitution was adopted, the first ten amendments, known as the Bill of Rights, were adopted. The last of these amendments provided that, "The powers not delegated to the United State by the Constitution, nor prohibited by it to the States, are reserved to the States respectively, or to the people." Since the U.S. Constitution does not contain any statement regarding building construction or fire prevention, the task of fire prevention and control passed to the states.

Each state has taken its own approach. Many states have adopted one of the model codes, like the one developed by the insurance industry. In fact, the insurance industry code was the only building code in the country until 1927 when the *Uniform Building Code* was published by the Pacific Coast Building Officials Conference. Because of different concerns in the southern part of our country, officials in that area developed a competitive document, the *Southern Standard Building Code*. That document was first published in 1945. Five years later, the Building Officials Conference of America, which later became the Building Officials and Code Administrators International (BOCA), published the *Basic National Building Code*. That code is now used in much of the country. BOCA provides a family of model codes, including codes for plumbing, electricity, and fire prevention. The National Fire Protection Association (NFPA) also has a fire prevention code. These codes are revised on a regular basis.[28]

The means by which the states adopt these codes and standards varies widely. Only a few states have statewide codes for building construction and fire prevention. Many states delegate code enforcement to the localities. Some states specify the code to be adopted if one is to be adopted, but leave to the local jurisdictions the choice of adoption and enforcement. Some states have codes that apply only to state-owned or state-financed construction. Many states leave to the local jurisdiction all matters of fire protection, both fire suppression and fire prevention.

Having a statewide code has many advantages. Although it is difficult for building and fire officials to administer several codes, imagine the problems of a housing or commercial developer. Every project requires some extra research to determine the code requirements applicable for the particular site. The cost of this activity is eventually passed on to the consumer. Efforts to reduce this confusion are ongoing and some wish to carry the logic of a standard code one step further: An effort to consolidate the several existing fire codes into one national fire prevention code is presently under way.

Codes and standards are published by organizations with an interest in standardizing or regulating some activity. There are several hundred such organizations in the United States. Usually the standard or code is actually written by a volunteer committee of experts who represent government, affected businesses, the insurance industry, the public sector, and citizens or consumers. These committee members draft a **consensus document** that is usually voted upon (approved) by the organization's entire membership. That document is then made available to the public for adoption.

Codes are generally written as **minimum standards**, meaning that the code provides the minimum level of acceptance needed to meet the code requirements. Unfortunately, given the nature of our society, the minimum is usually all we get.

> ### WHAT IS A FIRE PREVENTION CODE?
>
> A **fire prevention code** is a legal document that sets forth the requirements for life safety and property protection in the event of fire, explosion, or similar emergency. Its purpose is to minimize the risk of loss of life and property by regulating the use and storage of materials that might be on the property; by regulating the installation, maintenance and testing of fire detection and suppression systems; and by requiring access for fire apparatus.

Codes can be of two types. Some codes provide **specifications** by which a building is to be built, describing in detail the material to be used, the size and spacing of these components, and the methods of assembly. Other codes define the overall objectives or the **standard of performance** to be accomplished and leave to the designer how to accomplish these requirements. Most of the codes are moving to the performance-based approach. Performance codes are generally favored because they encourage the design and adoption of new materials and practices.

■ Note

An effort to consolidate the several existing fire codes into one national fire prevention code is presently under way.

consensus document
the result of general agreement among members who contribute to the document

minimum standards
as used in codes and standards, it indicates the least or lowest accepted level of attainment that is acceptable.

specifications
spell out in detail the type of construction or the materials to be used

standard of performance
a defined level of accomplishment or achievement

But a code or standard is no more than just a document. A code or standard has no real legal status until it is adopted by a government agency. It only applies to those citizens who are subject to the jurisdiction of that government agency. Codes passed by one state do not apply to the citizens of another state.

LOCAL CODE ENFORCEMENT AUTHORITY IS LIMITED IN SOME STATES

Government police powers, the fundamental power of the state to restrict the personal freedoms and property rights of individuals for the protection of the health, safety, and welfare of the public, is an inherent power of the states, possessed by them before the adoption of the United States Constitution and reserved to them under the tenth amendment. However, the tenth amendment makes no mention of powers reserved to the municipalities. That is because there are none.

Local governments have only those powers delegated by the state governments. In the late 1800s Chief Justice John Forrest Dillon, of the Iowa Supreme Court, wrote what was to become known as **Dillon's Rule**. As local governments in other states contested their authority in the courts, a majority of the states adopted Dillon's rule.

Simply stated, under Dillon's rule, local governments have only those powers expressly conferred by the state constitution, state statutes, or home rule charter, and those powers implied in, or incidental to, the powers expressly granted, and those powers essential to the declared purposes of the local municipality.

As a result, many states have limited the authority of local municipalities to make decisions regarding the adoption and enforcement of fire and building codes.

Dillon's Rule

a legal ruling from Chief Justice Dillon of the Iowa Supreme Court whereby local governments have only those powers expressly granted by charter or statute

The Role of the Federal Government in Fire Prevention

We noted earlier that the federal government has delegated to the states those powers not expressly assigned to the federal government. The power to deal with fire and building related matters is an example of this delegated power. However, in an interest to help the states, there are areas where the federal government has taken the lead. The Department of Health and Human Services and the Department of Labor, particularly the Occupational Safety and Health Administration (OSHA) have a significant role in the design and use of buildings where people live and work. The Consumer Product Safety Commission (CPSC) has an impact on the things we buy. By controlling the materials used in clothing and furniture, the CPSC is able to reduce the consequences of fire in these materials. Examples of this activity include control of the materials used in upholstering furniture and in children's sleep wear.

BUILDING CODES, FIRE CODES, AND CITIZENS' RIGHTS

Because of the general public interest in safety, we accept the government's intrusion into our personal lives in regulating the construction and use of certain structures. Building codes set forth legal requirements for the construction and occupancy of buildings to protect life, health, and the general welfare.

Building codes deal with the construction and major renovations of buildings and deal with such topics as fire safety, means of egress, light and ventilation, sanitation, and general structural design. Codes usually provide for height and area limitations of structures, with credit offered for more fire-resistive construction and for installed fire suppression systems. In many communities the building code is administered by the building department.

Fire prevention codes, on the other hand, deal with the use of the building *after* it is occupied. Fire codes are usually administered by a fire official. For example, the building code provides information regarding the number and location of exits. The fire code provides language regarding the proper maintenance of these important life safety features. Fire codes also provide for maintaining installed fire protection equipment.

Several final comments regarding codes and standards:

- They tend to be rather intimidating, but if you take the time to read and learn a little of their content, you will find the answers to many commonly encountered questions.

- You are not expected to memorize the codes, but feel free to carry the code book during inspections. If that is not convenient, look the questions up after the inspection.

- Most code- and standard-writing organizations offer staff assistance and usually answer questions on the phone.

Summary

Every fire officer and firefighter in the fire service has an opportunity to make a significant impact on the fire-loss situation in their community. Fire prevention can save lives and property just as much if not more than making an aggressive interior attack or making a dramatic rescue. Your actions will not make the front pages, but they will make a difference to the community you serve.

Review Questions

1. What are the two ways a fire department can protect its community from fire loss?

2. The Commission on Fire Prevention and Control identified three problems that inhibit fire prevention activities. What were they?

3. Name four ways fire loss is reported. What are recent values for each of these?

4. What are the leading causes of fire?

5. What is the leading cause of fire deaths?

6. Where do most of these deaths occur?

7. How has the smoke detector reduced this statistic?

8. What are the three "E's" of fire prevention?

9. How are fire prevention codes adopted?

10. How can company officers boost fire prevention efforts in the community they serve?

Additional Reading

America Burning, The Report of the National Commission on Fire Prevention and Control (Washington, DC: U.S. Government Printing Office, 1973).

America Burning Revisited, (Washington, DC: U.S. Government Printing Office, 1990).

Bender, John F., "Fire Prevention and Code Enforcement," *Fire Protection Handbook,* Eighteenth Edition (Quincy, MA: National Fire Protection Association, 1997).

Cote, Arthur, "Building and Fire Codes and Standards," *Fire Protection Handbook,* Eighteenth Edition (Quincy, MA: National Fire Protection Association, 1997).

Fire in the United States. Published regularly by the U.S. Fire Administration. The eighth edition, covers data from 1983 to 1990. The ninth edition covers the period 1985 to 1994.

Hall, John R., "America's Fire Problem," *Fire Protection Handbook,* Eighteenth Edition (Quincy, MA: National Fire Protection Association, 1997).

Powell, Pamula, "Fire and Life Safety Education," *Fire Protection Handbook,* Eighteenth Edition (Quincy, MA: National Fire Protection Association, 1997).

Robertson, James C., *Fire Prevention,* Fourth Edition (Englewood Cliffs, NJ: Prentice Hall, 1995).

Regular reports regarding Annual Fire Loss, Loss Fires, and Firefighter Injuries and Fatalities appear in the *NFPA Journal,* the membership publication of the National Fire Protection Association.

Notes

1. *America Burning,* The Report of the National Commission on Fire Prevention and Control. U.S. Government Printing Office, Washington, DC, 1973. p. 7.

2. Ibid., p. x.

3. Ibid., p. 8.

4. *America Burning Revisited.* U.S. Government Printing Office, Washington, DC, 1990. p. 93.

5. You should be able to get information like that presented here that describes your community's fire problem. Ask your fire prevention staff to send you their latest fire loss report. Many departments publish this data on an annual basis. In addition, *Fire in the United States* provides data for each state. The report is available from the U.S. Fire Administration.

6. *Facts about Fire in the U.S.* United States Fire Administration, Emmitsburg, MD, 1995. p. 6.

7. *Fire in the United States 1985–1994,* ninth edition, U.S. Fire Administration, Emmitsburg, MD, p. 1.

8. Ibid., p. 33.

9. Ibid., p. 35.

10. "Large Loss Fires", an annual report from the National Fire Protection Association, Quincy, Massachusetts, is published in the November/December issue of the association's publication, *NFPA Journal.*

11. *Fire in the United States 1985–1994,* p. 43.

12. Ibid., p. 37.

13. Ibid., p. 22.

14. *America Burning,* p. 14.

15. *Socioeconomic Factors and the Incidence of Fire,* United States Fire Administration. (Washington, DC, Federal Emergency Management Association, 1997), pp. 2–3.

16. *Fire in the United States 1985–1994,* p. 5.

17. Ibid., p. 4.

18. Ibid., p. 8.

19. Ibid., p. 78.

20. Ibid., p. 78.

21. Ibid., p. 61.

22. *Fire Death Rate Trends: An International Perspective,* United States Fire Administration. (Washington, DC, Federal Emergency Management Association, 1997), pp. 1–2.

23. *America Burning,* p. 11.

24. Heat, fuel, and oxygen represent the three sides of the traditional fire triangle. Using more current terminology, we substitute the term *energy* for fuel and *oxidizer* for oxygen.

25. *Facts about Fire in the U. S.,* p. 6.

26. *Legal Aspects of Code Administration.* Building Officials and Code Administrators International, Inc., Country Club Hills, Illinois, 1984. p. 9.

27. Arthur Cote and Percy Bugbee, *Principles of Fire Protection,* Second Edition (Quincy, MA: National Fire Protection Association, 1988), p. 2.

28. Ibid. p. 10.

Chapter

10

The Company Officer's Role in Understanding Building Construction and Fire Behavior

Objectives

Upon completion of this chapter, you should be able to describe:

- Basic types of building construction used by the fire service.
- The strengths and weaknesses of each building type.
- Basic fuel loading in structures.
- Fire ignition, growth, and development in structures.
- Fire control and extinguishment.

INTRODUCTION

In the previous chapters we discussed the preparation of officers in the fire service to perform the nonemergency tasks of the position. We change our course here a bit and look at the operational aspects of the job. No doubt you are an experienced firefighter. Many experienced firefighters have successfully moved from firefighter to fire officer with little or no formal training. To their credit, they were quite successful. But just think how much better they would have been if they had been trained and given a chance to prepare for that assignment.

This situation is especially true for new officers in operational situations. A new officer, riding on an engine or in a truck company arriving at the scene of a working fire or other emergency, may be overwhelmed. Some of this is due to a lack of experience but some may be due to a lack of training.

At the scene of emergencies, the first arriving officers are usually concerned with limited resources and time, and many unknown factors. Working in these conditions is difficult and dangerous; leading others under such conditions can be extremely challenging. In such adverse conditions the company officer is expected to be calm and decisive, to issue clear orders, and to keep track of all the activities. The company officer is responsible for the safety of others. The company officer is responsible for at least some aspect of the management of the incident, and in many situations, may be the incident commander.

As you move to the officer's position in the organization, the roles change quite a bit. As an officer, you are now the manager and leader of the company. We have talked about management and leadership in previous chapters. All of that information applies in your role as a leader during emergency activities as well.

In the next four chapters we discuss the company officer's role in identifying and mitigating problems at the scene of emergencies. This chapter focuses on information needed for fighting fires in structures; we saw in Chapter 9 that structural fires are frequent, and the loss is often high. This chapter deals with building construction and fire behavior. We will get to tactics in Chapter 13.

■ Note

Working in emergency conditions is difficult and dangerous; leading others under such conditions can be extremely challenging.

TODAY'S FIREFIGHTING REQUIRES EXPERIENCE, KNOWLEDGE, AND SKILL

Modern firefighting requires more than just experience; it requires knowledge and skill. With the gradual decline in the number of serious fires encountered, good officers will learn from every opportunity, including actual working incidents, participating in training and drills, attending seminars, and reading about the subject as much as possible.

Many years ago, the chief of the London Fire Brigade, Massey Shaw, visited the United States. Chief Shaw later wrote the following:

When I was last in America, it struck me very forcibly that although most of the chiefs were intelligent and zealous in their work, not one that I met even made a pretense to the kind of professional knowledge that I consider so essential. Indeed one went so far as to say that the only way to learn the business of a fireman was to go to fires. A statement about as monstrous and as contrary to reason as if he had said that the only way to become a surgeon would be to commence cutting off limbs without any knowledge of the human body or of the implements required.

There is no such short cut to proficiency in any profession and the day will come when your fellow countrymen will be obligated to open their eyes to the fact, that, as a man learns the business of a fireman only by attending fires he must of necessity learn it badly. Even that which he does pick up and may seem to know, he will know imperfectly and be incapable of imparting to others.

I consider the business of a fireman a regular profession requiring previous study and training as other professions do. I am convinced that where training and study are omitted and men are pitchforked into the practical work without preparation, the fire department will never be capable of dealing satisfactorily with great emergencies.[1]

Chief Shaw's comments may seem disrespectful to some, but in the time frame in which they were offered, they may have been amazingly perceptive. Before you take sides here, we want to point out that the fire service *has* made a lot of progress in the development of the very professional knowledge that Chief Shaw found deficient.

Regardless of one's experience, training, and education, even the best officers can fail, through no fault of his own. Emanual Fried, in his classic text *Fireground Tactics,* offered the following advice:

Command on the fireground is a demanding task for the officer. To improve the ability to handle the situations we have to concede certain basic faults that we find in ourselves.

Admit these possibilities:

you are going to get excited

you are going to yell

you are going to make mistakes

you are going to lose buildings.[2]

> ■ **Note**
> Many factors contribute to the company being ready to deal with emergency situations, including your knowledge of the job.

In spite of these challenges, you and your fire department are in the business of saving lives and property during fires and other emergencies. You respond to a variety of events hoping that you will be able to contribute something toward the successful outcome. Fortunately, most events do have successful outcomes. This does not happen by chance. Many factors contribute to the company being ready to deal with emergency situations, including your knowledge of the job.

Your Role in Keeping Ready

The basic unit of emergency response in the fire service is the company. It may be an engine company, ladder company, or other functional unit. A company may respond alone from a volunteer station in a small community or with several other companies in a crowded urban environment. A company may work independently at small-scale events, or as part of a multiunit response assignment at significant events. Regardless of the location of the community or the size of the event, the company's role is critical to the successful outcome of the event.

Readiness of the company's type or location, personnel and equipment are also important. The assigned personnel must be physically fit and well trained to perform any task they may face. They must be able to work together effectively, combining the strengths of the individual members to form a team. The team must be prepared to work effectively to achieve the tasks needed at the emergency scene.

As a company officer, your role at the scene of emergencies is to lead or direct the activities of others to accomplish the organization's goals. To do this, you must have the knowledge and skills to understand what is happening. You must also understand what action is needed to mitigate the situation. You must be able to effectively communicate your ideas to others, and finally, you must be a manager and weigh the risks involved, considering personnel safety, the limited time and resources available, and your own understanding of the situation.

Firefighting Is Still Our Business

Most textbooks and training programs still reflect the traditional nature of the fire service: They focus on fire-related emergencies. In reality, fire departments provide a host of other services to their community, including the delivery of emergency medical care, the mitigation of incidents involving hazardous materials (see Figure 10-1) and various rescue capabilities usually tailored to the local community's needs. As a result, we are seeing a decrease in the percentage of fire-related emergencies as a part of the number of total calls received by the fire department.

There is another factor that affects the department's firefighting activities. Where good fire prevention activities are in place, there are fewer fires. And where automatic fire detection and suppression equipment are present, timely fire notification and control lead to smaller and safer fires. This is good news for the fire service and for the community. However, it does present the fire service with a challenge: The basic fire department component, the company, is expected to be proficient in doing a greater number of things, while having fewer opportunities to practice the basic skills needed for their original task, effective fire suppression.

All of this suggests that there is much to be done to prepare the company for the challenges it faces. Firefighters and fire officers should take advantage of every opportunity to plan and train for the various situations they may face. They must be familiar with the community they serve and the activities that reg-

■ **Note**
Regardless of the location of the community or the size of the event, the company's role is critical to the successful outcome of the event.

■ **Note**
Firefighters and fire officers should take advantage of every opportunity to plan and train for the various situations they may face.

Figure 10-1
Firefighters are called upon to deal with many emergencies.

ularly take place in that community. They must be familiar with the special hazards that exist, hazards that pose threats to themselves and the citizens they are obligated to help during an emergency.

UNDERSTANDING THE COMMUNITY'S NEED

Fire departments should be assessing their communities' risk factors. This is the first step in identifying the strengths and weaknesses of the department in dealing with emergencies. The company has a role in this process as well, and should have a significant interest in the outcome. Regardless of whether the community has one fire company or a hundred, some part of the community should be a particular fire station's primary concern. Members of the company responsible for that area should be interested in the risk factors in that area. They should look at specific situations where they may be called upon to save lives and protect property and should also be thinking about the overall community consequences.

We will review two items that were probably covered in your basic training. This information is not offered as an insult to your knowledge, however, it is possible that you may have forgotten some of this important material. It is also possible that some new information has been added that might make your job a little easier and a whole lot safer.

life risk factors
the number of people in danger, the immediacy of their danger, and their ability to provide for their own safety

property risk factors
an assessment of the value and hazards associated with property that is at risk

community consequences
an assessment of the consequences on the community, which includes the people, their property, and the environment

physical factors
an assessment of the conditions relevant to population, area, topography, and valuation of a given area

access factors
an assessment of the department's access to and into a building

occupancy factors
an assessment of the risks associated with a particular structure based on the contents and activities therein

structural factors
an assessment of the age, condition, and structure type of a building, and the proximity of exposures

resource factors
an assessment of the resources available to mitigate a given situation

survival factors
an assessment of the safety hazards for both civilians and firefighters in a particular occupancy

---

RISK FACTORS AND COMMUNITY CONSEQUENCES

Life risk factors are affected by the number of people at risk, their danger, and their ability to provide for their own safety.

Property risk factors are affected by the characteristics of building construction, exposures, occupancy, and available resources.

Community consequences are related to the impact of the event upon the community, both during and after the event is concluded. Some of these consequences are immediate; some last for generations.

For company officers, a general knowledge of your area of responsibilities is essential for effective operations. This knowledge includes:

Physical factors: geographical size, population, valuation, response time, and topography of the community.

Access factors: access and barriers to all areas.

Occupancy factors: the nature of the businesses that occupy the buildings.

Structural factors: age, type, and density of structures.

Resource factors: fire department resources, and for fire fighting purpose, water supply capabilities.

Survival factors: stairwells and other penetrations to allow for rescue, fire spread, and potential falling hazards for firefighters.

BUILDING CONSTRUCTION

Given that a substantial amount of firefighting is done in structures, you should have a basic understanding of how buildings are built and how they react during structural fires. After all, when a building is on fire, it is being destroyed. You have to make quick decisions on how fast this destruction is taking place in order to make sound decisions regarding entering the building.

Some buildings are destroyed faster than others during fires. The fire prevention and building codes establish a reasonable tradeoff between economy and safety in the design of buildings. The best and safest buildings are built of fire-resistive materials, but this type of construction is expensive. As a result, compromises are allowed, based on the size and use of the structure.

Building Types

The fire and building codes recognize five general classes of construction:

fire-resistive construction

a type of building construction in which the structural components are noncombustible and protected from fire

Fire-Resistive Construction **Fire-resistive** construction is the best you can hope for. A fire-resistive building built of fire-resistive materials and the structural components of the building, that is the principal parts that hold the building up, are protected from direct exposure to any fire that may be ongoing in the building. While a fire-resistive building is fire resistive, no building is fireproof! Occupants bring combustible materials into all buildings to add to the quality of life and work. Most of these materials will burn.

The structural components in fire-resistive buildings are generally made of reinforced concrete or protected steel. Protected steel means that all structural components are covered with some form of fire-resistive material. Floors and walls are designed to limit fire spread. Generally these features will provide one to four hours of protection during normal fire conditions. Many fires have occurred in fire-resistive buildings. The contents of the building may be destroyed but the structure remains sound. This structural integrity aids evacuation, firefighting, and the likelihood that the building can be used again.

While this would seem the best option, even fire-resistive buildings present concerns to fire officers. We have already mentioned that although the building may be fire resistive, the contents can still burn. For buildings that were built in the last 50 years, fluorescent lighting and air conditioning have allowed huge work areas in all types of occupancies. Even in some large high-rise office buildings, entire floors are essentially one room. When one of these "rooms" becomes involved in fire, the fire can grow quickly and spread to all four walls of the building. If the fire breaks a window, especially a large window that extends to the ceiling, the fire can easily extend up the outside of the building to reach the floor above.

■ **Note**

The fire and all of its by-products also look for interior routes of extension.

The fire and all of its by-products also look for interior routes of extension. Any open penetration in concrete floors, any hole left during construction or renovation, or even a space around a duct or cable is sufficient to allow fire extension into adjacent areas or even the floor above. Several recent fires in office buildings have shown that not only are these statements true, but firefighting in such an environment can be extremely difficult and dangerous (see Figure 10-2).

noncombustible construction

a type of building construction in which the structural elements are noncombustible or limited combustible

Noncombustible Construction A second type of construction is called **noncombustible** (see Figure 10-3). In this case, the term accurately describes the materials used in *constructing* the building, however, the building may be susceptible to collapse during a fire. In noncombustible construction, the walls are typically of masonry or concrete materials while the roof is supported on unprotected steel beams or trusses. Because this type of construction allows for long spans without support, it is very popular in large one-story commercial and industrial facilities. Although the steel roof provides adequate strength to support the load during normal conditions, the unprotected steel fails during fire conditions. Again, fire officers should be aware of any building using this type of construction, and plan accordingly.

~20 min Building

Figure 10-2   *In Type I construction, the structural elements are noncombustible and protected from fire.*

ordinary construction
a type of building construction in which the exterior walls are usually made of masonry, and therefore, noncombustible: the interior structural members may be either combustible or noncombustible

Ordinary Construction Ordinary construction, the third building class, is found in older cities throughout the United States. **Ordinary construction** is characterized by masonry exterior walls. It is often referred to as Main Street, USA, because of its common use in small commercial occupancies in many communities. In some parts of the country, these structures are referred to as "taxpayers." These structures typically have a retail store of some kind on the ground floor and living quarters above for the owner. Taxes and utilities are charged off to the business and the owner gets a pretty good deal on housing costs. However, now one can find ordinary construction in nearly every occupancy class.

In ordinary construction, roof and floors were supported by roof rafters and joists respectively. Although not as massive as heavy-timber construction, these building components were of substantial size, often measuring 3 by 10 inches. These components are set into pockets in the masonry wall. To reduce the likelihood of collapse during a fire, the ends of these rafters and joists are often "fire cut" so that they pull out of the pocket during collapse, and save the walls from falling.

These structures were well built and many still survive. Their long life provides opportunities for many changes of occupants and for many alterations. As a result, there are openings, voids, false ceilings, and other avenues for rapid fire

Figure 10-3 *In Type II construction the structural elements are noncombustible. In some structures, the ability of the building to withstand the consequences of a fire is enhanced by coating the structural elements as shown here.*

travel. Some of these alterations have compromised the structural strength that characterized this type of construction (see Figure 10-4).

heavy timber construction

a type of building construction in which the exterior walls are usually made of masonry, and therefore, noncombustible

Heavy Timber Construction The fourth type of construction is called **heavy timber**, or mill, **construction**. Mill construction got its name from the type of structures that were built using this method of construction, primarily in the northeastern states. These buildings are usually larger than those of ordinary construction. This construction method features masonry exterior walls and heavy timbers to support the floors and roof. Heavy timbers are defined as lumber that is at least 8 inches square. The combination of masonry walls and heavy timbers provides strength for the structure, as evidenced by the fact that many of these buildings have been in use for generations.

Unfortunately, that extended use has brought its problems. The early use of these buildings often involved heavy machinery, which may have stressed the building and left oil-soaked floors. Renovations may have compromised the original design that provided the strength needed to support these heavy loads. And typically these buildings were huge. A well-involved fire in a mill-sized building can easily overwhelm the best of fire departments.

Figure 10-4 *Type III construction, often referred to as ordinary construction.*

There may be a bit of false security in the protection provided by heavy timber construction (see Figure 10-5). While vastly superior to what is used in construction today, even heavy timber construction can fail, sometimes surprisingly quickly. During a recent fire training exercise, a large heavy-timber building collapsed in just 20 minutes, a lot faster than anyone had expected.

Woodframe The last classification of buildings on our list of five common types is called **woodframe construction**, and as you might have guessed, it describes the construction method where the structure is framed out with wooden components. Since the pioneers arrived here nearly 400 years ago, woodframe construction has been widely used because of its relative low cost and ease of construction.

Woodframe construction (as shown in Figure 10-6) has always presented a fire problem because the structure itself can be destroyed in fire. Having a structural fire (that is a fire that actually involves the structure) when the building is woodframe construction usually means that some reconstruction will be necessary to make the building usable again.

Older woodframe structures are likely to have been built in what was called the *balloon-frame* construction style. In this type of construction, the vertical members, called *wall studs,* ran the full vertical dimension of the structure. This facilitated construction and reduced cost. As lumber costs increased, the cost of

woodframe construction

a type of building construction in which the entire structure is made of wood or other combustible material

Figure 10-5 *Type IV construction, called mill or heavy-timber construction. As evidenced here, the use of heavy timber construction is not limited to old mill buildings.*

these long components suggested that an alternative would be more economical. That alternative is called *platform frame construction*. In this type of construction, the walls are erected one level at a time upon the platform of the floor.

The long vertical spaces in the walls associated with balloon-frame construction presented considerable hazards due to the rapid and often undetected vertical fire spread within the walls of the structure. Platform frame construction reduced these hazards.

The cost of wood continues to increase. Builders are looking for alternative ways to use wood and for alternatives for wood. Regardless of the outcome of this evolution, we will continue to see wood used in construction, not just for structural components, but for interior trim and decoration as well.

One way that the amount of wood used in construction can be reduced is by using some form of "lightweight" construction (see Figure 10-7). The use of 2 by 10 inch structural components to support the floor has given way to a light-

Figure 10-6 *Type V construction or woodframe can be totally combustible.*

Figure 10-7 *Lightweight truss construction is commonly used today.*

■ **Note**
The lighter weight structural component is far more vulnerable to failure when exposed to fire.

weight truss that contains a fraction of the material used in the solid component. While the intent was to reduce the amount of wood, and hence reduce the cost of construction, the lighter weight structural component is far more vulnerable to failure when exposed to fire.

> ## LET'S STEP BACK AND SEE WHAT ALL OF THIS MEANS
>
> Type I is called fire-resistive. It is built to resist the impact of fire. The structure is made of noncombustible materials and they are protected from damage due to fire.
>
> Type II is called noncombustible meaning that the building is non-combustible. The building itself will not burn, but unlike Type I, the components may be exposed to fire and as a result, they may fail. For example, the steel truss roof components in the roof system of a Type II building will not burn, but since they are largely unprotected, they may fail if exposed to fire.
>
> Type III, ordinary construction, and Type IV, heavy timber construction, feature walls that are usually masonry and are noncombustible. However, the floor and roof system are usually made of wood and will burn.
>
> Type V, called wood frame, is completely vulnerable to fire.

In construction Types I and II, the building will not burn; in Types III and IV the walls will not burn; and in Type V everything burns.

The type of construction, fuel load, and smoke and fire behavior can usually be predicted for a given building. When planning for fire attack, carefully consider how each of the factors will impact on your operations. Building construction has a major impact on many of the factors in your planning and fighting any fire that may occur. For example, Table 10-1 identifies the types of buildings we have been discussing and a fire-related concern.

Fuel Load

fire load
stuff that will burn

fuel load
the expected maximum amount of combustible material in a given fire area

Nearly every building has combustible contents. **Fuel load**, or **fire load** as it is often called, is the quality of combustibles in the building. In a typical building, the contents include all of the furnishings, inventory, and the combustible wall and floor coverings. These are found in all classes of buildings. For purposes of determining how much fire these materials will produce, this load is usually expressed in pounds per square foot.

In addition to the contents, the structure itself presents a fire load except for Type I and II buildings. The structural fire load includes the combustible portions of the walls, floors, and roof of the structure. We will use this information later when we look at water flow requirements.

■ **Note**

Although fire is a serious threat, smoke kills far more people than fire.

Smoke and Fire Travel

Although fire is a serious threat, smoke kills far more people than fire. Smoke and gases produced by a fire move through the building well in front of any flames.

Table 10-1 *Fire-related problems associated with various building types.*

Building Type	Fire-Related Problems
I. Fire-resistive	Large buildings. Failure of the concrete exposing structure members. Retention of heat and smoke in the structure.
II. Noncombustible	Roof failure.
III. Ordinary	Fire travels in concealed spaces. Buildings are old and often altered.
IV. Heavy timber	Heavy fire load from the interior structural members and contents.
V. Woodframe	Combustible structure. In the case of lightweight construction, possible rapid failure and collapse.

When looking at buildings for purposes of preplanning for a fire emergency, consider the consequences of the problems associated with smoke, heat, and gases, and ways to mitigate these problems to reduce the threats they present to civilians and firefighters.

There are ways to detect and control the smoke and heat from a fire. Automatic detection equipment can notify the occupants and the fire department that there is a problem. Some detection systems automatically take control of the building's ventilation system to facilitate the removal of smoke and other gases and reduce the likelihood of these gases being transferred to other sections of the structure.

Installed Fire Protection Equipment

Installed fire detection and suppression equipment, including automatic sprinklers and standpipe systems are present in many buildings. Non-water-based fire extinguishing systems may also be present. Installed fire protection equipment does not prevent fires, but such equipment greatly assists fire companies in detecting and fighting fires.

To get full benefit from this equipment, all fire department personnel should be knowledgeable about the existence of these system. Fire officers should understand how these systems work and how the fire department can support their proper operation by increasing water pressure and flow when needed. We briefly examine three types of systems: fire detection equipment, automatic sprinklers, and standpipes.

Fire Detection Equipment Fire detection equipment neither prevents fires from occurring nor extinguishes them once they have started, but fire detection equipment can provide timely notification for both the occupants and the fire department.

Timely notification of occupants will significantly reduce the number of lives lost in fires. Timely notification of the fire department will probably reduce the severity of the fire. Fire detection equipment ranges from simple battery-powered smoke detectors commonly found in residential occupancies to elaborate integrated systems with devices that recognize flame, heat, or smoke in large industrial and commercial facilities. These system can also manage air handling equipment, elevators, and other building systems. Firefighters should know where these devices are located and check them for proper operation when they are on the property.

automatic fire protection sprinkler
self-operating thermosensitive device that releases a spray of water over a designed area to control or extinguish a fire

Automatic Sprinklers   **Automatic fire protection sprinklers** are one of the most reliable forms of fire protection in commercial, industrial, and other classes of occupancies. A sprinkler system automatically provides a water spray over the area where the fire starts, and simultaneously alerts both the occupants and the fire department that the sprinkler system is in operation.

Sprinkler systems are one of the greatest allies of the fire department. When a significant number of sprinklers are operating, the water pressure in the sprinkler system will drop, reducing the flow of water. Fire service personnel should know how to take appropriate action to augment the water supply to boost the pressure and maintain adequate waterflow. Fire personnel should realize that sprinklers will control a fire; manual firefighting may be required to extinguish it. Operating companies should avoid premature sprinkler shutdown during fire operations.

■ **Note**
Sprinkler systems are one of the greatest allies of the fire department.

Sprinkler systems do not prevent fires but they do have an impressive record of protecting property from fire loss. The greatest threat to a sprinkler system is for the water supply to be turned off. Without a water supply, a sprinkler system cannot operate. Fire service personnel should know where sprinklers are located in their jurisdiction (see Figure 10-8), and check them regularly to be sure that they are ready for proper operation.

standpipe systems
plumbing installed in a building or other structure to facilitate firefighting operations

Standpipe Systems   Unlike sprinklers, **standpipe systems** (see Figure 10-9) do not put out fires but they greatly help firefighters extinguish fires. Standpipes are simply large-diameter pipe plumbing systems that allow firefighters to get water to the seat of the fire easily and quickly. They are found in large buildings and tall buildings, both environments that present the potential for labor-intensive firefighting operations. Again, firefighters should know where these systems are located. They should also know how to connect a water supply to the system to provide adequate pressure and water flow for safe and effective firefighting operations.

Other Building Systems

We have already mentioned the elevators, ventilation, and firefighting systems found in large buildings. Others include electricity and gas, other transportation

Figure 10-8 *Fire suppression personnel should know the location of all sprinkler systems in their response area. They should check them regularly to make sure that they are ready for proper operation.*

Figure 10-9 *In addition to sprinklers and standpipes, fire personnel should be aware of and check all installed fire protection equipment.*

fire behavior
the science of the phenomena and consequences of fire

piloted ignition temperature
the minimum temperature to which a substance must be heated to start combustion after an ignition source is introduced

systems for people and materials within the building, and any special products that may be stored or used within the structure, especially those that may be flammable. You should understand these systems and how to control them so that you can use them safely and effectively, or secure them, as appropriate.

FIRE BEHAVIOR

Fire officers should also thoroughly understand fire behavior. **Fire behavior** is a term used to define the way fire performs or reacts in given situations. Every fire is unique, but there are some general concepts that help us better understand fire behavior. Understanding fire behavior helps us to be more effective at fire suppression. Understanding fire behavior is also important from a safety perspective, both for the occupants of the structure as well as the firefighters.

We need three things to start a fire: fuel, heat, and an oxidizer, represented by the three sides of the fire triangle (see Figure 10-10). The process is interactive. The fuel must be heated to its **piloted ignition temperature**, and the continuation of the process is dependent upon the heat being fed back into the process. This heat, in turn, helps warm the fuel. This process continues until all of the fuel is consumed, the oxidizing agent (usually as oxygen in the air) is reduced to a level where it can no longer support combustion, the heat is removed at a rate that prevents the feedback process from continuing, or the process is altered by the addition of some chemical.

All fires eventually go out. Some go out because all of the fuel was consumed; some go out because you altered one of the other conditions listed in the preceding sentence. All are used as fire extinguishment techniques. The fire tetrahedron (shown in Figure 10-11) better explains how fires burn and how fires are extinguished.

Figure 10-10 *The fire triangle helps explain how fires start. To have a fire, fuel, heat, and air must be present.*

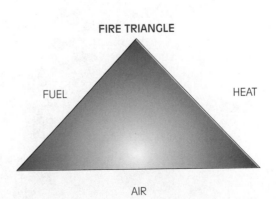

FIRE TRIANGLE

FUEL　　　　　　　HEAT

AIR

Figure 10-11 *The fire tetrahedron helps explain how fires burn and how they are extinguished. A fourth side, called a continuous chain reaction, recognizes the presence of an ongoing complex chemical process. Fires can be extinguished by removing one or more of the four sides.*

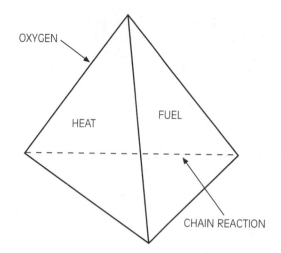

Fire Science

The processes involved in fire ignition and growth are essential for fire to survive and our understanding of these processes is essential for effective fire extinguishment. The following brief summary is not intended to be an exhaustive treatment of the subject of fire behavior, however, a brief review may help in your understanding of the information in this section.

Fire usually develops from a small flame to where there is significant involvement of the fuel. When this process takes place within an enclosed building, it usually passes through three stages.

First stage—called the **incipient phase**. A fire may smolder for a period of time that ranges from seconds to hours before there is sufficient heat to produce open flame. During this time there is little heat buildup but some smoke is evident.

Second stage—called the **free-burning phase**, as the fuel is burning free of nearly any constraint. Flames become visible along with an increased rate of heat generation and fuel consumption. Since the heat expands the air, there is a slight increase in air pressure within the structure. Where there are openings in the structure, smoke and heat are pushed from the building. Hot gases heat the fuel above the fire, accelerating its ignition. The process continues and accelerates. Because of the dynamic forces at work, we see rapid fire extension during this phase.

Third stage—called the **smoldering stage**. Assuming that the structure is still intact, the fire will have consumed a portion of the oxygen in the air, and the fire is reduced to where it just smolders again. However, unlike the first phase, the room is filled with hot flammable gases.

incipient phase
first stage of fire growth, limited to the material originally ignited

free-burning phase
second phase of fire growth, has sufficient fuel and oxygen to allow for continued fire growth

smoldering phase
third stage of fire growth; once the oxygen has been reduced, visible fire diminishes

The change from a free-burning fire where there is 21% oxygen to a smoldering fire with an oxygen-deficient atmosphere requires only a slight reduction in the oxygen concentration. Anyone who has ever adjusted the burners on the gas grill knows that it takes only a slight adjustment to significantly change the characteristic of the flame.

Fire Phenomena

When fire burns in structures, several fire-related phenomena occur. These include thermal stratification, rollover, flashover, and backdraft explosions.

Thermal Stratification When a fire burns in an open area, the heat rises and is dissipated into the atmosphere. When a fire occurs in an enclosed space, a very different process occurs. Part of this difference is due to the fact that the hot gases cannot escape as fast as they are created. While the feedback process that allows some of these hot gases to be reintroduced back into the fire presents some very complex concepts of fire dynamics, we know from our own experience that most of the heat from a fire rises and stratifies within an enclosed space. This natural process is called **thermal stratification**. The heat is at the top of the room and the cooler air is near the floor. This is why we tell people to crawl low in smoke.

thermal stratification
rising of hotter gases in an enclosed space

Figure 10-12 *In an outdoor fire, the heat dissipates into the air.*

When firefighters enter a space and start fire suppression operations, they can easily disturb the natural process that sends the heat to the highest areas in the room. When this occurs, firefighters endanger occupants and themselves alike. Since most fire operations deal with ordinary combustibles, water is used as an extinguishing agent. It works well, however, the inappropriate application of water disrupts the natural process of thermal balance.

Rollover When materials burn, they consume oxygen and give off combustible gases. These combustible gases are hot and therefore, less dense. These fuel-rich gases rush upward in the column of heat above the fire and fill the area. As additional gases are heated they also rise and try to displace the gases that are already there.

In an open environment, these heated gases escape (see Figure 10-12). However, in an enclosed environment such as found in structures, the gases spread out and travel along the ceiling (see Figure 10-13) well in advance of the front of the actual fire. In areas where there is limited opportunity for these hot gases to expand, such as in a long hallway, they may precede the main body of fire by as much as 20 feet. When these hot, fuel-rich gases meet fresh air, they ignite and burn. This dynamic process appears to roll along the ceiling and hence the name, **rollover**. This situation has this potential for placing firefighters in considerable danger.

rollover
reignition of gases that have risen and encountered fresh air, and thus a new supply of oxygen

Figure 10-13 *In a fire inside a structure, the heat banks down, accelerating the chain reaction, and placing civilians and firefighters in grave danger.*

flashover
a dramatic event in a room fire that rapidly leads to full involvement of all combustible materials present

backdraft explosion
a type of explosion caused by a sudden influx of air into a mixture of burning gases that have been heated to the ignition temperature of at least one of them

fire extension
the movement of fire from one area to another

Flashover Flashover is another fire phenomena that presents significant risks to firefighters. As a fire burns in a enclosed space, the contents of the room are gradually heated. Everything thing in the room has an ignition temperature and as the room temperature approaches the ignition temperature of the furnishings, they will ignite. **Flashover** can occur within a few minutes after ignition. Many factors determine when flashover will occur. Unfortunately, you seldom know all of these many factors in a given fire situation.

Numerous demonstrations of flashover have been captured on film and many firefighters have been trained in a device called a "flashover trainer." While not always predicable, the process is very real. However, the results are very real and are predictable. Anyone in the room at the moment of flashover has little chance of survival. Firefighters have experienced and survived flashover conditions. Those who survived their experience had and were using *all* the components of their personal protective clothing. In spite of their protection, most were injured, and all tell of the tremendous heat they experienced.

Backdraft Of all the fire-related events described here, **backdraft explosions** are the most forceful. As a fire develops in an enclosed space, it consumes oxygen and produces hot combustible gases. We have already described how this process produces what is known as flashover and rollover. As this process continues, it moves into the third phase of fire where the atmosphere within the room or building reaches a point where there is insufficient oxygen to support the open combustion process.

As this process continues, the entire space becomes filled with hot combustible gases. One side (the oxidizer) of the fire triangle has been removed and this has interrupted the complete combustion process. Eventually the flames diminish and the materials smolder, producing hot flammable gases. When firefighters open a door or window, they allow oxygen to enter the space. When the right combination of heat, fuel (as a combustible gas), and oxygen are present in the space, the triangle is complete once again. Since the fuel is in a gaseous form, and since it is at or above its ignition temperature and spread fairly uniformly throughout the space, a rapid ignition of these gases occurs, sometimes with explosive force.

We will add two additional terms to this list. While not really considered fire phenomena, they are closely related to the behavior of fire and to effective fire control. Understanding the two processes discussed here will enhance your effectiveness in fireground operations.

The first term, fire extension, describes the spread of the fire through the structure. During overhaul activities, firefighters are checking for **fire extension**. To be done effectively, this activity requires a knowledge of both building construction and fire behavior.

Fire follows any available path to find the fuel needed to sustain its own life. Fire travels within walls and ceilings, through voids and shafts, through any place where fuel and oxygen can satisfy the fire's ravenous appetite. Whenever

■ **Note**

Whenever openings are made in a building that contains a fire or suspected fire, whether to fight the fire or to check for fire extension, firefighters should have a charged hose line available to extinguish any fire that is found.

ventilation
a systematic process to enhance the removal of smoke and fire by-products and the entry of cooler air to facilitate rescue and firefighting operations

■ **Note**

Ventilation alters most of the fire-related phenomena described previously and reduces the probability of backdraft and flashover.

openings are made in a building that contains a fire or suspected fire, whether to fight the fire or to check for fire extension, firefighters should have a charged hose line available to extinguish any fire that is found. This applies to opening doors, windows, cutting holes in roofs, and opening walls or pulling ceilings to check for hidden fire extension.

Ventilation, the other term, is a key tactical operation that affects how the fire behaves. While often considered as an afterthought, ventilation must be considered along with rescue and fire fighting. Ventilation greatly alters the conditions within the structure. Ventilation removes heat, helps removes combustible and hazardous fire gases, improves visibility, and directs the fire's advance. Ventilation alters most of the fire-related phenomena described previously and reduces the probability of backdraft and flashover.

For civilians who may still be present in the building, ventilation can easily make a difference in their survival. Ventilation improves the likelihood that civilians will be found during search and rescue operations, and that they will be savable. Ventilation also greatly improves the conditions firefighters face while fighting the fire.

Ventilation can be accomplished by natural or forced means. Natural ventilation can be accomplished as easy as by opening a window. On the other hand, forced ventilation takes special equipment. A lot of attention has been given recently to the merits of using positive pressure ventilation (PPV) in conjunction with fire suppression operations. Ventilation activities during firefighting operations, especially in large occupancies, must be carefully coordinated in order to avoid the hazards associated with the fire-related phenomena just described. Ventilation is a key tactical activity that greatly affects the outcome of any fire suppression effort.

QUESTIONS THAT FACE THE FIREFIGHTER AT EVERY FIRE

- Where is the fire?

- Where is the fire likely to spread?

- Is there an immediate likelihood of a flashover?

- Is there an immediate likelihood of a backdraft explosion?

- Is there a likelihood of a collapse?

- Is there any building feature what will help slow the fire's advance?

- Are there any installed fire protection features to assist in fire suppression efforts?

- Are there civilians present in the building? How will our activities affect them?

What Makes the Fire Go Out?

Most structural fires are extinguished with water. Water has many desirable properties that make it an effective extinguishing agent, one of which is its ability to absorb heat. Many factors affect both the ways and the efficiency with which water absorbs heat from a particular burning combustible. The application rate and droplet size of the water impact on its effectiveness. The form and type of material burning have a significant impact on the amount of heat being generated as well as the ability of the water to cool the burning material and control the fire.

During firefighting it appears that water cools the burning materials. In reality, effective fire suppression occurs when the heat from the involved materials is transferred to the water being applied. When applied effectively, a large portion of that water is converted to steam. Under ideal conditions, nearly all of the water is converted to steam in this process.

Let's talk science for a minute. The amount of heat required to raise the temperature of one pound of water one degree Fahrenheit is called a **British thermal unit** (BTU). You might think that a British thermal unit is a metric unit, but it is not. We use BTUs for many purposes in life but we use it to our advantage here, not so much to show that water works well as an extinguishing agent, but rather to estimate how much water is needed.

Another scientific term we use is **specific heat**, a term used to describe the heat-absorbing capacity of a substance. As we have already noted, water is very effective in absorbing heat. In extinguishing fires, we apply water to absorb the heat generated by the fire. What we have is a heat transfer process: We reduce the temperature of the burning fuel and raise the temperature of the water.

If you were to raise the temperature of the water from 60°F to 212°F when applying it to burning materials, 152 BTUs would be absorbed for every pound of water applied. Water weighs about 8.3 pounds per gallon. Thus, each gallon of water has the potential for absorbing 1,266 BTUs when heated from 60° to 212°.

$$212 - 60 = 152 \text{ BTUs/pound}$$

$$152 \text{ BTUs/pound} \times 8.3 \text{ pounds/gallon} = 1,266 \text{ BTUs/gallon}$$

If you continue to heat the water, it will turn to steam.

Another advantageous property of water is that in the process of converting to steam, water will absorb many more BTUs. We know from research that every pound of water will absorb 970 *additional* BTUs as it turns from water into steam. Scientifically this phenomena is called the **latent heat of vaporization**—the quantity of heat absorbed by a substance when it changes from a liquid to a vapor. Again, multiplying by 8.3 pounds per gallon, the latent heat of vaporization for a gallon of water is 8,080 BTUs. Adding 1,266 and 8,080 we get 9,346 BTUs per gallon.

$$970 \text{ BTUs/pound} \times 8.3 \text{ pounds/gallon} = 8,080 \text{ BTUs/gallon}$$

Combining the results of the two:

British thermal unit
the amount of heat required to raise the temperature of one pound of water one degree Fahrenheit

specific heat
the heat-absorbing capacity of a substance

latent heat of vaporization
the amount of heat required to convert a substance from a liquid to a vapor

$$1,266 \text{ BTUs/gallon} + 8,080 \text{ BTUs/gallon} = 9,346 \text{ BTUs/gallon}$$

One gallon of water will absorb 9,346 BTUs of energy. Big deal you say. You will see the significance of this in a moment. If you continue to heat the steam, it will continue to expand, as evidenced by billowing of steam during firefighting operations.

We now look at the other side of the equation: how much heat do you have? Except for the kind of fires that fire protection engineers and scientists build for research purposes, you usually do not really know how many BTUs are being generated during a fire. But that research allows you to estimate the heat that is produced in the fires you typically encounter.

We do know that when materials burn, they generate heat. Scientists use the term **heat of combustion** when measuring the quantity of heat or energy released per unit of weight. For example, research indicates that about 8,000 BTUs are released from burning a pound of wood. Other products have other values, some ranging from two to three times more than wood. In general, flammable liquids and gases burn more readily than solids.[3] Some gases burn very hot. Not surprisingly they are used as fuel gases. These include methane, propane, and natural gas. However, for purposes of our illustration, we assume that the fuel is wood.

Research has shown that the average load of fuel in typical residential structures is about 4 pounds per square foot. Of course this loading factor is much higher in warehouses and in some industrial and commercial facilities. While a lot of this may be of interest to the scientist, we can benefit from this knowledge as well. Most structure fires are extinguished by using water to absorb the heat created by the fire. From basic training we recall that reducing the temperatures of the items that are burning below their ignition temperature will cause the fire to go out. In order to control the fire, the quantity of water applied to control the fire must be able to absorb the heat being produced. If you can estimate the amount of heat being produced, you can estimate the amount of water needed.

In real life you do not know exactly what is burning and even if you did, you would not have time to do the math. What you do want to know is how fast you have to apply the water to make an effective fire attack. The rate of water needed to control the fire is called the **theoretical fire flow**. There are several formulas for determining the fire flow, but one of the easiest ones to use is the one currently taught by the National Fire Academy (NFA). It is based on the area in square feet of the burning structure. The area is determined by multiplying the length by the width. The NFA formula is:

$$\text{Fire flow (in GPM)} = \frac{\text{Area in square feet}}{3}$$

For example, take a building 20 by 30 feet, or 600 square feet. Divide the area, 600, by 3 and you have 200. The answer is in gallons per minute, so the answer is 200 gpm. Remember that we are talking about flow *rate,* not the total

---

heat of combustion
the amount of heat given off by a particular substance during the combustion process

■ **Note**
Research has shown that the average load of fuel in typical residential structures is about 4 pounds per square foot.

■ **Note**
In order to control the fire, the quantity of water applied to control the fire must be able to absorb the heat being produced.

theoretical fire flow
the water flow requirements expressed in gallons per minute needed to control a fire in a given area

quantity of water required.[4] Knowing the fire flow helps in preplanning and firefighting operations.

This brief explanation of the NFA formula has not addressed how to deal with additional floors for fire involvement, exposures, and other important issues associated with fire flow. We will discuss this topic further in Chapter 13.

How can this help you in preplanning and firefighting? If you estimate the size, fuel, and level of involvement at the time you start fire suppression operations, you can

- Determine number and size of lines to provide the needed fire flow
- Determine the pumping capacity to supply these lines
- Determine if existing water supply resources are adequate
- Determine the number of personnel and companies need to initiate a fire attack

This information will serve you in two ways. It helps you preplan for fire attack. You have to make assumptions about the size of the fire at the time you start fire suppression operations, but you can make some estimate of the needed fire flow while preplanning for the emergency. From this you can determine how many and what size hose lines are needed, the pumping capacity required, and the water supply and the number of personnel needed.

A second application of the NFA formula facilitates **size-up** of the situation after arrival. Lacking a preplan, you can use the NFA formula to deal with the same questions regarding fire flow and resources that are previously listed. If you *estimate* the size of the involved area and level of involvement, you can better determine number and size of lines needed, if you have adequate pumping capacity to supply these lines, if the existing water supply resources are adequate, and if there are adequate personnel to initiate a safe fire attack. In most cases the number of fire fighting personnel present is the critical element. Where a hydrant is available, water supply, pumper capacity, and available hose lines generally exceed the capabilities of the personnel of the first arriving companies.

size-up
mental assessment of the situation; gathering and analyzing information that is critical to the outcome of an event

THE IOWA STATE UNIVERSITY FORMULA

Mr. Keith Royer, a longtime fire training instructor at Iowa State University, feels that numbers from the NFA formula are excessive. He researched fire behavior staring in the late 1950s and concluded that a formula based on the *volume* of the room rather than the area of fire involvement is more appropriate. His assumptions address interior operations where fire behavior may be limited by the volume of the air (oxygen) available, rather than the fuel.

Royer noted that 1 gallon of water will absorb the heat in 200 cubic feet of the type of hot air typically found in a room where fire is occurring. He also noted that a gallon of water will expand to approximately 200 cubic feet when converted to steam. To be conservative he reduced these factors by 50%.

For interior fire fighting operations, Royer's formula is volume of the room divided by 100 equals fire flow.

$$\text{Volume (in cubic feet)}/100 = \text{GPM}$$

For example, an area of 600 square feet with an 8-foot ceiling would have a volume of 4,800 cubic feet. According to Royer, a fire in that space could be controlled with 48 gpm! To work effectively, the application of the water and ventilation of the involved area must be carefully controlled and coordinated.

This limited amount of water is needed for fire fighting in a single room. It does not consider the need for backup lines, exposure protection, fire fighting in other involved areas, and other needs. Royer's comment regarding comparisons between his formula and those that suggest that more water is needed: "Not much burns when it's under water!"

Royer notes that using water effectively has several advantages. First, it reduces damage. It also makes for easier fire fighting. Using less water helps maintain the thermal balance in the room; the hot gases remain warm and will rise and dissipate. These are the gases that contain all of the products of the fire's decomposition process, and the moisture from the steam from the application of water. We want that cloud to rise, not settle down over the firefighters and their work area.[5]

Things have an interesting way of reinventing themselves every generation or so: Research is presently underway in Europe on this very topic.

Summary

The company officer must have a thorough understanding of fire behavior and building construction. These areas of knowledge are fundamental to the task of fire suppression. Understanding and being able to anticipate the forces that control the growth and extension of a fire also facilitate a more effective and safer extinguishment operation. Company officers should keep these items in mind as they plan, train, and operate at burning buildings.

In addition to knowing about the hazards of fire, planning is important too. Planning allows companies to prepare for certain expected operations by better identifying and analyzing the problem. Planning also allows for better identification of our resource needs and limitations, and to make assignments for first arriving units, based on probability and risk factors. We examine these topics in greater detail in Chapter 12.

Review Questions

1. What is meant by life risk factors?
2. What is meant by property risk factors?
3. What is meant by community consequences?
4. How do the each of the following factors affect firefighting operations?
   a. physical factors
   b. access factors
   c. structural factors
   d. survival factors
5. Define the five building classifications.
6. Identify a fire-related hazard associated with each type of building.

7. Define the following:
   a. rollover
   b. flashover
   c. backdraft
8. What is the National Fire Academy's fire flow formula?
9. How does the NFA fire flow formula help in prefire planning and fire suppression operations?
10. Why is an understanding of building construction and fire behavior essential for the company officer?

Additional Reading

Brannigan, Francis L., *Building Construction for the Fire Service,* Third Edition (Quincy, MA: National Fire Protection Association, 1992).

Brannigan, Francis L., "Building Construction Concerns for the Fire Service," *Fire Protection Handbook,*

Eighteenth Edition (Quincy, MA: National Fire Protection Association, 1997).

Cote, Arthur E., Jr., *Fire Protection Handbook* (Quincy, MA: National Fire Protection Association, 1997).

Davis, Richard, "Building Construction," *Fire Protec-*

tion Handbook, Eighteenth Edition (Quincy, MA: National Fire Protection Association, 1997).

Fitzgerald, Robert, "Fundamentals of Fire Safe Building Design," *Fire Protection Handbook,* Eighteenth Edition (Quincy, MA: National Fire Protection Association, 1997).

Fried, Emanual, *Fireground Tactics* (Chicago: Marvin Ginn Corporation, 1972).

Klinoff, Robert W., *Introduction to Fire Protection* (Albany, NY: Delmar Publishers, 1996).

Lloyd Layman, *Attacking and Extinguishing Interior Fires* (Quincy, MA: National Fire Protection Association, 1955).

Nelson, Floyd W., *Qualitative Fire Behavior* (Stafford, VA: International Society of Fire Service Instructors, 1991).

Quintiere, James G., *Principles of Fire Behavior* (Albany, NY: Delmar Publishers, 1997).

Notes

1. Lloyd Layman, *Fire Fighting Tactics* (Quincy, MA: National Fire Protection Association, 1953), p.8. Reprinted with permission from National Fire Protection Association, Quincy, MA 02269.

2. Emanual Fried, *Fireground Tactics* (Chicago: Marvin Ginn Corporation, 1972), p. 357.

3. Looking at some of the types of materials we commonly see involved in fire, we can get some idea of the approximate amount of heat being generated. For example, wood and paper yield about 8,000 BTUs/pound, polystyrene yields 18,000 BTUs/pound, and gasoline yields 19,000 BTUs/pound.

4. For additional information regarding the NFA fire flow formula, see the Preparation course student manual for the NFA training program entitled Managing Company Tactical Operations. Additional, information about this course is contained at the end of Chapter 12 of this book.

5. Keith Royer and Floyd Nelson, "Water for Fog Fire Fighting." *Fire Engineering,* August, 1959 and "Using Water as a Fire Extinguishing Agent," *Fire Engineering,* September, November and December, 1992. See also *Qualitative Fire Behavior* by Floyd W. Nelson, listed in the additional reading section.

The Company Officer's Role in Fire Investigation

Objectives

Upon completion of this chapter, you should be able to describe:

■ The common causes of fire.

■ The significance of arson as part of this nation's fire problem.

■ How to perform a preliminary fire investigation.

■ How to recognize evidence that would suggest that a fire was deliberately set.

■ How to secure the incident scene and preserve evidence.

INTRODUCTION

A major part of any active fire prevention program should be the investigation of fires. Fire investigations help identify the community's fire risk by determining the frequency and causes of the fire. This information can be used to provide the basis for fire prevention programs and for identifying the resources needed for fire suppression. Where a fire may have been deliberately set, a good fire investigation is the first of several steps needed to bring the perpetrator to justice. Fire officers have a significant role in this process.

All Fires Should Be Investigated

Every fire should be investigated to determine its origin and cause. We often equate fire investigation with arson and arson investigation. Although arson investigation is a part of the fire investigator's job, fire investigation has other purposes. A majority of fires (70 to 80 percent overall) are accidental in nature. (Statistics in your particular response area may suggest a different value.) Accidental fires should be investigated to determine their cause so that patterns of human behavior or equipment failure can be discovered and corrected.

Although most fires are accidental, statistics indicate a significant arson problem in the United States. We may tend to become complacent about the causes of the fires we are attending, but we should never lose sight of the fact that arson occurs in every community. We should always consider the possibility of arson at any fire.

accidental
refers to those fires that are the result of unplanned or unintentional events

We use the term **accidental** here to include all types of fires that were not deliberately set. Some would argue that accidental is too kind a term; that while accidents do happen and mechanical devices do fail, human carelessness and stupidity are also the cause of all too many fires. In some areas, all accidental fires are considered suspicious until a final fire cause determination is made or arson is ruled out by an investigator.

Fire investigators usually try to eliminate all possible accidental causes before considering arson. Arson is the result of an incendiary or deliberately set fire, but not all deliberately set fires are arson; it is not a crime to deliberately light a fire in your fireplace. In many areas, it is not a crime to deliberately set fire to trash or brush. The crime of arson requires criminal intent as well as the act itself. Where you cannot find reasonable evidence of accidental cause or carelessness, you should consider arson. It is a good approach to take and one that most investigators are taught to follow.

A company officer is often in a unique position for conducting fire investigations. They are likely to be among the first emergency responders at the scene and therefore have an opportunity to see the fire while it is still burning. For smaller events, they may be the senior fire department official at the scene, so they must complete the fire incident report. To complete the report they have to determine the origin and cause of the fire. Where arson is involved, the fire officer's observations and actions play a significant role in any legal proceedings that follow. With these thoughts in mind, let us consider the common causes of fire.

HAVE YOU HEARD MS. VINA DRENNAN TALK ABOUT THE COST OF CARELESSNESS?

Vina Drennan understands carelessness when it comes to fires; she is the widow of New York City Fire Department Captain John Drennan who, along with two other firefighters, died in a fire caused by a citizen's careless act. (A bag of trash was left on top of a gas stove.) Captain Drennan survived the fire but died several agonizing weeks later as a result of his injuries.

Shortly after his death someone said to Ms. Drennan, "I'll bet you would like to get your hands on those who were responsible for the fire." She was surprised at the suggestion. She responded by saying, "Oh no, they didn't mean it, they were just careless. Just careless . . . we tolerate a lot of carelessness in America."

She has traveled all over the country and talked about John Drennan's life. She also talks about fires; she has learned a lot about fires in the last several years. She compares the attitude toward fire and the fire loss in America with other countries, noting that we do not compare very well. She tells those who will listen that Captain Drennan loved his job, and he died in that job because of someone's carelessness.

She challenges all of those who hear her to get the message out: "*Tell the public that carelessness can kill!*" If we can reduce the number who suffer injuries from fires and if we can reduce the number of fatalities from fires, then we can find even greater meaning in John Drennan's life.

What Is Fire?

In Chapter 10 we reviewed fire behavior. We noted that a fire will not occur unless the three essential elements that comprise the fire triangle are present. One of these is heat. Contemporary fire science textbooks refer to this side of the fire triangle as **energy** (see Figures 11-1 and 11-2). Heat, or energy, warms the fuel to cause vapors to form and provides the ignition source in the fire triangle. Heat also promotes fire growth and flame spread. There are many commonly found ignition sources including chemical reactions, electrical energy, and mechanical energy.

The second essential element of fire is fuel. Fuel can be anything that is combustible. Fuel can be present as a solid, a liquid, or a vapor. Fuel for accidental fires includes items that are all around us, ranging from household furnishings to trash.

Oxygen is the third requirement. Oxygen is naturally present in the air. In fact, fire is usually defined in simple scientific terms as the process of **oxidation** of a fuel by atmospheric oxygen. Fire is often defined as a rapid, self-sustaining

energy
the capacity to do work; in the fire triangle, heat represents energy

oxidation
a chemical reaction in which oxygen combines with other substances causing fire, explosions, and rust

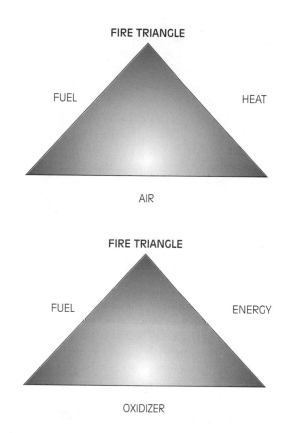

Figure 11-1 and Figure 11-2 *In Chapter 10, we noted that fuel, heat, and air are needed to start a fire. To better understand the chemistry and physics involved in the combustion processes, air is replaced with the more scientific term oxidizer; likewise heat is replaced with the term energy.*

oxidation process, usually accompanied by the evolution of heat and light in varying intensities.

COMMON CAUSES OF FIRES

Accidental Causes

In Chapter 9 we discussed the common causes of fire. Those same causes are discussed here briefly, with the focus on looking at their origin and cause.

■ **Note**

Most heating equipment is safe; what causes problems is usually not the equipment, but rather the way the equipment is used.

Heating Equipment Heating equipment is a leading cause of fires. By design, heating devices get hot. When sufficient quantities of that heat get close enough to combustible materials, you have a fire. Most heating equipment is safe; what causes problems is usually not the equipment, but rather the way the equipment is used. Fires occur when heating equipment is misused, improperly maintained, or placed too close to combustible materials, like bedding and wet clothing.

Chimneys are designed to vent the hot and hazardous by-products of the fire to the outside. When the chimneys are not properly maintained, those by-products accumulate in the chimney. Eventually, when they get hot enough, they ignite and cause a chimney fire. When the chimney is not properly maintained, the hot gases that are intended to go up the chimney can escape through the walls of the chimney. If they escape into the structure, they may ignite nearby combustible materials.

Hot fireplace ashes are another major concern in many parts of the country. Regardless of the type of fire that produced the ashes, ashes that remain warm can set fire to any nearby combustible material. Many people have carelessly set fires by placing hot fireplace ashes in a plastic can or paper bag in their garages or on their wooden patio decks.

Firefighters everywhere attend fires like these and can tell stories about how fires they have seen were started. But not many homeowners know what you know about fires, and some of those who do know do not seem to care. So, we keep having the same kinds of fires. Nearly every fire department in the country attends fires involving heating equipment during the winter season.

Cooking Cooking is the leading cause of injuries in residential fires, and fires resulting from cooking activity are destructive in terms of property damage (see Figure 11-3). Like heating equipment, cooking equipment is designed to provide a safe means for heating. When used properly, cooking equipment is not hazardous. Unsafe practices, poor housekeeping, inappropriate attire, and lack of

■ **Note**

Leading accidental causes of fire include: heating equipment, cooking and cooking equipment, the careless use and disposal of smoking materials, and electrical distribution systems.

Figure 11-3 *Fires in the kitchen can have serious consequences. Photo courtesy of the Fairfax County Fire and Rescue Department.*

maintenance can cause fires associated with cooking equipment. In most cases, it is not really the equipment that is at fault, it is human error.

Smoking Materials Smoking and the careless discarding of smoking materials are the source of many of the nation's fire fatalities. Such fires often involve a delayed ignition of upholstery materials, such as in a bed, chair, or sofa. The results are a slowly developing fire that eventually produces enough heat and toxic vapors to kill everyone present. When undetected, these fires are often significant in terms of their intensity, and because they frequently occur after bedtime, the results are fatal.

Electricity Electricity is all around us. When properly used, it is a safe form of energy. However, that energy has tremendous potential for causing damage. Electrical appliances and electrical distribution systems cause fires. Safeguards are built into appliances and electrical distribution systems to protect us from fire and personal injury. However, the safeguards do not always protect individuals from careless acts.

One of the major protection devices in any electrical system is the electrical panel. It is a good place to check after any fire. You should try to determine if the electrical circuit was energized at the time of the fire. Examine the panel itself as well as the area around the meter to determine if there was any tampering before the fire or any evidence of arcing or other malfunction during the fire.

When you are looking at wiring inside a structure after a fire, it is often difficult to determine if the fire caused damage to the wire or if the wire caused the fire. The nature of the damage is a function of the cause of the fires, the extent of the exposure of the wiring to heat, the duration of the fire, and the protection on the wire. In a typical residential fire involving a room and its contents, the wiring does not get hot enough to melt. On the other hand, wiring can be damaged by the heat generated by an excessive amount of electrical current. In this case, you may see melting of insulation in areas of the structure where there was no fire damage and evidence that the wire was warm enough to deform due to the heat generated by the electrical current.

Arcing, as a result of poor connections, can damage wiring. Arcing usually leaves a distinctive mark on the wire either in the form of beaded ends or one or more cavities in the wire itself. The arcing usually causes sparks, which may ignite nearby combustible materials. Finding one or more clues associated with electrical distribution systems and electrical equipment may help you determine the cause of the fire. Electrical problems are often missed as a cause of fire. Fires are also unfairly blamed on electrical problems.

Natural Causes

There are other causes of fires. Many wildland fires are the result of lightning. Lightning can strike anywhere, even in the middle of congested areas. Storms

■ **Note**
In a typical residential fire involving a room and its contents, the wiring does not get hot enough to melt.

■ **Note**
Arcing usually leaves a distinctive mark on the wire either in the form of beaded ends or one or more cavities in the wire itself.

often bring lightning strikes to structures, downed powerlines, and other problems. Fires in the wildland environment endanger any structure that is present. Even in the urban or suburban environment, fires can start as a result of exposure to other fires, or from flaming materials carried downwind from the original fire.

ARSON

■ **Note**

Arson contributes to more than 500 fire-related deaths each year, is a leading cause of fire loss (measured in dollars), and one of the leading causes of fires (measured in numbers of fires).

Arson is a pervasive problem affecting many types of property. In 1994, arson fires accounted for 28% of all fires in the United States.[1] Arson remains a leading cause of fires in every type of structure, in vehicles, and in the outside environment. Many arson fires occur in residential properties. These fires contribute to 90% of all arson deaths, 65% of all arson injuries, and 40% of the total dollar loss from arson.[2]

Although we tend to focus our efforts on arson involved in structural fires, a majority of all fires occur in vehicles and outside property (see Figure 9-1). The number of fires in these categories tends to be high, but the other statistics (loss of life, injuries, and dollar loss) are relatively small at between 10% and 20% of the total, and as a result, the impact of individual events tends to be overlooked. Overall arson contributes to over 500 fire-related deaths each year, is a leading cause of fire loss (measured in dollars), and is one of the leading causes of fires (measured in numbers of fires)(see Figure 11-4).

Figure 11-4 *Arson fires involve nearly all types of property. This one caused extensive damage in a high school. Photo courtesy of the Fairfax County Fire and Rescue Department.*

> ### ARSON IS A SERIOUS PROBLEM IN THE UNITED STATES
>
> In each of the past ten years there have been over 500,000 arson fires in the United States. Arson is a leading cause of fire and annually kills hundreds of Americans, injures thousands more, and causes over $3 million in property damage.[3]

■ Note

Motives for arson include spite and revenge, fraud, intimidation, concealment of another crime, vanity, pyromania, emotional disfunction, civil disorder, and the acts of juveniles.

Individuals set fires for many reasons. Arson has traditionally been thought of as part of a fraud to collect money, but it is increasingly being used to hurt people. Many arsonists are not "hired torches" who burn buildings for commercial gain, but spiteful individuals who deliberately set fires to hurt people. Their acts range from revengeful attacks on an individual or a family, to the recent bombings of the World Trade Center in New York and the Alfred P. Murrah Federal Building in Oklahoma City (see Figure 11-5). Other motives include vandalism, intimidation, concealment of a crime, other economic motives, vanity, pyromania, and emotional dysfunction including juveniles in crisis.

Figure 11-5 *The Alfred P. Murrah Federal Building. Photo courtesy of Michael Regan.*

motive
the goal or object of
one's actions; the
reason one sets a fire

MOTIVE

The reason a person sets a fire is known as the **motive**. There are several reasons or motives for arson-related fires. Many fires are set to obtain money. The usual scenario is that property is deliberately destroyed by fire in an effort to collect money from an insurance company or others. The motive here is fraud: an unfair or dishonest act to obtain something, in this case, to obtain money.

Another common motive for arson is revenge. Revenge fires are usually associated with rage related to a broken romantic relationship or particularly hostile labor-relations problems in the workplace. Because of the attitude of the arsonist, these fires are frequently started with a flammable substance or even an explosive device, with devastating results.

WHAT IS ARSON?

Arson is a legal term. Under common law, arson was defined simply as the malicious burning of someone else's house. Today, the definition of arson varies from state to state, but generally the definition has been extended to include any property including one's own property, and to designate four levels or degrees of arson:

- The burning of dwellings
- The burning of buildings other than dwellings
- The burning of other property
- The attempted burning of buildings or property

THE VALUE OF FIRE INVESTIGATION

In the state laws that regulate your fire department's operations, there should be a statement regarding fire investigation. The following is typical:

The local fire official shall make an investigation into the origin and cause of every fire occurring within the limits of the jurisdiction.

Taken literally, that statement suggests that you would call the fire chief to *every* fire. For most fires, calling an investigator, much less the chief, is not practical or necessary. One of the reasons we fail to call out an investigator is we may have to wait for him or her to arrive, and may have to wait even longer until the

■ Note
Every fire officer should become proficient in determining the "origin and cause" of every fire.

■ Note
Experienced fire officers should be able to determine the origin and cause of nearly all fires they attend.

■ Note
If in doubt, ask for the fire investigator and ask as soon as possible.

investigation is completed. Another reason is that we are often quick to assume the fire was accidental.

In many jurisdictions, fire investigators do not respond to fires that are considered accidental or that result in minor property damage. In such locations, it is important that every fire officer become proficient in determining the "origin and cause" of every fire.

The average fire is attended with one or two engine companies. One of the officers is usually assigned the task of gathering the important information and filling out the fire report. Part of the report asks for information about the cause and origin of the fire. Experienced fire officers should be able to determine the origin and cause of nearly all fires they attend. These reports are important in that they help the community, state, and to some extent, the nation, better identify our true fire problem. Therefore we encourage every fire department and every fire officer to complete the report that supports the gathering of data. Most of the states participate in the National Fire Incident Reporting System (NFIRS). Reporting continues to get easier as new computer technology is introduced.

Although officers should be able to determine the origin and cause of most fires, as soon as it become obvious that a fire investigator is needed, she should be requested. That determination might be made as early as the arrival of the first-in companies. Local policies and protocols suggest exactly how this should take place. Our recommendation is, if in doubt, ask for the fire investigator and ask as soon as possible. The investigator's arrival during fire suppression or at least during overhaul will assist in determining the cause of the fire, help build a better case if arson is involved, and may even allow suppression companies to leave sooner they might have otherwise.

GUIDELINES FOR CALLING THE FIRE INVESTIGATOR

It is suggested that an investigator should be called in the following situations:

- Any event that involves an obvious incendiary fire
- Any fire that results in death or serious injury
- Any event that produces burn injuries, especially those involving direct flame contact, fireworks, or is the result of an assault
- Any event that involves an exploding or incendiary device
- Any event that involves property damage in excess of $50,000
- Any event that produces damage to government property
- Any vehicle fire that is not the result of an accident
- Any event in which the officer in charge is unable to determine the cause

In any event in which the fire officer has concerns about these or other issues, the officer should talk with a fire investigator by phone or other means before leaving the scene.

Although we are still assuming that the fire is accidental 85% of the time, we should still accurately determine the cause of the fire all of the time. In cases where the fire is determined to have been accidental, timely and accurate fire cause determination is important for the owner, the insurance company, and the fire department. In addition, timely and accurate fire investigations help identify faulty equipment, unsafe habits, and patterns of activity. For example, a series of similar events may look like an unrelated series of *accidents* until one notices a pattern of time, addresses, or causes.

■ **Note**

Timely and careful gathering of evidence will aid in building a case that may bring someone to justice.

Although most fires are accidental, you should be thinking about the possibility of arson on every call. In cases where you have a reason to believe that the fire may have been deliberately set, timely and careful gathering of evidence will aid in building a case that may bring someone to justice. It may take some effort to note the things identified in the following paragraphs while they are occurring, but these observations may be useful during the size-up process, regardless of the cause of the fire, and they may be very useful if it is later determined that the fire was the result of arson. In many situations, it is very difficult to put these facts together after the fire is over. With that thought in mind, let us look at the things that the first responders may be able to note.

OBSERVATIONS OF FIRST RESPONDERS

Firefighters, especially those arriving first at the scene of a working fire, have a unique opportunity to make observations regarding a fire's behavior. Although their primary purpose is to save lives, protect property, and extinguish the fire, a moment of time making observations about the fire and the surrounding conditions may help in both the suppression effort and in the fire investigation that follows. Since even the best of fire investigators is usually not present during these first few minutes, and since you cannot recreate these condition once an investigator is present, it is important that you retain enough of the details so that they can be recalled and shared with the investigator and included in your fire report.

■ **Note**

First responders have unique opportunities for making observations while en route to the fire, upon arrival, during the initial stages of suppression, and during overhaul activities.

First responders have unique opportunities for making observations while en route to the fire, upon arrival, during the initial stages of suppression, and during overhaul activities. Some information can even be obtained during the first notification of the fire to the fire department.

During the Initial Notification Most fires are reported today by telephone. The first clue of a fire is usually a citizen's call to the fire department or emergency dispatch center. The caller may be an occupant or someone who just noticed the fire. Multiple calls pertaining to the same event indicate that it is probably significant. The information from the caller, as well as the information previously gathered in conjunction with the company's preplanning efforts, are integrated to form the basis for the response action. The calltaker should get the caller's name, phone number, and location; you may want to talk to her later.

While en Route   Weather, traffic conditions, time of day, and other factors should be noted and also incorporated into the response action. All of this information is important for the responders. Some of it may become important as you try to determine the cause of the fire.

Upon Arrival   The first emergency personnel to respond to the scene of a fire have a unique opportunity to help determine the origin and cause of the fire. These first responders can observe the amount of smoke and fire showing, the condition of the building, and activities in the immediate vicinity. The first responders may be able to note who is present at the scene upon their arrival, especially someone who is alone or demonstrating unusual behavior.

Upon arrival at the scene of a fire, you would expect people to be exiting the building and showing concern for those who may remain in the building. You should note the appearance and mental state of any occupants of the building. In most cases one or more of the occupants will meet the fire officer. You would expect people to be dressed appropriately for the time of day. Finding the occupant in nightclothes at 4:00 A.M. is normal; finding the occupant of a residence dressed for travel at that hour should raise some questions.

Fires also generally attract spectators. For all sorts of reasons, people will stop and watch a fire and many will linger to watch the fire department's suppression activities. This is normal behavior, too. We usually do not have time to observe the crowd of spectators that gather at the scene of a fire, but noting any who may be demonstrating unusual behavior may be very useful. Concern and curiosity are normal behaviors, but any unusual behavior, especially indications of excitement or pleasure should be noted. If there has been a series of fires in a neighborhood and a particular person has been present at several of these events, this person's appearance should be carefully noted and pointed out to the investigator or police.

We should also be concerned about those who might be leaving. It is difficult to note details of a person's appearance, activity, or even get a description of his vehicle during this time, but simply noting that the event occurred has its value. Persons who may have been working on the premises should be carefully noted. Construction activity, especially in cold weather, may have been the cause of the fire. Careful observations help you note and later recall the details that may be useful to the fire investigator (see Figure 11-6).

Noting the appearance of the fire itself is one of the most important observations that first responders can make. Information regarding the color and location of flames, and the color, quantity, and location of smoke provides valuable information for both fire suppression and for fire investigation. Note if and where the fire has **vented** itself, the number of rooms and floors involved, was it one fire or did there appear to be several separate fires burning simultaneously,

■ **Note**

Noting the appearance of the fire itself is one of the most important observations that first responders can make.

vented
opened to the atmosphere by the fire burning through windows or walls through which heat and fire by-products are released, and through which fresh air may enter

Figure 11-6 *Fire suppression personnel should be prepared to describe their observations and activities to the fire investigator. Photo courtesy of the Fairfax County Fire and Rescue Department.*

any evidence of collapse or pending collapse, and the presence of explosions or other unusual sounds. Pay particular attention to the presence of sound from any fire alarm systems.

The appearance of the property itself should be noted. The question, "Was the building occupied?" really has two meanings. Clearer understanding might occur if we were to ask, "Was the building in use at the time of the fire?" and "Was anyone inside the building at the time of the fire?"

Unoccupied buildings, that is, those abandoned by their owner, become a frequent shelter for homeless persons. In this case the answer to the first question (Was the building in use?) might be no, while the answer to the second question (was anyone inside?) might be yes. These "unoccupied buildings" are often the scene of significant fires, especially during colder weather, when the occupants build fires in the structures to keep warm. A homeless person is still a human being.

A completely separate issue deals with the security of the building. For buildings in which there were people present at the time the fire ignited, you would expect the doors, and possibly even the windows to be unlocked and maybe even open, depending upon the weather. Conversely, for buildings that are in use, but unoccupied at the time of the fire, you would expect to find these openings secured. Where automatic fire doors are installed, you should expect to find them closed. Deviations from these conditions should definitely be noted and reported both to the incident commander and the fire investigator.

The observations made during the early stages of the event are critical. Whether it be food on the stove or a fully involved structure as a result of arson, there are always indicators present that help in the investigation.

■ Note
Observations made during the early stages of the event are critical.

During the Initial Stages of Suppression Once entry is gained and fire suppression activity is started, there is often another opportunity for the first responders to note the conditions present. Note any unusual fire conditions, including the fire's behavior during fire suppression, as these should be the cause of concern both during fire suppression and later during the fire investigation. From experience you learn how fire acts when appropriate fire suppression actions are started. In all properties, you should listen and note the sound of the fire alarm system. If there was a sprinkler system in the building, you should be able to determine if the sprinkler system was operating and if the sprinklers were able to help control the fire.

During Overhaul For first responders, overhaul activities may present the best time to determine the cause of the fire. Although the purpose of overhaul is to be sure that the fire is completely extinguished, at the same time you can search for and gather clues to help determine the cause of the fire. It is important for those conducting the fire investigation to maintain a good command organization and to wear proper protective clothing during overhaul operation, but most of the urgency needed for effective fire suppression activities is no longer required. With a more deliberate pace, you can work safely, carefully, and systematically to overhaul and look for the origin and cause of the fire.

During overhaul, it may be necessary to pull down ceilings and wall materials. If the investigator is present, he or she may want to look around before this activity takes place. When this occurs, the overall investigation may be completed sooner. Any effort to move and remove the contents of the room to facilitate overhaul impedes the investigator's efforts. If the investigator has time to look at this material before it is moved and discarded, she will have a much better picture of the fire scene, and may see evidence that would otherwise be missed. In the final analysis, if you can help the investigator save some time and effort, everyone may be able to go home a little sooner. If circumstance preclude an investigator being at the scene, then the company officer and others present must do some of the work of the investigator. The first thing to look for might be whether any items appear to be missing. In most cases, pictures, personal property, and other items that one would expect to find in the places where you would expect to find them should still be in place.

■ **Note**
Cooperation during the investigation and overhaul phases provides better results and may allow everyone to go home sooner.

Sometimes the fire investigator may be several hours away. This situation presents the fire department with some very difficult choices. Although the fire suppression personnel may be most concerned with completing the overhaul of the fire and getting back into service, they should realize that their overhaul activities will likely destroy most of the evidence the investigator needs to determine the cause of the fire.

Firefighters should also be able to note the construction features that allowed for fire, heat, and smoke to travel through the structure. Open stairways, open windows, laundry chutes, and concealed voids all present opportunities for fire travel. Where burn patterns are apparent, do they indicate normal fire spread? Normally these patterns help determine the nature of the fire as well as its path of travel from the point of origin. Note and preferably photograph as much evidence of the fire's behavior as possible. These methodical and careful observations will assist in making a complete investigation.

The information gathered at the time of the alarm, and from observations made during the response, size-up, and suppression activities can be combined with this post fire information to help determine the cause of the fire. Each has an important part in the overall investigation, and for the most part, only first responders are in a position to gather the information.

FIRST RESPONDERS' OPPORTUNITIES FOR MAKING OBSERVATIONS

During the initial notification of the event:

    Identification and location of the caller

    The caller's tone, level of excitement, and accent

    Background noises

While en route to the fire:

    Additional information that may be provided by caller, the weather, time of day, etc.

    Delays due to highway construction, trains, etc.

Upon arrival:

    Any persons present and what they are doing

    Any vehicles present and whether they are leaving

    Note the fire conditions: location and intensity of the fire, color of the flames and smoke

Observation during size-up:

    Operation of any alarm or suppression equipment

    Any unusual observations

    Methods of escape of any occupants

During the initial stages of suppression:

    Were furnishings and inventory in place?

    Was the fire alarm sounding?

    Was the sprinkler system operating?

During overhaul activities:

    Recheck items noted above

    Determine origin and cause

    Determine path of fire travel

LEGAL CONSIDERATIONS OF FIRE INVESTIGATIONS

Company officers should know several important concepts about the legal aspects of fire investigation. Remembering these will help keep you out of trouble and, at the same time, reduce the chance that you interfere with the fire investigation. This information may seem to be more appropriate for the fire investigator, but every firefighter and fire officer should realize that actions taken *before* the investigator arrives are significant in setting the stage for any possible criminal action that may follow.

■ **Note**

Every firefighter and fire officer should realize that actions taken before the investigator arrives are significant in setting the stage for any possible criminal action that may follow.

Let us briefly review the legal rights of firefighters to enter the property of others. The Fourth Amendment of the U. S. Constitution and other laws hold that all citizens have the right to be secure in persons, houses, papers and effects, against unreasonable searches and seizures. The fire service has some rights too. In general, members of the fire service have the right to enter the property of another to extinguish a fire. The logic of this law is that the fire may spread to other property, and that the concerns of the community and its citizens outweigh the rights of any one individual. This same right generally means that the fire department has the right, within reason, to enter an adjacent property as needed to check for extension or facilitate fire extinguishment.

In addition, the fire department also has a responsibility to determine the origin and cause of every fire. As a result, a conflict may arise between the owner's constitutional rights of security in his property, and the community's needs, represented by the legal responsibilities of the fire department to conduct an investigation. These conflicting issues raise significant legal questions:

- How long can the fire department remain on the property after the fire is extinguished?
- What is required for the fire department to reenter private property to start or conclude a fire investigation?

More than 20 years ago, a fire in Michigan provided an opportunity to answer these questions. To briefly summarize the events, the fire started about midnight in January. At 2:00 A.M. the fire chief arrived and found evidence of arson. Efforts to conduct a thorough investigation at that hour were hampered by darkness, cold weather, poor visibility, and other dangerous conditions. At about 4:00 A.M., and with suppression activity complete, the fire department departed the scene.

Around 8:00 A.M. that same morning, the assistant chief, who was the primary fire investigator for the department, went to the property to determine the cause of the fire. Several weeks later a representative of the state police entered the property to take photographs and seize evidence. Other searches followed. The owner was eventually tried for arson. The owner tried to suppress the evidence, saying that the searches were in violation of the protection provided by the U. S. Constitution. The case eventually made its way to the U. S. Supreme Court.

MICHIGAN V. TYLER

In 1978 the U. S. Supreme Court laid down the rules for postfire searches in a landmark case known as *Michigan v. Tyler.* In summary, the Court ruled that the fire department has the right to enter property under emergency conditions and has the right to remain on the property until the emergency is over to extinguish the fire. The Court further ruled that the fire department could remain on the scene for a reasonable time for purposes of determining the origin and cause of the fire. Once the initial stage of the investigation is complete, the fire department should leave the property. If a member of the fire department wishes to return to the property, it should be under the auspices of an administrative search warrant or a criminal search warrant, as appropriate or with the freely given consent of the owner. Evidence obtained under other conditions is inadmissible.

As we can see, the court held that people have a right of privacy in their property, and that evidence obtained during warrantless searches will not be admissible in an effort to convict someone of arson. The major function of a warrant is to provide the property owner with sufficient information to reassure him that the law enforcement agency's (in this case the fire department) entry onto his property is legal.

In summary, the court held that entry into a burning building for fire suppression is acceptable and that once inside, firefighters have the right to determine the cause of the fire and to seize evidence that is within plain view. Evidence obtained under these condition is admissible.

Three Legal Ways to Search the Property of Another

Just as any firefighter has the right to enter the property of another to suppress fire, the person in charge has the right to enter the property to determine the cause of the fire. But as the Supreme Court noted, there are conditions that must be satisfied.

There are three methods by which the fire department may reenter the property *after* the fire is extinguished. The first is with the owner's consent. In many cases, simply explaining what needs to be done and asking permission to enter is all that is required to allow you to reenter the property and conduct an investigation. It is recommended that this consent be documented with a "consent to search" form. Most fire departments have a standard form for this purpose.

Another alternative is to obtain an **administrative search warrant**. Note that the stated intent is to determine the origin and cause of the fire, an **administrative process**. So far there is no reason to presume any criminal action. On the other hand, suppose that the evidence gathered during the department's initial

administrative search warrant
a written order issued by a court specifying the place to be searched and the reason for the search

administrative process
a body of law that creates public regulatory agencies and defines their powers and duties

search warrant
legal writ issued by a judge, magistrate, or other legal officer that directs certain law enforcement officers to conduct a search of certain property for certain things or persons, and if found, to bring them to court

probable cause
a reasonable cause for belief in the existence of facts

■ **Note**
Company officers should remember that they have a right to complete an investigation, and that they should have continuous presence on the property until that investigation is complete or an investigator has released them.

investigation, or at a subsequent investigation, does in fact suggest that there is a possibility of arson.

Now the rules change. In this situation, the fire department will need a **search warrant**. There is reason to suspect that a crime has occurred and this is now a criminal process. A criminal search warrant requires **probable cause** and spells out the conditions of the search, the area or areas that may be searched, and the material that may be seized. Evidence obtained under other conditions is inadmissible.

Hopefully at some point along this path the fire department has been able to obtain the assistance of a fire investigator (see Figure 11-7) or a police officer who is more familiar with this process. The point of our brief discussion here is to make you aware of these basic rules so that your actions do not jeopardize the rights of the citizens or the opportunity to obtain legal evidence that might be used in court. Company officers should remember that they have a right to complete an investigation, and that they should have continuous presence on the property until that investigation is complete or an investigator has released them.

Figure 11-7 *Fire investigators must work carefully to locate, gather, and preserve evidence.*

Figure 11-8
Maintaining security of the fire scene is important for the safety of the firefighter and for the preservation of evidence. Photo courtesy of the Fairfax County Fire and Rescue Department.

Maintaining security of the scene is important throughout these activities (see Figure 11-8).

Recording the Information

It is important to record the observations made at a fire. We do not expect to be questioned about every fire we attend, however, we might be questioned about an occasional fire. Unfortunately, we do not know in advance which one it will be, so it would be wise to make a few notes about every fire.

Why would you be questioned about the fire, or about your actions at a particular fire? The owners might be interested in your actions. The insurance company might be interested in what you saw upon arrival and what you did. Certainly, if there is any evidence of arson, you will be asked about what you saw and what you did. Months may pass before someone starts asking you to recall the events of a particular fire. In the meantime, you may have been to other fires in similar structures. How will you be able to recall the facts regarding a particular event?

First responders should note the address of the fire, the type of building, the location and intensity of the fire, the color and intensity of the smoke, and action taken. Your report should also note if anything was unusual within the structure.

■ **Note**
If there is any evidence of arson, you will be asked about what you saw and what you did.

How did your firefighters get in, and what did they see after they gained access? These are simple questions that can be easily answered shortly after the call is complete. Trying to recall the events of a particular fire a few weeks or months later may be impossible.

In most fires, there are normal answers to these questions. In those cases where arson is involved, the answers may suggest unusual conditions that warrant closer examination. Some good questions to assist in the process:

- Who discovered the fire?
- When was the fire first discovered?
- When was the fire reported?
- Who provided the first report of the fire?
- Who extinguished the fire?
- Who provided scene security?
- Who has pertinent knowledge regarding the fire?
- Who had a motive for setting the fire?
- What happened during the fire?
- What actions were taken by firefighters?
- What damage occurred?
- What do the witnesses know?
- What evidence was found?
- Where did the fire start?
- Where did the fire travel?
- Where were the occupants?

The Fire Report

Careful recording of the observations made during the course of any fire may be useful later, especially if the circumstances led to a trial. The information should include information regarding the response, size-up, suppression, and overhaul phases of the firefighting activity. Many of the questions will be prompted by the requirements to complete an official report. For most departments, this is on an official form that prompts many of these questions.

However, the most important part, and for some the most difficult part, is the written narrative statement. It should answer any questions not required in the form, explain any unusual circumstances, and otherwise complete any missing information. It is generally acceptable to express these statements using a personal style. "Upon arrival, I saw. . . . Firefighter Smith and I forced the front door. We advanced a 1 3/4-inch hose line. . . ." Usually it is easier to write and easier to read this sort of narrative if it is written in the first person and if the facts are reported in the order in which the events occurred.

Figure 11-9 *Writing a complete and accurate fire report is an important part of the company officer's job.*

When writing your report, you should avoid using terminology that might not be fully understood by people who are not in the fire service. Also avoid opinions. As Joe Friday said, "Just the facts." After you have written the first draft, proofread your work and clean up the grammar and spelling errors. Drawings are also useful and should be attached to the report. Complete the fire incident report as fully and as accurately as possible (see Figure 11-9). When you are writing this report, you have no knowledge of its potential use. It is wise to assume that every report could be reviewed by the department or used in court.

Documenting the Fire Scene

The fire scene can be documented by notes, diagrams, and photographs. We have already discussed the fire report. Notes are also important. Written notes can provide as much detail as necessary to help recall the facts at a later time. Some investigators use a tape recorder to facilitate the note-taking process, transcribing the recorded information to written documentation as soon as possible.

Photography provides a visual image of the fire scene. Photographs can be used to assist the investigator in recalling what was seen and may be used as evidence in court. If possible, the photographs should include scenes taken during fire suppression operations, and of the bystanders at the scene. As soon as the fire is over and conditions permit, an organized series of pictures should be taken to document the entire scene.

■ Note

An organized series of pictures should be taken to document the entire scene.

Just as the investigator follows a path toward the point of origin, so should the photographer follow a path that starts with external views, followed by internal views. The pictures should tell a story and work in a logical way toward the point of origin. This can also be done with a video recording. Regardless of the cause of the fire, or the quality of the written report, a picture may well be worth a thousand words if questions later arise regarding the department's actions, where items were located, the extent of the damage, evidence of other crimes, and so forth. Diagrams are also useful in documenting the fire scene. Here the entire fire scene is captured on one piece of paper. Diagrams can be useful when used in conjunction with photographs, especially if the location of each photograph is noted on the diagram as the picture is being taken.

Regardless of the method by which the documentation is obtained, it is important that the person be competent in using the equipment involved. Competency does not imply that one is an expert, rather that one knows how to operate the equipment and has a reasonable expectation of getting the desired results. Photographing a fire scene can be quite difficult. Challenges include the lack of good lighting, poor visibility, and the tendency of everything involved in the fire to look black. All of these can be easily overcome with training and experience.

Fire investigators take care of these activities when they are on the scene, but in many situations, you may not have the luxury of an investigator. The fire may be considered accidental or too small to justify an official investigation, or an investigator may not be readily available. In such cases, the origin and cause of the fire must be investigated and documented by the company officer.

■ **Note**

The fire may be considered accidental or too small to justify an official investigation, or an investigator may not be readily available. In such cases, the origin and cause of the fire must be investigated and documented by the company officer.

Courtroom Demeanor

In cases involving arson, it is not unusual for fire officers and even firefighters of the first arriving company to be asked to report their observations in the courtroom. Since the trial might take place many months after the actual fire, it is important to review the facts prior to appearing in court. In this case, the photographs, diagrams, and notes that you made during and shortly after the fire prove their value.

The choice of wearing a uniform or civilian attire is often decided by the custom of the local court. Find out what is appropriate. Regardless of the attire, it is important that you appear professional. This means that you are well groomed in every regard from having a neat haircut to having well-shined shoes. The following are also suggested:

- Sit upright with your feet flat on the floor.
- Keep hands in a natural position and try to avoid gesturing.
- Speak to the jury if you can, rather than to the attorney.
- Refer to the defendant courteously as "the defendant" or Mr. or Ms. _____.

- Answer questions with confidence gained by being familiar with the facts.
- Take time to be sure you understand the question. If unsure of the question, ask to have it repeated or rephrased. After all, you are not on trial.
- If the opposing attorney objects to a question, wait until the judge rules on the question before answering.
- Tell the truth at all times.
- If you do not know the answer, say so.
- Do not volunteer information.

Summary

Company officers should be aware of the causes of fire and the impact of arson in their community. They should be able to determine the cause of most fires and recognize the signs of arson. They should be able to conduct a basic investigation to properly determine and document the origin and cause of the fire. They should understand both the rights of the property owner and the requirements for conducting a lawful investigation.

This chapter has only scratched the surface of the topic of fire investigation. It is an area that deserves the attention of every fire professional. You are encouraged to read additional information on this important topic, and to take training and educational programs whenever the opportunity presents itself.

Training for firefighters and fire officers should include courses in determining the origin and cause of fires and arson recognition. (NFPA standards for both firefighter and fire officer certification include this requirement.) The fire science program at most community colleges includes a course in fire investigation. Many states offer training for firefighters and officers and certification programs for fire investigators.

The National Fire Academy, recognizing the need for additional training in these topics, provides a series of training programs to help you learn about fire investigation and arson recognition. A listing of these programs, starting with their basic awareness-level training program, is shown below:

Course Name	Course Length
Arson Detection for First Responders	2 days
Fire Cause Determination for Company Officers	6 days
Fire/Arson Investigation	2 weeks
Management for Arson Prevention and Control	2 weeks

Contact your training officer or state fire training director for additional information about these training programs.

Review Questions

1. What are the most common causes of fire?
2. What is arson?
3. What is meant by the term "motive"?
4. What are some common motives for arson fires?

5. What information should be noted by the first arriving personnel at the scene of any fire?

6. What information should be noted by fire personnel during fire suppression operations?

7. What information should be noted during overhaul?

8. What are the ways you can enter private property to conduct a fire investigation?

9. Why is it important to complete a fire incident report for every event?

10. What is the company officer's role in fire cause determination?

Additional Reading

Barracato, John, *Arson!* (New York: W. W. Norton and Company, 1976).

Brannigan, Francis L., "Building Construction Concerns for the Fire Service," *Fire Protection Handbook,* Eighteenth Edition, (Quincy, MA: National Fire Protection Association, 1997).

Custer, Richard L. P., "Fire and Arson Investigation," *Fire Protection Handbook,* Eighteenth Edition (Quincy, MA: National Fire Protection Association, 1997).

Davis, Richard, "Building Construction," *Fire Protection Handbook,* Eighteenth Edition (Quincy, MA: National Fire Protection Association, 1997).

DeHann, John D., *Kirk's Fire Investigation,* Fourth Edition (Saddle Brook, NJ: Brady, 1996).

Fitzgerald, Robert, "Fundamentals of Fire Safe Building Design," *Fire Protection Handbook,* Eighteenth Edition (Quincy, MA: National Fire Protection Association, 1997).

Mercilliott, Frederick, *Arson in The First Degree* (West Haven, CT: University of New Haven Press, 1995).

NFPA 921, *Guide to Fire and Explosion Investigations* (Quincy, MA: National Fire Protection Association, current edition).

Notes

1. *Fire in the United States 1985–1994.* (Emmitsburg, MD: United States Fire Administration, 1997), p. 184.

2. *Fire in the United States: 1983–1990.* (Emmitsburg, MD: United States Fire Administration, 1993), p. 219.

3. *Arson in the United States,* United States Fire Administration. (Washington, D.C., Federal Emergency Management Association, 1997), p. 1.

Chapter

12

The Company Officer's Role in Planning and Readiness

Objectives

Upon completion of this chapter, you should be able to describe:

- Preemergency planning activities.
- Training and education needs for company members.
- The company officer's role in maintaining company readiness.

INTRODUCTION

Good preemergency planning is essential for safe emergency scene management. The plan should provide information about building layout, access, construction features, occupancy, and a host of other important items that are needed before deciding to enter the building to attack the fire directly. The advantages of preemergency planning are often overlooked. We continue to read of firefighters who are injured and killed in structural fire situations where there was no preemergency plan, or if there was a preplan, the plan was not followed. Preemergency planning should address the very topics that may likely become an issue during any significant emergency in the building. These include layout, contents, construction, and type and location of installed fire protection equipment.

The information and plan have a variety of uses. The primary purpose is to provide information during the response and initial firefighting operations in a particular building, or other situation, but you can also use the same information for training purposes. For example, when the preplan calls for special evolutions that are not undertaken on a regular basis, this should suggest that these evolutions become a regular part of your training program. The preplan itself should be a part of the training program. A regular review of the plan by every member of every company that will respond to a working event at that address will help everyone to better remember the features and hazards of a particular address.

Most of the material in this chapter deals with fire-related situations, in part because this book was written to address the requirements listed in NFPA 1021. You should be undertaking the same planning and training efforts for dealing with all of the emergency situations that you are likely to encounter (see Figure 12-1). Fortunately, many of the tools used in planning for fire emergencies can be

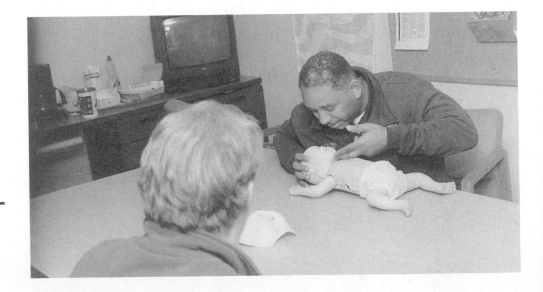

Figure 12-1 *Company officers should help their firefighters maintain basic skills.*

used to address the other problem areas as well. As a minimum, your planning and training programs should certainly address the medical emergencies, hazardous materials events, and rescue situations that you are likely to encounter in your particular community.

PREEMERGENCY PLANNING

preemergency planning
preparing for operations at the scene of a given hazard or occupancy

Preemergency planning, preplanning, and preincident planning are all terms that mean essentially the same thing. **Preemergency planning** is a process of preparing a plan for emergency operations at a given occupancy or hazard. Preemergency planning enhances effective and safer operations, helps you save lives and protect property.

Preemergency Planning Activities

Preemergency planning consists of these interrelated activities:

- the preemergency survey
- the development of information resources that would be useful during the event
- the development of procedures that would be used during an emergency

■ **Note**
Preemergency planning often starts with a survey of the property.

Preemergency planning includes the gathering and storing of information to facilitate the strategic dispatch and placement of personnel and equipment to deal with fire, rescue, and hazardous materials emergencies.

preemergency survey
the fact-finding part of the preemergency planning process in which the facility is visited to gather information regarding the building and its contents

Preemergency planning often starts with a survey of the property. This activity is called a **preemergency survey**. Preemergency surveys are essential for developing preemergency plans. Preemergency plans, in turn, are essential for developing strategies for safe and effective fireground operations.

In Chapter 5, we described the problem-solving process. We said that the first step in the problem-solving process is to recognize that we have a problem. Succeeding steps were to define the problem, collect data, analyze that information, develop alternatives, select the best alternative, implement, monitor, and correct as needed.

■ **Note**
Look at the fire scene as a problem that needs to be solved. The first two steps in the preemergency planning process are to identify the problems and gather the facts needed to find a solution.

Look at the fire scene as a problem that needs to be solved. The first two steps in the preemergency planning process are to identify the problems and gather the facts needed to find a solution.

Preemergency planning is a problem-solving process. You want to know as much as you can about the facilities and the activities that are within your area so that you are ready to deal with any emergencies that do occur. You are in a unique position of being able to define the problems, gather information, and develop plans for dealing with the emergency situations that might be encountered.

Preemergency planning information should be recorded and stored so that it is readily accessible to you in emergencies. The preemergency plan or simply, "preplan," allows you to make informed decisions while en route to and at the scene of an emergency. Preplanning benefits everyone involved with the event and contributes to more effective and safe incident management.

Always review your plan during training activities or at a real emergency. You will find that real incidents do not always turn out exactly like your plan, but the fact that you have a plan gives you a far better starting point than if you had nothing at all.

After you have a training exercise or an event in which the plan is used, you should review the plan and ask others for feedback. This is similar to the controlling or evaluation phase of the management cycle. From the feedback, you should be able to make some improvements to your plan. That entire process is what this chapter is all about.

USING FIRST-DUE COMPANIES FOR PREPLANNING ACTIVITY

Most fire departments have standard policies for the preemergency planning activity, including how to conduct a preemergency survey. Your department can gather this information in several ways, but the most logical way is to assign the company or emergency-response unit that is most likely to be the first to arrive at the address during an emergency to make the survey. This response unit, usually referred to as the first-due company, is likely to be there first during any emergency, and will probably have the greatest impact on mitigating any event that does occur. It is logical that they should be tasked with planning the actions that should be taken.

This procedure has several other advantages. First, it allows all of the firefighters from the company to visit the site and become familiar with the layout of the building under favorable conditions. A second benefit is that it permits all of the firefighters to have a chance to look at the process and storage hazards that may be present, and how these would impact on any fire or other emergency that might occur. Preemergency surveys are also an opportunity to let the building owner, occupants, and even the public know that the fire department is interested in protecting their property, just like it says in the fire department's mission statement. Preemergency surveys are valuable in many ways.

■ **Note**

Using first-due companies for preplanning activities allows personnel to become familiar with the property and allows the public to see the firefighters' interest in protecting the property

---

Where to Start

If your fire department has been involved with this activity for a number of years, your major task is simply to maintain the system, keep the existing information up to date and pick up new structures as they are built and occupied. On the other

■ **Note**
Given that you cannot plan for every situation, you should plan for those that present the greatest risk.

target hazards
locations where there are unusual hazards, or where an incident would likely overload the department's resources, or where there is a need for interagency cooperation to mitigate the hazard

hand, if your department has not taken advantage of preplanning, you may be asking, "Where do I start?"

Look first at target hazards. It would be nice to plan for all of the emergencies that might occur at all of the facilities within your jurisdiction, but it is unlikely that this will happen. Given that you cannot plan for every situation, you should plan for those that present the greatest risk. We will call these facilities **target hazards**. Target hazards are those occupancies that present high risk to life safety and property.

Target hazards usually include:

- occupancies such as health-care facilities, including convalescent homes
- large public assembly facilities, including airports, theaters, libraries
- multiple residential occupancies, including apartment buildings, motels, and dormitories
- schools and educational centers
- occupancies such as jails and prisons
- occupancies that present difficult challenges—high-rise buildings, and large buildings of any classification
- occupancies where access by fire department vehicles precludes the normal method of operations
- occupancies where there may be high property value
- occupancies where there is a high likelihood of a fire
- mercantile and business occupancies
- industrial facilities and hazardous storage facilities
- large unoccupied buildings
- buildings under construction or demolition

Even though residential occupancies (single-family homes, individual apartments, etc.) collectively represent the locations for the majority of fire calls, specific single-family homes are generally exempt from the type of formal preemergency planning discussed here. However, all departments should have standard procedures for dealing with fires and other emergencies that would likely occur in residences in the area they serve.

■ **Note**
The preemergency survey is the fact-finding portion of the preemergency planning process.

Preemergency Surveys

Preemergency surveys are an essential part of developing preemergency plan-of-action procedures to mitigate fire, hazardous materials incidents, and medical service calls in advance of an emergency situation. The preemergency survey is the fact-finding portion of preemergency planning. Preemergency surveys provide valuable information to facilitate an effective response assignment and provide responding personnel with timely information regarding the features and hazards of the facility. In addition to ensuring the effective use of resources during an emergency, such surveys provide documents to help train new personnel and help all personnel become aware of the hazards that may result in an accident or other unexpected event during emergency operations.

The first step in the survey process is to make an appointment to visit the facility at a mutually convenient time. Try to reduce the interference your intrusion may bring. Shopping facilities should be avoided during peak shopping days. Restaurants should be avoided during their busiest rush hours.

Once on the site, the company officer should look for the facility owner or manager. Once that person is found, you should introduce everyone (including all of the members of your company) and clearly state the purpose of your visit. Establish a pleasant but businesslike relationship. While conducting these surveys, remember that you are a guest on someone's property and that good manners and good appearances are important.

Ask to have someone go with you and be sure that they have keys to allow your access into as many areas as possible. This does several things. First and most important, for liability reasons, you do not want to be walking around the facility unescorted. Second, you may need to have access to areas that are generally kept locked or you may want to ask questions about a particular area or operation. Having someone along with you facilitates the entire process.

The next step is to develop a systematic plan for seeing all of the facility. The complexity and size of the facility determines how long it will take to make the survey.

It is important that you get a good feel for the layout of the building. Sometimes a picture, or in this case, a drawing is worth a thousand words. Actually two drawing are useful. One of these is called a **plot plan** (see Figure 12-2), an overhead view of the entire property showing the structure's placement, access, and other major external features. A second drawing is the **floor plan** (see Figure 12-3). This drawing shows the major internal structural features of the building. In buildings of more than one story it may be necessary to have floor plans for each floor, unless all of them are alike.

If possible, try to obtain a copy of these drawings. This will save you considerable time in taking and recording dimensions. During the survey, note any variations between the drawing and the building as it appears. Be sure to return the documents if you are asked to do so.

Allow enough time to complete the survey properly. In most cases your company will be in service and available for calls. However, many jurisdictions have established procedures whereby companies that are involved with preemergency surveys are not called unless they are really needed.

Even when a preplan already exists, it is worthwhile to visit the site once a year to make sure that the plan is still up to date, to see if there have been any structural or other alterations that might affect the plan (see Figure 12-4), and to allow new members of your company to see the building for the first time.

plot plan
a bird's-eye view of a property showing existing structures for the purpose of preemergency planning, such as primary access points, barriers to access, utilities, water supply, and so on

floor plan
a bird's-eye view of the structure with the roof removed showing walls, doors, stairs and so on

■ **Note**

The survey information
and the preemergency
plan should give first
consideration to life
safety.

life safety
the first priority during
all emergency opera-
tions that addresses
the safety of occu-
pants and emergency
responders

fire control
activities associated
with confining and
extinguishing a fire

property conservation
the efforts to reduce
primary and secondary
(as a result of firefighting operations) dam-
age

What to Look for During the Preemergency Survey

Reflecting on your mission statement, your first obligation is to save lives. The survey information and the preemergency plan should give first consideration to **life safety**. We usually do this by making sure that everyone is safely evacuated from the building and accounted for.

Your second priority should be to attempt to save property loss by stopping the fire, usually referred to as **fire control**. Your final concern should be to consider what you can do to reduce the loss that will result from the fire and the fire suppression efforts. Information regarding this aspect of the operation, usually referred to as **property conservation**, should be addressed during the prefire planning.

Just as your first priority during an emergency focuses on life safety, your first priority during the survey should be to look for hazards that may impact on

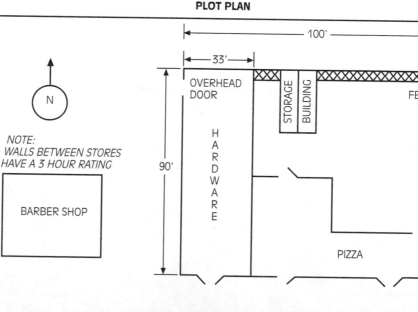

Figure 12-2 *Sample plot plan.*

FLOOR PLAN

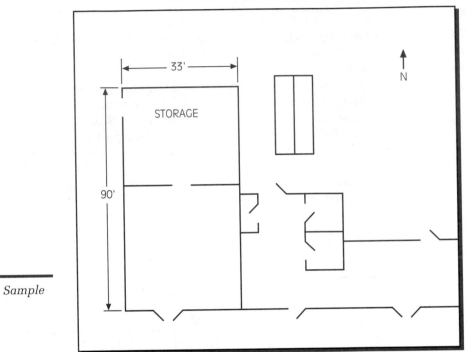

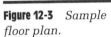
Figure 12-3 *Sample floor plan.*

Figure 12-4 *A photograph of the building also helps. This is the same building shown in Figure 10-3, but a year later, after it was completed and occupied.*

life safety. This means you should look for the location of means of egress for everyone in the building, making sure they are clear of obstacles and properly lighted.

Life Safety Concerns You should be aware of the number of people who would be present and what they would likely be doing. Dealing with ambulatory children in the middle of the day is quite different than dealing with nonambulatory adults in the middle of the night. Age, mental and physical conditions, and time of day are all important life safety considerations. They should be addressed in your preplanning activities.

There are several alternatives available to protect the civilians who may be in the building. While the most common approach is evacuation, there are several situations where evacuation is not the appropriate answer. An alternative may be to relocate the occupants to another part of the building, to an **area of refuge**. This is a likely situation in a large high-rise building.

Another alternative may be to have the occupants remain in place, as would likely occur in a hospital. Obviously, in either of these situations where the occupants are in the building, life safety considerations must be incorporated into the building's design and into your preemergency planning.

Building Condition You should also take interest in the structure itself (see Figure 12-5). We addressed this topic at length in Chapter 11, but certainly the building

> ■ **Note**
> You should be aware of the number of people who would be present and what they would likely be doing.

area of refuge
portion of a structure that is relatively safe from fire and the products of combustion

Figure 12-5
Preplanning includes a visit to the site on a regular basis.

■ **Note**

The building type, age, and condition are all important for the occupants' and firefighters' safety, as well as the way you fight the fire.

type, age, and condition are all important for the occupants' and firefighters' safety, as well as the way you fight the fire. The design and condition of the roof are important when planning strategies for safe firefighting. Think about how the fire will likely spread and how you stop that fire. These should also be addressed in good preemergency planning efforts.

Building Contents   Ask the owner or the manager about high value items. Ask about where the files are stored and where other items that would prevent the occupant from resuming operations are located. Make note of these items. You will want to give priority to these items during the salvage operations that would follow any fire.

Building Systems   You should also note the location and function of elevators, material handling equipment, and other mechanical devices that may become hazards during fire conditions. These hazards may present problems for safe evacuation, as well as means by which the fire may rapidly spread. Give attention to storage and process operations involving flammable materials.

Check the ventilation system. Determine if it will help spread the smoke, heat, and fire. Determine if the ventilation system would be used to control these same factors. Look for and make note of openings in the building that would facilitate ventilation. Most modern buildings have skylights and other fixtures that allow the roof to be opened without cutting. Timely ventilation will significantly help in the evacuation of the occupants, permit safer and more effective firefighting, and greatly reduce property loss due to the spread of heat and smoke throughout the building.

■ **Note**

You should always be interested in any installed fire protection equipment.

You should always be interested in any installed fire protection equipment (see Figure 12-6). Clearly understand the scope and function of any installed alarm systems. Your survey should capture appropriate information on how sup-

Figure 12-6   *During site visits, be sure to note the fire department connections.*

pression personnel can safely reset these systems after false activations. And if sprinkler systems and standpipes are included, be sure that you understand the pipe arrangement and how to augment the existing water supply. Where any of these features are present, they should be noted in the preplan. Tactical planning considerations should work in harmony with the installed fire protection equipment, not compromise its intended efforts.

INFORMATION NEEDS FOR PREEMERGENCY PLANNING

Preemergency planning is a process of preparing a plan for emergency operations at a given building or hazard. Most preemergency planning is directed to a specific location or occupancy. When the target of the planning process is a fixed location, the following information should be incorporated into the plan information:

- Building number and street address
- Other names commonly used to describe the occupancy
- Occupancy load during the day and night
- Building description (number of floors, dimensions, or square feet of floor space)
- Hazards to personnel
- Installed fire detection equipment
- Installed fire suppression equipment
- Water supply
- Access to utility cutoffs
- Priority salvage area
- Special considerations

■ Note
Certain vital information should be incorporated into preemergency plans.

Water Supply Take careful note of the water supply that may be available, both on site and within the immediate vicinity of the building. It seems elementary to mention that water supply is a critical item, but we point out that the best fire departments with the best of preemergency information and plans will need water to put out a fire. Where a hydrant system is available, the hydrant's capabilities should be verified (see Figure 12-7). Where water may come from other sources, plans should be made and resources identified for moving the water to the scene of the fire.

Exposures There may be as many as six exposures in a fire. Each should be considered in preemergency planing. Often an exposure presents greater risks than

Figure 12-7 *If in doubt, have the hydrant flow tested to be sure that it can provide adequate water during a time of need.*

the fire building itself. In such cases, the planning effort should identify these risks. It may be that with limited resources, the exposures deserve the full attention of the fire department, rather than the target building. These factors should be carefully addressed in any preemergency planning documents.

Contents Hazards Some buildings contain chemicals and other hazards that when heated and mixed together, present significant safety, health, or environmental problems. Again, preplanning should allow you to identify these hazards. If a significant fire should occur at one of these buildings, it may be better to allow the fire to burn and focus personnel energies on protecting the exposures and the environment. More than one fire suppression effort has resulted in creating a lake of highly contaminated water. If this contaminated water reaches the community's drinking water, there could be long-term problems associated with the incident. The overall mission of the fire department is to protect the community. Saving the community's water supply for the next generation may have far greater value than saving part of a building that will likely be torn down after the fire is over.

    The more you know about these factors, the better you will be able to deal with the emergency problems that occur. While the resource and time constraints may remain, many of the unknown factors can be eliminated. Elimination of unknown conditions will make the working conditions safer and more effective.

■ **Note**
Your preemergency plan should address the community's overall risk. Sometimes it may be more prudent to let a fire burn and focus efforts on protecting exposures and the environment.

Use a checklist to identify and record important information. Start by getting the correct address and the name of the building. Get the name of the owner, manager, and other key representatives, and their phone numbers both during working hours and nonworking hours; it may be important to contact one of these individuals when the facility is closed.

A program for conducting preemergency surveys does not replace the need for conducting regular fire prevention inspections. Both should be ongoing activities, done for different reasons. Preemergency surveys are best done by suppression personnel within their own response area.

Fire prevention inspections may be done by suppression personnel or others. In any case, fire code violations found during a preemergency facility survey shall not be overlooked. Severe violations should be corrected at once. Where immediate correction is not possible, immediately advise your fire prevention division of the situation. In all cases, hazards should be reported using the appropriate reporting forms. Copies of the report should go to the person responsible for the building as well as the fire department's fire prevention division.

CONVERTING YOUR SURVEY INFORMATION INTO A PLAN

Turning all of the information gathered during the preemergency survey into a preemergency plan is the second step in preemergency planning. Preemergency planning gathers information from a variety of sources. We have the preemergency survey, of course. The information gathered during the survey, information about the department's resources, and the department's standard operating procedures suggest appropriate actions for planning the initial strategies that would be needed during a real emergency. The topics listed on the previous pages will provide a guide for these planning efforts.

You may also have information from the fire prevention division of the fire department. They may be inspecting the facility on a regular basis due to the nature of its occupancy. They may have issued a permit to store certain material or conduct certain operations, either of which you should know about.

Upon completion of the survey the company should work together to develop a plan (see Figure 12-8) that will address the most likely scenarios that would occur at that address. While we often think of the threat of fire, we should also consider EMS and hazardous materials control, and rescue situations as well.

The planning effort should address the same factors that were addressed during the survey. In addition, the plan should address the resources needed to deal with the emergency, and how these resources will be initially organized. For example, the plan should include placement of first-arriving companies, identify water supply sources, and provide a starting point for fireground strategy. Many buildings, because of their age or condition, are not worth the risk of a single firefighter to conduct interior firefighting operations. Where these buildings are identified, the initial mode of operations should clearly indicate that all firefighting activity will be from outside the structure.

Figure 12-8 *Get as much participation as possible in the preplanning activity.*

PREEMERGENCY PLANS SHOULD ADDRESS

- Life hazards problems, for occupants and firefighters
- Availability of egress for the occupants
- Availability of access for firefighters
- Places where a fire would most likely start
- Factors that would influence the spread of the fire
- Factors that would influence the intensity of the fire
- Initial strategies for effective fire containment and control
- Factors that would limit the fire department operations
- Resource needs including those of other agencies
- Apparatus placement
- Location of the command post

> ■ **Note**
> Part of preplanning includes forecasting the resources that may be needed to mitigate the situation.

Resource Needs to Deal with the Emergency

Part of the preplanning process includes forecasting the resources that may be needed to mitigate the situation. Matching resources to the needs of routine operations is easily accomplished. Room-and-contents fires, vehicle accidents, and standard EMS calls make up the bulk of most fire department operations. The events can usually be controlled with first-alarm resources. But for large events, you will need additional resources. Your preemergency planning efforts should address the potential for needing additional resources. If additional resources are likely to be needed, add that information to the preemergency planning information. Many jurisdictions provide special dispatch assignments, based on the preemergency planning information.

For example, when planning for operations at target hazards, you should always be thinking of the potential of an event that will require resources beyond the normal first-alarm assignment. In shopping centers and high-rise buildings, the mere transportation of firefighters and their equipment to the fire requires a tremendous support team. Hazardous materials incidents also require considerable support. And during wildland operations, we see the need for armies of firefighters to extinguish the fire while remaining safe. Preemergency planning should look at these needs. Even during routine firefighting, we should have relief crews on scene, ready to step in when needed.

Once you have identified your resource needs, you should consider how these needs will be met and how rapidly these resources can be on scene. Even in large cities, getting second and third alarm companies to the scene can take some time. In smaller communities, especially those communities where volunteer firefighters provide fire protection, there may be appreciable time delays as the firefighters and their equipment travel considerable distances to the scene.

In addition to fire department resources, you should consider the resources of other agencies. For example, will the police department be able to help with traffic and crowd control? What resources may be available on site and from other industrial or commercial facilities in the immediate vicinity? What about resources from federal and state agencies?

Interagency Cooperation

Identifying those resources is only the first step. Resources from agencies outside of your own should have an opportunity to review your plans and understand their role. The time to develop a clear understanding of the lines of communication and authority are during the planning stages, not during the time of an actual emergency. There should also be a clear understanding of who is authorized to make the request for assistance, what will be sent, and a resolution of questions regarding liability issues that may arise during an actual emergency.

Interagency cooperation must be carefully planned to reduce conflict, confrontation, and confusion during an emergency. Effective cooperation includes a study of each agency's capabilities, regular sharing of information, and joint training exercises so that everyone understands the capabilities and limitations of all of the other agencies.

■ Note
Once you have identified what your resource needs are, you should consider how these resource needs will be met and how rapidly these resources can be on scene.

■ Note
In addition to fire department resources, you should consider the resources of other agencies.

■ Note
Interagency cooperation must be carefully planned to reduce conflict, confrontation, and confusion during an emergency.

PREEMERGENCY PLAN INFORMATION SHOULD INCLUDE

- Address
- Owner
- Occupancy or use
- Means of access and entry
- Personnel hazards
- Fire behavior predictions
- Locations of stairs and elevators
- Ventilation system
- Installed fire protection systems
- Exposures
- Special hazards
- Resource needs
- Company assignments
- Estimated fire flow
- Water supply
- Predicted strategies

The National Fire Protection Association (NFPA) publishes several documents that aid the preemergency planning process. NFPA 170, *Fire Safety Symbols,* provides a good listing of standard symbols used in preplanning and other fire-related applications.

PREPLANNING FOR OTHER EVENTS

Over the last several pages we have discussed the need for preemergency planning at target hazards. We usually think of target hazards as fixed sites with specific addresses. You should also be thinking about the typical incidents that would benefit from the same preemergency planning process. In this case, we are thinking more in terms of typical or recurring events rather than specific hazards.

During some of your preemergency planning efforts, you may lack a specific address, but the general characteristics of the occupancy allow you to develop a preemergency plan for the type of hazard, rather than a specific address. Examples include various types of residential structures common to the first-due assignment area, high-rise buildings, shopping centers, service stations, and schools.

In addition to events that occur at structures, departments should be prepared to deal with other situations that are likely to occur within their jurisdic-

■ **Note**
In addition to events that occur at structures, departments should be prepared to deal with other situations that are likely to occur within their jurisdiction.

tion. Nearly every fire department has roads and highways, railroads, pipelines, and waterways. Each of these modes of transportation presents opportunities for fire and rescue situations that will involve the fire department.

The preplanning for these events often leads to a department-wide **standard operating procedure**, or SOP. The SOP is usually a several-page document that follows an established format and lists the various hazards associated with the type of hazard, and establishes standard procedures for first arriving companies. Unlike the preemergency plan information for site-specific hazards in which you will have time to look at information while en route to the emergency scene, many fire departments expect officers to be able to recall from memory the general procedures to follow at these selected target hazards.

RECORDING, STORING, AND RETRIEVING THE INFORMATION

The National Fire Academy suggests a very simple form for recording the information for the preemergency plan called a **Quick Access Prefire Plan** (see Figure 12-9). Usually a follow-on page of typed information, a drawing of the plot plan, and a drawing of the floor plan accompany this basic information.

There are a variety of ways of storing preemergency planning information. Regardless of the method selected, the main purpose of having the information is so that it is available when and where it is needed. If an emergency arises at a particular address, the preplan should be reviewed by companies while they are en route to the emergency. The information is also needed by companies operating at the scene.

Each company should have access to preemergency plans for their first-due response area in their apparatus. An additional set of the preemergency plans should be kept at the station near the watch desk for use by companies that may be reassigned to fill in at the station. These companies would likely be called up if additional alarms are needed. This same copy can be used for training and drill purposes.

A computer provides a compact way of storing a great deal of information and retrieving it rapidly. The information may be stored at the dispatch center and transmitted to a company when it is needed, or the information may be stored on a portable computer located in the vehicle. Although there are many other modern ways of storing and using this information, many fire departments still rely on the reliable three-ring notebook.

Good information in a readily available format permits timely access and allows for better analysis and the quicker start of solutions to mitigate the problem.

Where there is a lot of stored information about a particular facility, some departments keep the information on the facility site in a **lock box.** The lock box is located in an accessible location on the property and can contain preemergency planning information, material safety data sheets, the names and phone numbers of key personnel, and most importantly, keys that allow emergency responders to enter the facility. Obviously, the fire department must have access to the lock box

standard operating procedure (SOP)
an organized directive that establishes a standard course of action

Quick Access Prefire Plan
a document that provides emergency responders with vital information pertaining to a particular occupancy

■ Note
Each company should have access to preemergency plans for their first-due response area in their apparatus.

lock box
a locked container on the premises that can be opened by the fire department containing preplanning information, material safety data sheets, names and phone numbers of key personnel, and keys that allow emergency responders access to the propert

Quick Access Prefire Plan

Building Address: *1020 Jeanne Street*

Building Description: *1-story, ordinary construction; 2 occupancies in building*
firewall between occupancies

Roof Construction: *Wooden 2" x 10" rafters, plywood, composition roof covering*

Floor Construction: *Concrete slab*

Occupancy Type:	Initial Resources Requred:
Commercial	*2 Eng., 1 Truck or Squad, 1 Amb., BC*

Hazards to Personnel:
Pesticides, flammable/combustible liquids

Location of Water Supply:	Available Flow:
Hydrant across Jeanne Street	*1200 GPM*

	Estimated Fire Flow			
Level of Involvement	25%	50%	75%	100%
Estimated Fire Flow	375	750	1125	1500

Fire flow of largest fire area–hardware store and 2 exposures

Fire Behavior Prediction:
Rapid horizontal spread within one occupancy

Predicted Strategies:
Confinement, ventilation, extinguishment

Problems Anticipated:
Poor rear access, limited horizontal ventilation

☐ Standpipe:	☐ Sprinklers:	☐ Fire Detection:
None	*None*	*Smoke Detectors*

Figure 12-9 *The National Fire Academy's Quick Access Prefire Plan.*

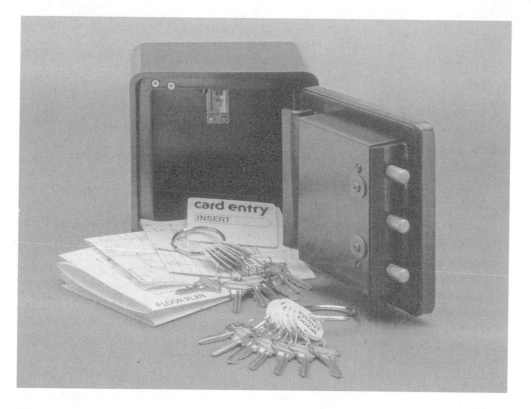

Figure 12-10 *A lock box provides information as well as a means for entry. Photo courtesy of The Knox Company.*

(see Figure 12-10). The lock box has many advantages, especially in that it allows for items such as keys to be left at the site in a secure place. Having the detailed preemergency planning information there may also be advantageous, especially when such information requires storage space beyond the capabilities of first-arriving units. If a lock box is used, the basic information (address, etc.) should be available to companies while en route.

Training and Practice

Training is a vital part of preemergency planning. If personnel do not know of the plan or do not know how to implement it during a time of need, the effort to gather the information and plan the effort has been largely wasted. Training and practicing for dealing with emergency situations should be a continuing process.

COMPANY TRAINING

The remainder of this chapter discusses fire department training at the company level. Training is the key to fire department effectiveness. No factor has more

■ **Note**
Preemergency plans should be exercised through regular training and practice.

effect on a fire department's operations than its training program. Since the fire department's main function is to save lives and protect property, members of the department must be qualified to successfully and efficiently perform a wide range of tasks in carrying out these important functions.

The realistic financial constraints of the world in which we work combined with the mission of the fire service requires the greatest efficiency from every agency and from every member of these agencies. With decreased fire activity, most fire firefighters are not called upon to save lives and protect property on a daily basis. However, all firefighters are expected to maintain a high level of readiness, should that need arise. Efficient use of training time provides significant opportunities to increase the readiness of the fire department and its personnel.

During the recent past, fire service training has changed immensely. For many years, teaching a few basic manipulative skills was considered to have met the requirements of an adequate fire service training program. But the role of the fire service is constantly changing and becoming more complex. New tasks are being undertaken and new challenges to existing procedures are being faced everyday.

Today, the training process starts when an individual joins the fire service and should continue throughout that individual's career. The fire service training program must prepare the member to meet the known challenges of today as well as be ready to meet the future.

Educators tell us that training precedes learning. Learning is the process of acquiring new knowledge and skills. To be effective, a training program must provide a process that encourages firefighters to learn information and develop skills that can be retained and used. Effective fire department training programs are essential for the safe and efficient operation of any fire service organization and for its members.

All company officers in a fire department should fully support the fire department's training program and should take personal interest in making sure that the department's training activities are enthusiastically carried out within their area of responsibility.

In this regard, company officers are responsible for training members assigned to them. Typically, most of the department's training program is conducted at the company level under the direction of the company officer. Such training activity strengthens the bonds among personnel while developing proficiency in individual and company evolutions. Company officers should periodically evaluate the abilities of their personnel to both determine the effectiveness of the training and to provide a valid basis for individual performance evaluations.

Company officers should observe the performance of their personnel during emergencies to see that they are using the same techniques that were presented in training activities. Critiques of company operations at fires and other emergencies can help identify the company's performance as a team. Company officers are probably the best qualified individuals in the department to assist the department's overall training efforts by identifying topics, procedures, and conditions where training is required.

■ **Note**

The realistic financial constraints of the world in which we work combined with the mission of the fire service requires the greatest efficiency from every agency and from every member of these agencies.

■ **Note**

To be effective, a training program must provide a process that will encourage firefighters to learn information and develop skills that can be retained and used.

The overall effectiveness of the department's training program should also be continually evaluated by the department's senior staff. The department and company training program, as well as the results of that training, should be reviewed on a regular basis. The results of that review should suggest topics for future training activities.

New Personnel

New personnel should receive comprehensive training during their recruit school. During the period following the successful completion of that school, they are usually required to serve as probationary firefighters. This period is important because the opportunity and motivation for individual learning is at its greatest during these first few months. It is also a time to teach good habits and reinforce the training provided by the training academy. The training accomplished during this period will become the basis for that which follows throughout an individual's career. New personnel should be trained consistent with the performance objectives outlined in NFPA 1001, the *Standard for Fire Fighter Professional Qualifications.*

Figure 12-11

Company personnel should maintain basic skills while learning new ones.

Individual and Company Skills

NFPA 1001 provides a list of the requirements for basic firefighter skills. These are not one-time events, a rite of passage so to speak. All firefighters should be able to pass any part of that test on any given day. To maintain this level of proficiency, company-level training activities should also include a regular review of the individual tasks required of all firefighters during their probationary period, because the basic knowledge and skills are often quickly lost without regular use.

Regular training on all the skills required of the firefighter (see Figure 12-11) may well exceed the time available, so some selectivity in training topics is usually required. Company officers should be able to recognize the needs of their company and emphasize particular areas of training. These training needs can usually be recognized by observing the performance of individuals and companies during training evolutions and at actual emergency situations.

Multiple-Company Evolutions

Multiple-company training evolutions provide an opportunity to put it all together, to test the system as it will operate during a real situation. The pace of activity is sometimes slowed to permit monitoring the progress and directing the training activities, but the events should all occur and should be conducted under realistic conditions whenever possible.

Multiple-company training evolutions permit firefighters to use their skills in realistic activities under conditions that can approach those that can be expected during a real emergency (see Figure 12-12). In such evolutions, company

■ Note

Company officers should note the performance of their personnel during training and actual emergency situations, and provide training in those areas in which skills do not meet expectations.

Figure 12-12 *Multi-company training evolutions provide an opportunity to exercise the Incident Management System.*

officers take their place as leaders in the operation. Their leadership ability as well as their prior efforts in providing training for their personnel will be quickly apparent. Although these activities provide a test for the readiness of all individuals and their units, remember that these activities represent a training exercise: This is a time for learning, a time to identify and correct problems that may exist.

Multiple-company training evolutions also provide an opportunity to train and test the fire department's incident command and communication systems. Such training activities provide an ideal opportunity for officers to move into positions of increased responsibility in managing emergency incidents and in controlling increased numbers of resources. These activities also provide a realistic opportunity for using and evaluating the department's communication's resources.

Multiple-company training evolutions provide a unique opportunity to work with companies that would probably be encountered on a mutual aid response. These training sessions provide an opportunity for the mutual aid company to become familiar with the neighboring department's procedures, response area, and high-risk targets. This is a good time to test those preplans and discover incompatibilities in equipment and communications capabilities. As a company officer, you do not have to wait until a more senior officer suggests the need for such evolutions; you should be able to make it happen on your own initiative.

Suggested training scenarios for individuals, company, and multiple-company training are included in Appendix B. These are just suggestions: You should have additional ideas for your company.

Live-Fire Training

Live-fire training provides unique opportunities for developing skills and self-confidence at all levels. Live-fire training permits those who have completed their basic training to operate under realistic conditions (see Figure 12-13). This training provides an ongoing opportunity for maintaining and improving acquired skills for all personnel. Live-fire training provides a means where the firefighter can display many combinations of skills, and at the same time, develop an appreciation of the necessity for personal safety during actual emergency evolutions. In order to ensure safety for all of the students and trainers during live-fire training exercises, all participants should have achieved a basic level of training by having completed the requirements for Firefighter I certification.

When and where possible, training-facility burn buildings should be used instead of acquired structures for all live-fire training events. When acquired structures are used, the building must be carefully inspected and prepared for the training evolution. Consideration shall be given to the presence of hazards, including utilities, pressure vessels, closed tanks, and interior finishes that may contribute to the rapid spread of fire. These hazards should be removed before the training starts. Consideration shall also be given to the age and condition of the structure itself, the presence of live and dead loads, and the probable effect that

■ **Note**
This is a good time to test those preplans and discover incompatibilities in equipment and communications capabilities.

■ **Note**
Live-fire training provides unique opportunities for developing skills and self-confidence at all levels.

■ **Note**
When acquired structures are used, the building must be carefully inspected and prepared for the training evolution.

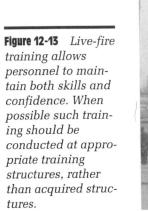

Figure 12-13 *Live-fire training allows personnel to maintain both skills and confidence. When possible such training should be conducted at appropriate training structures, rather than acquired structures.*

the fire will have on the integrity of the structure. When evaluating acquired structures, the following items should be considered:

- Floors, railings and stairs shall be made safe.
- Walls and ceilings shall be intact or patched.
- Debris that might contribute to an unsafe condition shall be removed.
- Low-density combustible fiberboard and unconventional interior finishes shall be removed.
- Extraordinary dead weights shall be removed.
- Means for controlled roof ventilation shall be provided.
- Chimney hazards shall be eliminated.
- Utilities shall be disconnected.
- Asbestos shall be removed.

In all cases, live-fire training evolutions should be conducted in accordance with the current edition of NFPA 1403, *Standard on Live Fire Training Evolutions.*

In addition to the safety considerations for their own personnel during live-fire training, fire departments should carefully consider the safety, health, and environmental consequences of their live-fire training activities on the community, both during and after the training activity.

Training for Officers

To the extent possible, fire departments should encourage the continuing education and training of their personnel, not only to meet the minimum qualifications for promotion to positions of greater responsibility, but also to provide these officers with the knowledge, skills, and abilities needed to be good managers and leaders.

Such education and training is available through many community colleges, state and regional fire schools, various training conferences and seminars, and of course, the National Fire Academy at Emmitsburg, Maryland. When such training programs are offered in-house, the subject matter of officer training courses should be designed to permit individuals to enhance their abilities and at the same time, meet the performance objectives outlined in NFPA 1021, *Standard for Fire Officer Professional Qualifications.*

Other Training

Specialized and advanced training should take place throughout a firefighter's career. Such training includes courses in emergency medical procedures, recognition and control of hazardous materials, specialized rescue procedures, driving of emergency vehicles, equipment operation, fire investigation, fire prevention, fire safety education programs, and customer service. Fire departments should be alert to the changing needs of the fire service, and to the needs of their personnel. As we said at the start of this section, training is the key to fire department efficiency.

To ensure consistency in the delivery of training among various components of the fire department and over a span of time, all training should be conducted according to standardized lesson plans. Such lesson plans are intended only to assist instructors in providing effective and standardized training and should be looked upon as providing the *minimum* amount of material to be presented. Lesson plans should never be considered as restricting new approaches for presenting material, improving teaching effectiveness, or maintaining student interest.

Training activities allow you to prepare your personnel and evaluate your equipment in what might be considered a practice session for the real event. No sports team or show company would dare perform without practicing. Why should you?

Company-level training includes classroom instructions, practice of evolutions, drills, familiarizations, inspections, and preincident planning, to name just a few. Several hours of each workday should be devoted to such activity.

Physical Training Is Important Too

In Chapter 8 we noted that the requirements of firefighting are physically demanding. Part of having your company ready for dealing with emergency

■ Note
Training activities allow you to prepare your personnel and evaluate your equipment in what might be considered a practice session for the real event. No sports team or show company would dare perform without practicing. Why should you?

situations includes that all company personnel be physically and mentally ready as well. Good physical condition increases every firefighter's ability to perform his or her duties and reduce the likelihood of injuries while performing those duties. Once again, we see an important role in this process for the company officer. As the company officer, you should demonstrate good leadership and provide, by personal example, the benefits of a good fitness program. The readiness, success, and safety of your company depend upon it.

One Final Comment

We started this chapter by drawing a parallel between the preemergency planning process and the problem-solving process. You have identified the problem, gathered information, considered alternative solutions, and implemented a solution. Remember the last steps in the process; after training activities or after a real event, take time to go back and review the preemergency plan to see how it worked. When appropriate, make the necessary changes. This process will enable you to have plans that really work, and that people will readily accept.

Summary

Company officers should develop a personal plan that will allow them to effectively lead their company to accomplish all of the tasks we have considered in this chapter. This will help the company become proficient in its tasks and ready to safely respond to help the citizens of the community it serves. The ability of the company to provide these essential services lies with the knowledge and dedication of the company officer. Understanding the problems and the solutions increases readiness and reduces the challenges. Readiness also involves being ready to perform. Constant planning, training, and physical fitness are essential for company readiness.

Review Questions

1. a. What is a preemergency survey?
   b. What is a preemergency plan?
2. Why is it best to use fire suppression personnel to develop these plans?
3. What information should be contained in a preemergency plan?
4. What features of the building should be noted in the preemergency survey?
5. What types of occupancies should be preplanned?
6. How can the preemergency planning information be made available during a time of need?
7. Should all members of the company be familiar with preemergency plans for occupancies in their area? Why?
8. Why is it import to have a continuing program of company level training?
9. What are the benefits of multiple company training activities?
10. What is the company officer's role in company readiness?

Additional Reading

Brannigan, Francis L., "Building Construction Concerns for the Fire Service," *Fire Protection Handbook,* Eighteenth Edition (Quincy, MA: National Fire Protection Association, 1997).

Davis, Richard, "Building Construction," *Fire Protection Handbook,* Eighteenth Edition (Quincy, MA: National Fire Protection Association, 1997).

Fitzgerald, Robert, "Fundamentals of Fire Safe Building Design," *Fire Protection Handbook,* Eighteenth Edition (Quincy, MA: National Fire Protection Association, 1997).

Freeman, Raymond, "Theory of Fire Extinguishment," *Fire Protection Handbook,* Eighteenth Edition (Quincy, MA: National Fire Protection Association, 1997).

Linder, Kenneth W., "Water Supply Requirements for Fire Protection," *Fire Protection Handbook,* Eighteenth Edition (Quincy, MA: National Fire Protection Association, 1997).

Managing Company Tactical Operations, published by the National Fire Academy, Emmitsburg, Maryland. This three-part training course focuses on preparation, preplanning, and managing typical structural firefighting activities. The first two parts, called "Preparation," and "Decisionmaking," are particularly relevant. Contact your training officer or your state fire training director for additional information.

Wilson, Dean, "Fire Alarm Systems," *Fire Protection Handbook,* Eighteenth Edition (Quincy, MA: National Fire Protection Association, 1997).

Chapter

13

The Company Officer's Role in Incident Management

Objectives

Upon completion of this chapter, you should be able to describe:

■ The factors to be considered during size-up.

■ The factors that determine how fire spreads in a structure.

■ The procedures to control, confine, and extinguish fires and protect exposures in structure fires, in outdoor situations, and where hazardous materials may be present.

■ The duties and responsibilities of officers using the incident management system at responses involving one or more units.

INTRODUCTION

As you pull up in front of Ms. Jane Doe's house, you see enough smoke coming out of the back of the house to know that this is no false alarm, nor is it someone's backyard barbecue. Ms. Doe is in her front yard, her hands clutching a wet towel. She is understandably upset. Her kitchen is on fire!

For you or any first-arriving officer, this is show time, time to put all of your training, all of your experience, and all of your fancy equipment to work. And for you, and for all of those overqualified, hard-charging firefighters you brought with you, this is why you joined the fire service. Will things go well? Will you stop the fire with no additional damage to Ms. Doe's house? Will you and your firefighters operate safely?

Strong leadership is needed at events like these. We know that every group has a leader, and in this case we expect the officer to be the leader, to be in charge, to be making decisions and giving directions. We want the company officer to be managing the event using a proactive management style.

Good scene management focuses on the priorities of the incident, provides for the safety of all concerned, and seeks to reduce further loss. The old philosophy, "grab a line and put the wet stuff on the red stuff," has long been outdated. Today, the company officer is expected to take command, provide leadership, identify problems, establish priorities, allocate resources, coordinate activities, and protect the safety of firefighters. When these things occur, company officers are in a proactive mode and are properly fulfilling the leadership role of the position they hold.

This chapter only introduces these concepts. We cannot cover all the aspects of incident management in one chapter, or even in one book. Clearly, considerable learning and experience must be integrated together to become an effective incident commander. But we can start that process here. Company officers should be able to command first-alarm resources. For small events they should be able to manage the event from start to finish. For larger events they should be able to set the foundation for what will follow, pending the arrival of a more senior officer. In either case, action taken during the first 5 minutes has a significant impact on the overall outcome of the event.

At an emergency event, the incident commander is responsible for the overall management of the incident (see Figure 13-1) and for the safety of everyone involved. At such events, the incident commander must assume command, evaluate the situation, initiate, maintain, and control communications, formulate strategy, and develop an organization to fit the needs of the event.

INCIDENT PRIORITIES

The Plain City fire department's mission statement includes a phrase about saving lives and property. Your decision-making process at the scene of any emergency should focus on this same statement: Save lives and property. These

■ **Note**
Good scene management focuses on the priorities of the incident, provides for the safety of all concerned, and seeks to reduce further loss.

■ **Note**
The company officer is expected to take command, provide leadership, identify problems, establish priorities, allocate resources, coordinate activities, and protect the safety of firefighters.

■ **Note**
Action taken during the first 5 minutes has a significant impact on the overall outcome of the event.

Figure 13-1 *The incident commander is responsible for managing the overall event and for the safety of everyone involved.*

become the guidelines for your activity: life safety, incident stabilization, and property conservation.

Your first priority is *life safety* for the civilians who may be at risk and for the firefighters under your command (see Figure 13-2). Life safety must always be the first consideration of emergency responders. The first responsibility of company officers is to safeguard human safety; to protect the safety of those under their direct supervision; and to take all reasonable risks in protecting the lives of the public they are sworn to protect.

In structural firefighting we use the term *rescue* to describe the activities associated with the civilians who may be at risk. Rescue includes protecting the occupants, removing those who are at risk, and searching for those who may not be able to help themselves.

The second priority is *incident stabilization*. You cannot be held accountable for what happened before you arrived on scene, but you are clearly accountable for what happens after you arrive. You must be sure that your information, your planning, and your actions all support an effective, coordinated, and safe operation. In firefighting operations, we call this fire control. In a hazardous material spill we attempt to contain the product. In both cases we want to keep from doing any additional damage.

The third priority is *property conservation*. Firefighters often think of property conservation in terms of salvage operations, often done after the fire is controlled or even extinguished. Property conservation should be addressed in

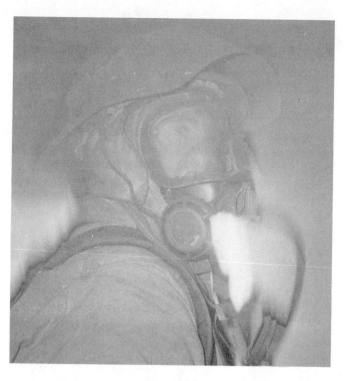

Figure 13-2 *The incident commander's first priority should be life safety, including the safety of the firefighters involved in the operation. Photo courtesy of Michael Regan.*

■ **Note**
Positive property conservation measures can have a significant positive impact on the property owner and the community.

benchmarks
significant points in the emergency event usually marking the accomplishment of one of the three incident priorities: life safety, incident stabilization, or property conservation

preemergency planning efforts and started upon arrival of fire suppression forces. Property conservation involves using the least destructive means for entering, using early and aggressive ventilation, using a coordinated fire attack with the proper selection of hose and nozzles, effective fire stream management, and throwing salvage covers early in the event before water has a chance to do any damage. Property conservation does not mean putting out the fire at the expense of losing the building. Ms. Doe will be pleased if you put her fire out. She will be delighted if you do it without destroying her house.

Positive property conservation measures can have a significant positive impact on the property owner and the community. Irreplaceable possessions can be saved, damage reduced, and the building reoccupied within a minimum period of time. If the building is a commercial occupancy, the rapid resumption of operations will mean that workers are back at work earning money and spending that money in the community.

Some departments have **benchmarks** to note the accomplishment of these three priorities. These benchmarks are transmitted to the incident commander who records the time. In some departments they are also transmitted to the communications center. This transmission officially records the time of the event, and provides a concise progress report for senior fire department staff officers who may be monitoring the event by radio.

For fire-related activities, these conditions are as follows:

Incident Priority	Transmission
Life safety = search, rescue	"all clear"
Incident stabilization	"under control"
Property conservation	"loss stopped"

Things are not always tidy during emergencies and there are times when the incident commander must overlap or mix activities to make these priorities happen. For example, during interior fire operations, it may be necessary to control the fire while searching and rescuing trapped civilians. In the auto accident scenario (see Figure 13-3), you may have to stabilize the vehicle before you start to

Figure 13-3 *Incident stabilization focuses on reducing injuries and loss after operations are started. Photo courtesy of Michael Regan.*

treat or remove victims. Both activities protect your own safety as well as the safety of the victims. In both examples, life safety (the first priority) is the focus of all of the activity.

At the scene of incidents, you have to quickly sort out a lot of information and give directions. This process is complicated but it can be simplified by using the standard decision-making process at all emergency events. Using a standardized approach helps you focus on the incident priorities we just mentioned.

There are other benefits of using a standardized approach. When incident commanders follow a consistent pattern, they reduce the tendency to overlook important activities. For others at the scene, your plans and your directions will more likely be understood when using a standardized approach. A good approach is to use the command sequence.

THE COMMAND SEQUENCE

■ **Note**
The command sequence is size-up, developing an action plan, and implementing the action plan.

The command sequence is a three-step process that helps the incident commander effectively manage the incident. These steps are size-up, developing an action plan, and implementing the action plan. In some ways this looks like the problem-solving process we discussed in earlier chapters. Here the incident commander gathers information, analyzes the information, determines and ranks the problems, defines solutions, and selects tactics. Finally, the incident commander issues orders to implement the tactics selected.

Understanding Size-Up

Before any action is undertaken, you should take a moment to analyze the problem and think about the action that you are going to take. That statement is true in nearly all of your endeavors, and certainly true for those leading personnel at the scene of emergency activities. The first part of this process is called size-up. Size-up is gathering and analyzing information that is critical to the outcome of the event.

WHAT'S SHOWING?

For fires, there are three possible situations: nothing showing, smoke showing, and fire showing. Each situation should cue the first-arriving officer on the appropriate action to take, and for the instructions to be given to other units. Your preemergency planning, training, and department SOPs should address these three commonly encountered situations.

Most of us think of size-up as what is seen through the windshield when you pull up in front of Ms. Doe's residence. It is all of that and a whole lot more (see

Figure 13-4 *Size-up involves more than what you see upon arrival.*

Figure 13-4). Size-up includes all of the information available. It includes the information gathered during your preemergency survey. It incorporates your preplanning efforts, based on the information gathered from those surveys. During size-up, you should also consider your department's standard operating procedures (SOPs) and the resources available to deal with the situation.

Hopefully you have good preemergency plans for your community. Given the time of day and the nature of the occupancy, the first-arriving officer should be able to make some estimate of the nature of the life hazard encountered. Even without a preemergency survey, officers should know enough about their first-due response area to know the type of construction, the access points, the water supply, the probable fire behavior, and the safety concerns that are likely to be encountered.

The officer's knowledge of the first-due area should also include information about the best response route, the natural barriers (rivers and terrain), the man-made barriers (railroad tracks and highways), and all of the significant hazards that may be encountered. Your mind should be like a little computer, constantly storing information about the area, and analyzing the information to deal with the problems that might be encountered.

Along with an assessment of the problem, the first arriving officer should have a good assessment of the resources that can be expected to help deal with the problem. In larger cities, Ms. Doe's fire will draw a crowd of eager firefight-

■ **Note**

Remember that size-up is the first step in the action plan. With good size-up information you are better able to develop an action plan.

ers and a parade of snazzy emergency vehicles that will cause her considerable embarrassment. On the other hand, in many communities where there is only one company on duty, or where the community's volunteer firefighters may be working, you may get only one unit to the scene within a reasonable time. Again, the time of day may be important, not just in terms of the conditions that are likely to be encountered, but also in terms of the available resources.

Depending upon the situation, there may be some nonfire department resources on scene as well. Many communities notify the police when the fire department is called out, and they often arrive ahead of the fire fighters. In many cases, they can undertake some of the size-up activities for you. Where this is done, establish a good working relationship. Help the police by offering them training on providing the information you need. Private ambulance companies often respond to structural fires, for the same reasons that a fire department emergency medical service (EMS) unit is assigned on working fires in many larger cities. Regional or state hazardous materials teams may be called when the local department is not prepared to deal with hazmat problems. You should know about these resources and how they can help at the scene of any emergency.

Along with all of this information, you will make your own observations while en route and upon arrival at the incident. When you see Ms. Doe's house, you can focus on the specific problems you face (one excited lady; one kitchen on fire). You now have a moment to make a rapid mental evaluation of the situation and of the relevant factors that may be critical to this unique incident.

■ **Note**

The first-arriving officer must assess the conditions calmly and be proactive rather than reactive to the situation.

For many officers, the first few moments at the scene create a lot of stress. For a new officer, every call is a significant event. With experience and maturity, most officers routinely deal with ordinary events. However, even seasoned veteranz sometimes encounter situations beyond their previous experience. Hopefully, they will be able to draw on their previous experiences and remain calm, even though the event seems overwhelming. In any case, the first-arriving officer must take a moment to assess the conditions calmly and be proactive rather than reactive to the situation. Preemergency plans and standard operating procedures need to be considered and used if possible. Hazards need to be identified and incorporated into the planning process (see Figure 13-5). In all situations, officers should organize their thinking into three areas:[1]

- What do I have?
- Where is it going?
- How can I stop it?

During fire-related events, the fire officer should be concerned about the following:

Building Construction The important thing to remember here is that each type of building presents us with a different set of problems in dealing with fire spread and fire control.

Figure 13-5 *The first arriving officer should take note of hazards that may be present.*

Location and Access Finding the building is part of your job. In addition, you need to know how to get past any fences and doors you may encounter. You should also have obtained information about exposures, water supply, terrain problems, and other features that impact on your operations (see Figure 13-6).

Figure 13-6 *Gaining access to the building can be a challenge.*

Occupancy and Contents For Ms. Doe's house and its 2,000 square feet of living space, the first-arriving officer can anticipate the size and contents of the structure just from knowing the neighborhood. Your preemergency survey and planning efforts really pay off for larger structures.

Location of the Event within the Structure You also need to know where the situation is located within the structure. Again, your prior knowledge will be useful here. If your problem is a fire, knowing its location helps you answer all three of the questions: "what have I got, where is it going, and how do I stop it?" Lacking this information will seriously adversely affect your ability to answer any of these questions.

Time of Day and Weather Conditions Of course you will know the time of day and the weather situation, but you should always be thinking about how these conditions will affect both the situation and how you can mitigate the situation. For example, a fire at night in any residential structure presents us with significant life safety concerns. A fire in a commercial building at night may present reduced life safety concerns but will likely have a delayed discovery. Your ability to get to that

Imagine the difference of pulling up to a well-involved fire in a structure of 20,000 square feet of floor space in a building that you have carefully surveyed and planned. Contrast that with your situation while standing in front of a similar building that you have never seen before (see Figure 13-7). You may have trouble getting past the front door.

Figure 13-7 *Certain structures have obvious hazards. What hazards do you see here?*

fire may also be delayed while you attempt to gain access. Again, you see some advantages to your preemergency planning activity.

environmental factors
factors like weather that impact on firefighting operations

You will also know about the weather and other **environmental factors** that impact on your operations. Extremely hot or extremely cold weather affects firefighting operations and firefighters. Cold weather can make life difficult, especially when the cold is mixed with precipitation (see Figure 13-8). Cold weather will probably slow your response and your operations once you are on the scene. Cold weather can also have an impact on any citizens affected by the event. Imagine evacuating all the occupants of an apartment house or office building while your fire fighting activities are taking place. In nice weather, they will have a chance to watch your operations and visit with one another. In cold weather, they may be at greater risk from the elements than they were from the fire. The same can be said for the those involved in some sort of transportation accident. We should not lose sight of the first priority: human life.

Figure 13-8 *Severe weather can hamper operations.*

Several acronyms can be used for keeping track of all of these factors. One of the traditional ones is *COAL WAS WEALTH*:

- **C**onstruction
- **O**ccupancy
- **A**pparatus and personnel
- **L**ife safety

- **W**ater
- **A**uxiliary appliances
- **S**treet conditions

- **W**eather
- **E**xposures
- **A**rea
- **L**ocation
- **T**ime
- **H**eight

Keeping track of all this information can be quite a challenge. Many departments use a tactical work sheet for larger events but there are benefits of having some sort of checklist for all events. Many officers keep a copy of some sort of work sheet folded up in the pocket of their turnout coat. A sample tactical work-sheet is shown in Figure 13-9. This sheet provides a checklist for size-up factors, a place to record assignments, and a space for a simple diagram.

Risk Analysis Although there are dangers associated with all activities, firefighting certainly ranks high as a risky activity. One of the considerations for the first-arriving officer is a risk/benefit analysis that determines the strategy to be initially used for mitigating the event. Where lives are at risk, any reasonable risk that may result in saving those lives is warranted. However, where lives are not at risk, you should never place firefighters at risk. We have been reading about this for years, yet we continue to read of firefighters taking risks to enter buildings that are scheduled for demolition.

Where the risk justifies entering the building, you should realize that just because you can safely *enter* the building does not mean that you can safely *stay* in the building. For many types of buildings, modern construction techniques are not designed to enhance structural integrity during fire conditions, and some modern buildings will collapse, in some cases, sooner than one might expect. Normally the arriving firefighters do not know how long the fire has been burning before they arrived, nor are they likely to know how the fire has impacted on the structural integrity of the building. The length of time firefighters can be expected to sustain an interior attack is usually very limited at best.

■ **Note**
Where the risk justifies entering the building, you should realize that just because you can safely enter the building does not mean that you can safely stay in the building.

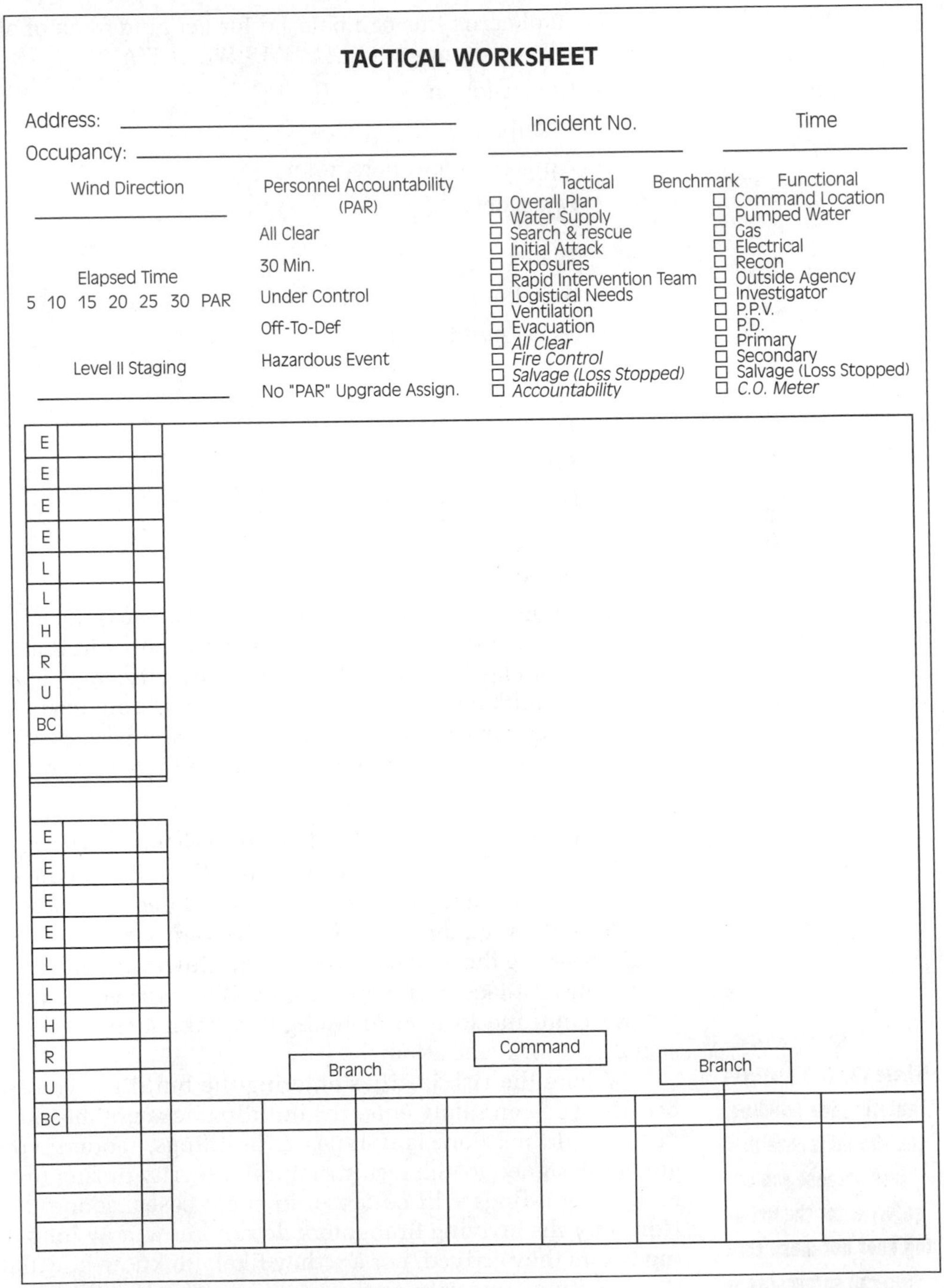

TACTICAL WORKSHEET

Address: _____   Incident No. _____   Time _____

Occupancy: _____

Wind Direction	Personnel Accountability (PAR)	Tactical Benchmark	Functional

Wind Direction _____

Elapsed Time
5 10 15 20 25 30 PAR

Level II Staging _____

Personnel Accountability (PAR)

All Clear

30 Min.

Under Control

Off-To-Def

Hazardous Event

No "PAR" Upgrade Assign.

Tactical   Benchmark   Functional

☐ Overall Plan   ☐ Command Location
☐ Water Supply   ☐ Pumped Water
☐ Search & rescue   ☐ Gas
☐ Initial Attack   ☐ Electrical
☐ Exposures   ☐ Recon
☐ Rapid Intervention Team ☐ Outside Agency
☐ Logistical Needs   ☐ Investigator
☐ Ventilation   ☐ P.P.V.
☐ Evacuation   ☐ P.D.
☐ All Clear   ☐ Primary
☐ Fire Control   ☐ Secondary
☐ Salvage (Loss Stopped) ☐ Salvage (Loss Stopped)
☐ Accountability   ☐ C.O. Meter

E
E
E
E
L
L
H
R
U
BC

E
E
E
E
L
L
H
R
U
BC

Command

Branch   Branch

Figure 13-9 *A tactical worksheet helps keep track of the information.*

If there are sufficient resources on hand to mount an aggressive interior fire attack, the save might be worth the risk. But in many cases, especially where resources may be limited or where the building's condition is doubtful, it may be best to focus resources on protecting the exposures. Decisions regarding interior operations are among the hardest you can expect to face. Remember that the fire is not your fault, and that you may not be able to save all of the burning buildings you are likely to encounter. Property that is already lost is not worth losing a firefighter (see Figure 13-10).

■ **Note**
Property that is already lost is not worth losing a firefighter.

Figure 13-10
Property that is already lost does not justify risking firefighters. Photo courtesy of the Fairfax County Fire and Rescue Department.

■ **Note**

There are three modes of operation at a fire: offensive, defensive, and transitional.

offensive mode
firefighting operations that make a direct attack on a fire for purposes of control and extinguishment

defensive mode
actions intended to control a fire by limiting its spread to a defined area

transitional mode
the critical process of shifting from the offensive to the defensive mode or from the defensive to the offensive

Selecting the Correct Mode There are three modes of operation at a fire: offensive, defensive, and transitional.[2] When in the **offensive mode**, operations are conducted in an aggressive interior attack mode, taking the attack to the fire. You must determine that the interior attack is worth the risk. You must also determine if you have sufficient resources to safely mount and sustain a coordinated attack. The offensive mode is a good choice: many fires are put out this way and lots of lives and property are saved with this mode of attack.

When fire conditions have advanced to the point where there is little chance of saving lives or property, or when there are insufficient resources on hand to safely mount and sustain an interior fire attack, you must resort to defensive operations (see Figure 13-11). In the **defensive mode**, operations are conducted from a safe distance outside of the structure, and may focus more on containing the fire rather than extinguishing it.

The third mode is called transitional. During the **transitional mode**, operations are changing from either an offensive to a defensive mode, or from a defensive to an offensive mode. Obviously the transitional mode it not something you would start with, but rather a phase you pass through as operations are shifted from one mode to another. Operations may have started in the offensive mode but

Figure 13-11 *At times the mode will be obvious.*

firefighters were unable to make any headway, so the officers prudently withdrew their forces from within the building. Conversely, the event may have started as a defensive operation, with firefighters knocking down some of the fire or providing a holding operation while sufficient companies arrived and prepared for a coordinated interior operation. In either case, the process of shifting from one mode to another presents its own risks and has to be carefully managed. Effective fireground communications become very important during these precarious moments.

Command Decisions As if you did not already have enough to think about, another critical decision that has to be made is in regard to your personal role in the initial operations. There are three modes of command: command, attack, or a combination of both. Once again, you have a choice to make: You can take command, elect to start combating the situation, or seek a hybrid solution of trying to do both.

In selecting the **command role**, the first arriving officer elects to maintain control of the incident and coordinates the activities of the first-arriving companies. In essence, the company officer becomes the first incident commander. As the incident commander, the company officer has overall control of the event, doing the size-up, directing personnel on scene, and giving instruction to arriving companies by radio. For smaller events this is standard procedure. The company officer may decide that one or two companies are sufficient to handle the event and release the remaining companies from the assignment. In such cases the first-arriving company officer will likely remain in charge of the event, complete the report, and perform all of the other duties of the incident commander.

The second option that the company officer has is to personally go into action. This option is referred to as the **attack role**. In this case, the first-arriving officer decides that the urgency of the situation warrants his or her immediate personal action, and that such action will likely have a significant positive affect on the outcome of the event. This option may be justified where an immediate rescue is required and can be safety undertaken, or where a quick fire attack may knock the fire down and prevent significant extension. One again, company officers are asked to make an on-the-spot risk/benefit analysis, and then take immediate action with limited resources based on that analysis. Times like these can be difficult.

When the officer selects the option of going into action, he should pass command to another officer who is or who soon will be on scene. There has been some misunderstanding about this process, especially in departments where it is not often accomplished. When the first-arriving officer selects to commit to the attack mode, he communicates this over the radio to the next-arriving company. The officer of the next-arriving company now becomes the incident commander. If the second arriving company is another engine company, its major task is usually establishing a reliable water supply. The officer of that company may be in a good position to become the incident commander. Obviously, too much of a good thing can be harmful, and most departments have a rule about passing the

■ Note

There are three modes of command: command, attack, or a combination of both.

command role
a situation in which the first-arriving officer takes command until relieved by a senior officer

attack role
a situation in which the first-arriving officer elects to take immediate action and to pass command on to another officer

command again. The general rule is that the next time that command is passed, it must be to a chief officer. This policy precludes the constant turnover of command as companies arrive.

In some situations the first-arriving officer may not really have the luxury of a choice, especially in smaller departments where the initial response might be limited to one or two engines and a handful of firefighters. Additional resources could be about a day away. In this case, the first-arriving officer is likely to be involved with a combination of both command and hands-on firefighting activities. This arrangement is obviously not good as it certainly places the company officer in a position where the best she can hope for is to try to perform one task without serious risk of compromising the other. In such cases, another officer should attempt to get on scene and relieve the first-arriving officer of the responsibilities of command.

COMMUNICATIONS IS THE GLUE THAT HOLDS THE COMMAND SEQUENCE TOGETHER

Effective management of emergency situations depends on proper use of command and communications. An effective command system defines how the incident is controlled and how the management of the incident is established and progresses. Communications ties the command system together and provides an opportunity for everyone to pass along and understand essential information. When using the communications system, messages should be brief, impersonal, and to the point.

■ **Note**
The initial report should address four questions: What do I have, what am I doing, what do I need, and who is in charge? Advise everyone else what you are doing, your location, and your immediate plans.

initial report
a vivid but brief description of the on-scene conditions relevant to the emergency

Brief Initial Report Part of the initial size-up is providing a good report of the situation you see. The **initial report** should paint a concise but vivid oral picture of the conditions you observe as well as a quick summary of your intentions and needs. Specifically, the initial report should address four questions: "What do I have, what am I doing, what do I need, and who is in charge?" Advise everyone else what you are doing, your location, and your immediate plans.

The first part is obvious and very important. In a sentence or two you should describe what is happening in front of you. If a building is on fire, describe the size and occupancy of the building and the evidence of fire conditions you can see. If you are arriving at the scene of a vehicle accident where you have more victims than you can likely handle with the resources on hand, the same rule applies: Describe what you see.

Advise everyone else what you are doing, your location, and your immediate plans. Let them know if you are entering the building, and if you are planning to operate in the offensive or defensive mode.

Next, give assignments to incoming units and ask for additional resources if needed. If you think that additional resources will be needed, *now* is the time to ask for them. Better to get them started and find that you really do not need them than to discover 5 or 10 minutes later that you need some help, and have to wait another 5 or 10 minutes until help arrives.

You may not have specific assignments for everyone. Let them know that. Let them know that you want them to continue in or go to a staging area. They can wait there, ready to go to work. This gives you a moment to assess the situation. On the other hand, if you have one of those "nothing showing" situations, you may want them to slow down the response a little. Most departments have standard procedures for doing this.[3]

Who is in charge? Here the *command statement* enters into the report. If the first arriving officer is assuming command, that should be clearly stated. The officer or the dispatch center identifies the command, usually by street or building name. If there is any question as to your exact location, you should clearly indicate your location so the arriving units will know where to find the command post.

A good initial size-up report provides incoming units with the information needed to mentally and physically prepare for the call. Most of the calls these days do not involve working fires. As a result, we fall into a pattern of "here we go again" for a report of a fire alarm or sprinkler activation. A good size-up report will jolt firefighters from the usual "here we go again" mindset into the reality that this is a real fire and we have some real work to do. It also alerts the command center and senior officers who may be listening that firefighters are about to go into battle. Here is an example.

> *I'm on scene of a two-story town house residence with smoke and fire showing from the front and side two of an end unit. Engine 2: pick up my supply line on Dawn Street and have your personnel report here for a backup line from Engine 1. I will assume command, designated as Huth Street Command. Please give me a second alarm and police units for evacuation.*

Most fire officers do a pretty good job on size-up reports at fire situations, but some do not do a very good job at EMS and hazmat situations. While both EMS and hazmat events present a host of unknowns, one should be able to describe what they see. Let us take an EMS call for example:

> *Engine 1 is on scene of a two-car accident at East Highway and Route 123. Both vehicles are extensively damaged. One car is on its side in the ditch and appears to have a small fire under the hood. It looks like at least six people are involved, and all are still in their vehicles. I am assuming command and will check the fire and assess the patients in the first car. Rescue 2, please take the second car. Send two police units for traffic control and two additional ambulances.*

■ Note
A good initial size-up report provides incoming units with the information needed to mentally and physically prepare for the call.

■ **Note**

Size-up must be ongoing throughout the event.

Ongoing Size-up At this point the action starts. However, size-up does not stop when the action starts. Size-up must be ongoing throughout the event. Many responses start without all the information, or firefighters find that the information is incorrect. As additional information is obtained, be sure that it is shared with those who need the information.

Two things are certain to change. The first is the situation. Hopefully your actions will have some positive impact on the situation. Report those changes on a regular basis. Many departments have a system in place that prompts a status report every 10 or 15 minutes. These automatic prompts also provide an opportunity to be sure that the incident commander has everyone accounted for. After 10 or 15 minutes of interior firefighting operations, you should see significant progress. If that progress is not apparent, it may be time to withdraw the troops and shift to a defensive mode. Sustained interior operations invite disaster!

■ **Note**

Along with a status of the event, the incident commander should keep track of the resources available for the operation.

Along with a status of the event, the incident commander should keep track of the resources available for the operation. Are additional units needed? Are additional firefighters needed to relieve those now engaged? Are support units needed to sustain the personnel and apparatus that are on scene? Will you need lights, food, shelter, toilets, compressed air to recharge the SCBA units, and so forth? Only those at the scene can anticipate these needs. A good incident commander thinks about these things (or has someone else doing it) and has them on scene before they are needed.

Developing an Action Plan

action plan
an organized course of action that addresses all phases of incident control within a specified period of time

An **action plan** is an organized course of action that addresses all phases of incident control within a specified timeframe. Note several key words here. First, it is an *organized* course of action. All of the operations should be conducted in an organized manner. This does not infer that it is necessary to have a meeting and vote on the actions you are taking; it just means that *someone* is in charge, and that someone has analyzed the situation, and issued orders that will permit safe and effective operations. Second, the action plan should *address* all phases of the emergency. Action planning does not stop until the last unit leaves the scene. Action planning, like the ongoing size-up, should be a continuing process.

The third key element in the definition of action plan is *within a specified timeframe.* With a little experience you should be able to estimate how long certain actions will take. Your action plan should allow for the fact that it takes several minutes from the time you, as the incident commander, give an order, until the action starts. It may take additional time for the results to be seen. Meanwhile, the fire is still burning or the victim is still trapped. You may need to think about your actions in slices of time, especially when coordinating the efforts of several companies.

> ## REALITY CHECK
>
> For some readers, this discussion about planning and applications of management science at the fireground may look like a lot of modern textbook stuff with no practical application. Timeout, folks! These ideas have been around for a long time, and they *do* have lots of practical application. Consider that Chief Layman's books were both published in the 1950s and Chief Emanual Fried's classic *Firegound Tactics* was published in 1972.
>
> Charles Walsh and Leonard Marks, both deputy chiefs of large fire departments, wrote a book in the 1970s entitled *Firefighting Strategy and Leadership*. Chapter 2 of their book is entitled "Action Plan." They compare developing an action plan at the scene of a fire or other emergencies to the same problem-solving process we discussed earlier.
>
> Please do not put off by the dates these books were published. That's the point: these ideas have been around for a long time. Many aspects of firefighting have changed remarkably in that period of time, and some aspects are still as they were described years ago. As a good officer, it is important that you not only be up to date, but that you remember the basic and timeless concepts of effective firefighting and fireground management.

Strategy Strategy is one of those mystery words we hear in classes on firefighting. It is a term we have borrowed from the military. Strategy is fundamental to planning and directing effective military operations. Along those lines, Paul Nitze, Secretary of Defense in the 1950s, said that ". . . strategy implies an organized authority capable of sustained action along lines of authority." He was talking about military strategy of course, but his words fit our situation well also. One can see how the word was brought to the fire service. For our purposes, **strategy** is the *overall* plan that is used to gain control of an incident.

strategy
sets broad goals and outlines the overall plan to control the incident

Chief Lloyd Layman, author of *Fire Fighting Tactics,* gave us many ideas about firefighting that are still with us. Chief Layman listed seven basic strategies of firefighting as follows:[4]

	rescue	
	exposures	
[size-up]	confinement	ventilation
	extinguishment	salvage

Notice that even in his time (the late 1940s), Chief Layman felt that the first step was to conduct a proper size-up. In the second column you see chief Layman's list of basic strategies related to fire fighting in a logical sequence. Rescue comes first, then protect the exposures, if needed. Next, focus on confining and extinguishing the fire. Ventilation and salvage are also important strategies but do not lend themselves to a fixed location in this sequence and so, Chief Layman wisely listed them separately. They should be used whenever needed.

Defining the strategy has many advantages. One of the advantages is that when the strategy is well defined, everyone understands the tasks at hand and can focus their efforts on making it happen. Without a strategic vision, everyone will contribute, but in a less organized manner. This leads to freelancing and confusion.

We find another military term to define the next step in the process. If strategy is the broad picture, then **tactics** will help you narrow your focus a bit and look at how to reach your strategic goals. On might say that tactics provide the highways that allow us to reach those goals. As the strategic goals were related to time, so are the tactics. They must also be focused on a specific location and measurable so that you can see if the desired results are happening.

Strategy and tactics lead us to an action plan. After thinking about the event (size-up), and planning (strategy and tactics), you finally get to make it happen. The first two steps should not be viewed as a waste of time and effort. We have simply said that where the firefighters' activities are preceded by the incident commander taking a moment to identify the problems and considering the various alternatives, we are more likely to have a satisfactory outcome. The action plan puts the planning and thinking phases into motion.

When assigning resources to carry out the action plan, the incident commander should have a realistic assessment of what can be accomplished with the resources on hand. This applies to the resources, collectively and individually. If the incident commander is blessed with three well-staffed engine companies and two truck companies, a lot can be accomplished. But even here, there are limitations. The crew from one of these companies can only be expected to deal with one problem at a time and be in one place at a time. If there are not enough resources, now is the time to ask for more.

Putting Your Action Plan to Work

To get things started, the incident commander communicates orders. One way to do this is to assign tactics to individual companies. The incident commander can direct a company to confine the fire or conduct a primary search. By assigning tactics, the incident commander allows the companies to determine the tasks that best accomplish the required action. The directions should be clear and concise. Assigning tactics usually reduces radio traffic.

Once the task is underway the incident commander should get some feedback about progress. If the news is good, the incident commander wants to hear about that. And if the news is bad, the incident commander wants to hear about that too. The incident commander may be the boss, but the boss depends upon timely reports to make decisions. Keep the boss informed.

A second way that the incident commander can make an assignment is to

■ **Note**

When the strategy is well defined, everyone understands the tasks at hand and can focus their efforts on making it happen.

tactics
various maneuvers that can be used to achieve a strategy while fighting a fire or dealing with a similar emergency

■ **Note**

The action plan puts the planning and thinking phases into motion.

tasks
the duties and activities performed by individuals, companies, or teams that lead to successful accomplishment of assigned tactics

mutual aid
assistance provided by another fire department or agency

assign **tasks** with very specific directions. This option will probably increase the amount of radio traffic and will require a little more of the incident commander's time and attention, but there are times when this approach is appropriate, for example, when the task is critical to the overall operation, or the incident commander wants the task accomplished in a way that departs from normal procedures, or when critical safety considerations are paramount. It also is appropriate to use this approach when the company involved is inexperienced or when the incident commander is not sure how the company might perform the task. The best example of this might be when a **mutual aid** company is assigned the task.

Good Training and SOPs Pay Off   Many departments use standard operating procedures (SOPs). These are departmental policies that provide a predetermined course of action for a given situation. With SOPs, the companies can get to work and perform their assignments with little or no direction from the incident commander (see Figure 13-12). SOPs can work well, but you should be aware that some companies will race ahead with their assigned task, without the incident commander having had an opportunity to complete the size-up or give some coordinating instructions. Incident commanders can avoid this problem by making assignments to companies as soon as possible, or by telling companies to wait for instructions. It may also be appropriate for the incident commander to tell

Figure 13-12 *Good training and SOPs allow companies to get to work with a minimum of direction from the incident commander.*

arriving units that because of some unusual circumstances, the SOPs are not appropriate and will not be used.

Basic communications skills are important at times like these. There is a lot going on, and things need to be resolved as rapidly as possible. Everyone is looking to the incident commander to give a few orders to put the operation into motion. Yelling, raising your voice, and speaking rapidly will certainly reveal to everyone listening that you are excited. At this time you want a calm, reassuring voice. As much as anything else, the incident commander's calm voice tells us that someone understands what is happening and has the solution well in hand. That is exactly what the incident commander is supposed to be doing. No one wants to hear an excited incident commander on the radio or see an excited incident commander running around the fireground.

Just as the incident commander gets the big picture and breaks it down into strategy, tactics, and tasks for companies, the company commander should do the same thing for the individuals who comprise the company. Each person needs to have some idea of the big picture and the role they will play in the overall outcome of the event. Each of them must also clearly understand his or her individual assignment. Only with understanding and coordination can you operate safely and efficiently.

Fire Attack   The effectiveness of the initial attack often determines the outcome of the event. The effectiveness of the initial attack's leader, company officer, chief, or whoever, is dependent upon that individual's knowledge of fire behavior, and his decisions with regard to the size, number, and placement of hoselines during that initial attack.

The most common fires encountered are considered free burning, that is, they are in the second of the three stages of fire. When possible, such fires should be fought by a direct, offensive attack. Direct attack is the surest and quickest way of controlling the fire and reducing the loss (see Figure 13-13). You should make every possible effort to attack the fire from the unburned side, and push the fire back to those areas where it has already caused damage. This is a fundamental concept of good firefighting. It reduces property loss and usually enhances the safety of the operation.

You may not be able to attack the fire from the unburned side in certain situations. These might include a commercial occupancy where access is limited to one end of the building, in multiunit residences (apartments, town houses, etc.) where you do not have access to the unburned side, and where such action might endanger occupants or firefighters.

Fire Control, Confinement, and Extinguishment   These three words are used quite a bit in connection with firefighting operations. Let us take a moment to agree on what they mean. Fire control includes both confinement and extinguishment of the fire.

Figure 13-13 *Offensive firefighting delivers the fire attack to the seat of the fire. This mode of operation is preferred when conditions permit.*

confinement
an activity required to prevent fire from extending to an uninvolved area or another structure

fire extinguishment
activities associated with putting out all visible fires

overhaul
searching the fire scene for possible hidden fires or sparks that may rekindle

Confinement means to cut off the fire from further advance, to keep it from spreading and doing any further damage.

Fire extinguishment means to extinguish all *visible* fire. Not all of the fire may be extinguished at this point; that will occur during **overhaul**. To be effective in fire control, one must understand fire behavior, building construction, and fire flow requirements.

Most firefighting is done with fire hoses and firefighters. In spite of all the advances we have made in the last few years, it still takes a minute or two to get a water supply established and a hose line in operation, directing water onto the fire. As a fire officer, you have to allow for this delay, and factor it into your thinking. If the fire has already passed you by the time your firefighters can open the nozzle, pushing the fire back to its area of origin will be quite a challenge.

During your planning, you should consider the number of firefighters available and how many hose lines they can put into operation. From this information you will have a good idea of how much water you can put on the fire. And you have to decide if this flow rate, or fire flow, is adequate to confine and extinguish the fire. Finally you need to coordinate the advance of the hose lines with other activities such as forcible entry, ventilation, search and rescue, and salvage.

Let's get back to Ms. Jane Doe's house for a moment. You have learned that she was in the house alone and was cooking. She went into the next room to answer the phone and while she was talking "something happened in the kitchen." The next thing she knew the smoke detector was going off and the kitchen was filling with smoke. You have seen this before, but Ms. Doe has not. She was smart enough to call 911 right away. Then she tried to get to the stove to turn things off. When she did that she instinctively grabbed a pan handle that was very hot.

What do you do? You get one of your medics to take care of Ms. Doe. Meanwhile, you calmly evaluate the fire situation, select to go for an offensive operation with an interior attack, mask up, take an attack line in though the front door, find your way back to the kitchen (not really hard to do because there is a big fire cooking lots of things now) and put the visible fire out with a few gallons of water (see Figure 13-14).

One of your colleagues was smart enough to go around to the back door and carefully open it, allowing some of the smoke and steam to vent outside through the doorway. You look around and find smoke throughout the house and considerable damage in the kitchen. While you will check further, it looks like the fire did not get into the structure itself. You have been on scene less than 5 minutes.

About this time your portable radio says: "Battalion 1 to Engine 1: I'm on scene. What have you got?"

Well, the battalion chief has finally arrived on his big white horse, or more likely in his big red Suburban. You walk out to the front yard to meet him (see Figure 13-15).

Figure 13-14
Firefighter in Ms. Doe's kitchen.

Figure 13-15 *"What have you got?"*

TAKING CUSTOMER SERVICE TO MS. DOE'S HOUSE

We have already discussed what we are doing for Ms. Doe and her fire. Let us look at what else we might do for her. Suppose that as you make your way to the kitchen you notice several family pictures sitting on a table in the hall. These will certainly be knocked over and likely broken as you and your firefighters advance a hose line or two. And you also notice a very nice antique chair that will certainly get in someone's way.

So what you ask; we have work to do in the kitchen. Yes, but property conservation includes little things that can mean a lot. Pick up those pictures and put them out of the way. You might even slide them into a drawer if you can find one close at hand. And pick up the chair and set it in the dining room where it will be out of your way and out of harm's way. After the fire is out, control the ventilation. Open windows or close doors to reduce the effects of the smoke that will smell up everything in the house for a month. Cover her pretty hall carpet with a plastic runner when you get a chance. Lots of little things that cost little but will mean a lot after you have gone (see Figures 13-16 and 13-17). Remember, we want her to be satisfied. But wouldn't it be nice if she was delighted with your work? She might even invite your company over for dessert after the kitchen is fixed.

Figure 13-16 *Fire departments can do much to help their citizens recover from the consequences of a fire. The first step is providing materials to assist during salvage operations. Boxes, bags, and plastic sheeting can help save a lot of property. Photo courtesy of Phoenix Fire Department.*

Figure 13-17 *Firefighters can also help citizens recover from the consequences of fire by moving their personal effects to a safe area. Photo courtesy of Phoenix Fire Department.*

What we have tried to do over these last few pages is set the stage for the chief's arrival, and get the show started. What seems like a lifetime has actually been only 5 minutes. However, those 5 minutes are the most critical during any emergency. A good company officer should be able to get on scene, conduct a size-up, give a report, and get the show started without having to wait for a more senior officer to arrive. In this case, you did the whole thing. Well done!

In situations like those at Ms. Doe's house, a good company officer should be able to coordinate the entire operation. The chief should show up and relieve the officer if needed, but if all is going well, let the company officer continue to run the show. That is how good company officers are trained to become good fire-ground commanders. On the other hand, if the situation is starting to escalate, or get out of hand, it may be time for the chief to take over. In either case, we want to use an Incident Management System.

THE INCIDENT MANAGEMENT SYSTEM

"NFPA 1500, the *Fire Department Occupational Safety and Health Program*, states that fire departments shall have written operating procedures for managing emergency incidents. The standard also requires training in incident management for all personnel involved in emergency operations. NFPA 1561, *Standard for Fire Department Incident Management System*, provides details regarding such a system. Incident Management Systems can provide better management of firefighters during emergency operations. Better management will enhance firefighter safety."

■ **Note**

The focus of IMS is on effective management of people and resources.

As we know it today, the Incident Management System (IMS) is an all-risk all-situation emergency management concept. The focus of IMS is on effective management of people and resources. To embrace this means that you must accept IMS as a workable concept in your organization. Next, you must plan, prepare, and train for the events that you are likely to face. You should remember the basic rules of effective management and use them when you are managing emergency activities.

IMS can be used for nonemergency activities too. Many training programs use IMS to manage large-scale training activities. There are other applications as well. Large-scale operations of any kind, from preparing for the Super Bowl in your city to having a fire department funeral, can benefit from the organizational concepts of IMS. IMS works across political and organizational lines. It can be used to manage any large-scale event.

For managing resources during emergency situations, IMS is designed to work from the time of alarm until the incident is concluded. The title "incident commander" may apply to an engine company officer or the chief of a major department, depending upon the situation. The structure of the IMS can be established and expanded depending upon the needs of the event.

Figure 13-18
Firefighter safety is always important.

IMS accountability, rapid intervention, risk management, safety factors, two in-two out, rehabilitation, safety officers. We hear these terms in our training programs and see them in the literature. Where do they come from and how do they fit into the business of firefighting? All of these terms deal with *safety* during emergency operations (see Figure 13-18). All are addressed in NFPA 1500 and were covered briefly in Chapter 8 of this book. NFPA 1500, NFPA 1561, and other documents provide vital information pertaining to enhancing safety during emergency operations. Company officers should be intimately familiar with these concepts to ensure the safety of their personnel and the ability to operate as an effective unit during large-scale events. Additional information about IMS is contained in Appendix C.

Summary

This chapter provided a brief discussion of the role of the company officer during emergency operations. Clearly one chapter does not do justice to this important topic. Company officers should be able to analyze emergency scene conditions, conduct size-up, develop and implement an initial action plan, and deploy resources to safely and effectively control an emergency.

While the focus of this chapter and the entire book is to meet the performance standards of NFPA 1021, company officers should be able to cope with EMS and hazardous materials events as well. Most firefighters today are certified as EMTs and as Hazardous Materials Responders at the Operations Level. Training for both of these programs should include managing incidents in these specialty areas. If you can integrate the IMS information on the preceding pages with what you have learned in those courses, you should be ready to manage routine events involving medical emergencies, hazardous materials incidents, and even kitchen fires at Ms. Doe's house.

Good luck.

Review Questions

1. What are incident priorities?

2. What is meant by command sequence?

3. What is meant by size-up?

4. What are the parts of size-up?

5. What are the three modes of firefighting?

6. What information should be included in the initial report?

7. What is an action plan?

8. What is meant by strategy; tactics?

9. What is IMS? When should it be used?

10. What is the company officer's role in incident management during the early phases of fires and other emergency activity?

Additional Reading

Brunacini, Alan, *Fire Command* (Quincy, MA: National Fire Protection Association, 1985).

Clark, William E., *Firefighting Principles and Practices, Second Edition* (Tulsa, OK: Fire Engineering, 1991).

Fried, Emanual, *Fireground Tactics* (Chicago: Marvin Ginn, 1972).

Layman, Lloyd, *Attacking and Extinguishing Interior Fires* (Quincy, MA: National Fire Protection Association, 1955).

Layman, Lloyd, *Fire Fighting Tactics* (Quincy, MA: National Fire Protection Association, 1953).

Managing Company Tactical Operations, developed by the National Fire Academy, Emmitsburg, Maryland, is a three-part training course that focuses on preparation, preplanning, and managing typical structural firefighting activities. The second course, "Decisionmaking," is particularly relevant to the material covered in this chapter.

The National Fire Academy also presents an excellent resident course entitled "Command and Control of Fire Department Operations at Target Hazards." Contact your training officer or state fire service training director for more information.

NFPA 1561, *Standard on Fire Department Incident Management System* (Quincy, MA: National Fire Protection Association).

Norman, John, *Fire Officer's Handbook of Tactics* (Tulsa, OK: Fire Engineering, 1991).

Purington, Robert G., "Fire Streams," *Fire Protection Handbook,* Eighteenth Edition (Quincy, MA: National Fire Protection Association, 1997).

Walsh, Charles V., and Leonard G. Marks, *Firefighting Strategy and Leadership,* Second Edition (New York: McGraw-Hill, 1977).

Notes

1. These questions seem to focus on firefighting, but remember that we are discussing a thought process that can be applied to all emergency situations.

2. Some suggest that there is really a fourth mode, the no-attack mode. There are two situations where no-attack might be the correct solution. Obviously, if there is no fire, you do not need to mount an attack. The other possibility is where the fire is so involved that saving the building is impossible, or where fighting the fire would present unacceptable risks to the firefighters or the environment, that you elect to let it burn.

3. SOPs, training, and good fireground discipline are important at times like these. The incident commander has quite enough to think about without having to decide where each apparatus should be placed. But the incident commander should remember that placement is important. Once pumpers and ladder trucks "go to work," it is hard to relocate them.

4. Chief Layman's book is entitled *Fire Fighting Tactics,* and his original list was called Basic Divisions of Fire-fighting Tactics. Chief Layman made some remarkable contributions to the fire service and was way ahead of his time in many areas. His choice of words (tactics instead of strategy) is understandable. Reprinted with permission from National Fire Protection Association, Quincy, MA 02269.

Appendix

A

NFPA 1021

Reprinted with permission from NFPA 1021, *Fire Officer Professional Qualifications,* Copyright © 1997, National Fire Protection Association, Quincy, MA 02269. This reprinted material is not the complete and official position of the National Fire Protection Association, on the referenced subject which is represented only by the standard in its entirety.

Standard for

Fire Officer Professional Qualifications

1997 Edition

This edition of NFPA 1021, *Standard on Fire Officer Professional Qualifications*, was prepared by the Technical Committee on Fire Officer Professional Qualifications, released by the Technical Correlating Committee on Professional Qualifications, and acted on by the National Fire Protection Association, Inc., at its Annual Meeting held May 19–22, 1997, in Los Angeles, CA. It was issued by the Standards Council on July 24, 1997, with an effective date of August 15, 1997, and supersedes all previous editions.

This edition of NFPA 1021 was approved as an American National Standard on August 15, 1997.

Origin and Development of NFPA 1021

In 1971, the Joint Council of National Fire Service Organizations (JCNFSO) created the National Professional Qualifications Board (NPQB) for the fire service to facilitate the development of nationally applicable performance standards for uniformed fire service personnel. On December 14, 1972, the Board established four technical committees to develop those standards using the National Fire Protection Association (NFPA) standards-making system. The initial committees addressed the following career areas: fire fighter, fire officer, fire service instructors, and fire inspector and investigator.

The Committee on Fire Officer Professional Qualifications met through 1973, 1974, and 1975, producing the first edition of this document. The first edition of NFPA 1021 was adopted by the Association in July of 1976.

Subsequent to the adoption of the initial edition, the committee has met regularly to revise and update the standard. Additional editions were adopted and issued by the NFPA under the auspices of the NPQB in 1983 and 1987.

The original concept of the professional qualification standards, as directed by the JCN-FSO and the NPQB, was to develop an interrelated set of performance standards specifically for the fire service. The various levels of achievement in the standards were to build on each other within a strictly defined career ladder. In the late 1980s, revisions of the standards recognized that the documents should stand on their own merit in terms of job performance requirements for a given field. Accordingly, the strict career ladder concept was abandoned, except for the progression from fire fighter to fire officer. The later revisions, therefore, facilitated the use of the documents by fields other than the uniformed fire services.

In 1990, responsibility for the appointment of professional qualifications committees and the development of the professional qualifications standards were assumed by the NFPA.

The Correlating Committee on Professional Qualifications was appointed by the NFPA Standards Council in 1990 and assumed the responsibility for coordinating the requirements of all of the professional qualifications documents.

The 1992 edition of NFPA 1021 reduced the number of levels of progression in the standard to four.

This new edition represents an effort on the part of the technical committee to update the standard based on several years of further use. With this edition, NFPA 1021 is converted to the job performance requirement (JPR) format to be consistent with the other standards in the Professional Qualifications Project. Each JPR consists of the task to be performed; the tools, equipment, or materials that must be provided to successfully complete the task; evaluation parameters and/or performance outcomes; and lists of prerequisite knowledge and skills one must have to be able to perform the task. More information about JPRs can be found in Appendix C.

The intent of the technical committee was to develop clear and concise job performance requirements that can be used to determine that an individual, when measured to the standard, possesses the skills and knowledge to perform as a fire officer. The committee further contends that these job performance requirements can be used in any fire department in any city, town, or private organization throughout North America.

1021–2   FIRE OFFICER PROFESSIONAL QUALIFICATIONS

Technical Correlating Committee on Professional Qualifications

Douglas P. Forsman, *Chair*
Oklahoma State University, OK [M]

Fred G. Allinson, Nat'l Volunteer Fire Council, WA [L]
  Rep. Nat'l Volunteer Fire Council
Stephen P. Austin, State Farm Fire & Casualty Co., DE [I]
  Rep. TC on Investigator, Professional Qualifications
  (VL to Professional Qualifications System Management)
Dan W. Bailey, USDA Forest Service, MT [E]
  Rep. TC on Wildfire Suppression Professional
  Qualifications
  (VL to Professional Qualifications System Management)
Boyd F. Cole, Underwriters Laboratories Inc., IL [RT]
  Rep. TC on Emergency Vehicle Mechanic Technicians
  Professional Qualifications
  (VL to Professional Qualifications System Management)
David T. Endicott, Prince William County Fire & Rescue
Service, VA [U]
  Rep. TC on Fire Fighter Professional Qualifications
  (VL to Professional Qualifications System Management)
Charles E. Kirtley, Colorado Springs Fire Dept., CO [RT]
  (VL to Professional Qualifications System Management)
Jack K. McElfish, Richmond Dept. of Fire and Emergency
Services, VA [E]
  Rep Int'l Assn. of Fire Chiefs
Michael J. McGovern, Washington State Council of Fire
Fighters, WA [L]
  Rep. Int'l Assn. of Fire Fighters

William E. Peterson, Plano Fire Dept., TX [M]
  Rep. TC on Inspector Professional Qualifications
  (VL to Professional Qualifications System Management)
Hugh A. Pike, U.S. Air Force Fire Protection, FL [E]
  (VL to Professional Qualifications System Management)
Bruce R. Piringer, Fire & Rescue Training Inst., MO [SE]
  Rep. TC on Rescue Technicians Professional
  Qualifications
  (VL to Professional Qualifications System Management)
Ted Vratny, Ted Vratny—Public Safety Communications
Consulting, CO [U]
  Rep. TC on Telecommunicator Professional
  Qualifications
  (VL to Professional Qualifications System Management)
Alan G. Walker, Louisiana State University, LA [E]
  Rep. TC on Fire Officer Professional Qualifications
  (VL to Professional Qualifications System Management)
Johnny G. Wilson, GA Firefighter Standards & Training
Council, GA [E]
  Rep Nat'l Board on Fire Service Professional
  Qualification
John P. Wolf, University of Kansas, KS [SE]
  Rep. TC on Accreditation and Certification
  (VL to Professional Qualifications System Management)

Alternates

John W. Condon, Nat'l Volunteer Fire Council, OR [L]
  (Alt. to F. G. Allinson)

Michael. W. Robinson, Baltimore County Fire Dept., MD
[E]
  (Alt. to J. G. Wilson)

Jerry W. Laughlin, NFPA Staff Liaison

Committee Scope: This Committee shall have primary responsibility for the management of the
NFPA Professional Qualifications Project and documents related to professional qualifications for
fire service, public safety, and related personnel.

Technical Committee on Fire Officer Professional Qualifications

Alan G. Walker, *Chair*
Louisiana State University, LA [E]

Steven T. Edwards, Maryland Fire and Rescue Inst., MD [SE]
Robert S. Fleming, Rowan College of New Jersey, NJ [U]
Daniel B. C. Gardiner, Fairfield Fire Dept., CT [U]
David H. Hoover, The University of Akron, OH [SE]
Franklin T. Livingston, Brownsburg Fire Territory, IN [E]
Steven D. Mossotti, Mehlville Fire Protection District, MO [L]

Chris Neal, City of Stillwater Fire Dept., OK [M]
    Rep. Int'l Fire Service Training Assn.
LeRoy Oettinger, Montgomery County Fire/Rescue, MD [U]
Philip Sayer, Galt Fire Dept./Sayer Farms Inc., MO [L]
    Rep. Nat'l Volunteer Fire Council
Donald W. Teeple, Colorado Springs Fire Dept., CO [L]

Alternates

George F. Malik, Chicago Fire Dept., IL [SE]
    (Voting Alt. to AFEM Rep.)

Jerry W. Laughlin, NFPA Staff Liaison

Committee Scope: This Committee shall have primary responsibility for documents on professional competence required of the fire service officers.

These lists represent the membership at the time each Committee was balloted on the text of this edition. Since that time, changes in the membership may have occurred. A key to classifications is found at the back of this document.

NOTE: Membership on a committee shall not in and of itself constitute an endorsement of the Association or any document developed by the committee on which the member serves.

Contents

NFPA 1021

Standard for

Fire Officer Professional Qualifications

1997 Edition

NOTICE: An asterisk (*) following the number or letter designating a paragraph indicates that explanatory material on the paragraph can be found in Appendix A.

Information on referenced publications can be found in Chapter 6 and Appendix B.

Chapter 1   Administration

1-1 Scope.* This standard identifies the performance requirements necessary to perform the duties of a fire officer and specifically identifies four levels of progression.

1-2 Purpose.   The purpose of this standard is to specify the minimum job performance requirements for service as a fire officer.

1-2.1   The intent of the standard is to define progressive levels of performance required at the various levels of officer responsibility. The authority having jurisdiction has the option to combine or group the levels to meet its local needs and to use them in the development of job descriptions and specifying promotional standards.

1-2.2   It is not the intent of this standard to restrict any jurisdiction from exceeding these minimum requirements.

1-2.3   This standard shall cover the requirements for the four levels of progression—Fire Officer I, Fire Officer II, Fire Officer III, and Fire Officer IV.

1-3* General.

1-3.1   All of the standards for any level of fire officer shall be performed in accordance with recognized practices and procedures or as defined by an accepted authority.

1-3.2   It is not required for the objectives to be mastered in the order in which they appear. The local or state/provincial training program shall establish both the instructional priority and the program content to prepare individuals to meet the performance objectives of this standard.

1-3.3   The Fire Fighter II shall meet all the objectives for Fire Officer I before being certified at the Fire Officer I level, and the objectives for each succeeding level in the progression shall be met before being certified at the next higher level.

1-4* Definitions.

Approved.*   Acceptable to the authority having jurisdiction.

Authority Having Jurisdiction.*   The organization, office, or individual responsible for approving equipment, an installation, or a procedure.

Comprehensive Emergency Management Plan.   Planning document that includes preplan information and resources for the management of catastrophic emergencies within the jurisdiction.

Fire Department.   An organization providing rescue, fire suppression, and other related activities. For the purposes of this standard, the term "fire department" shall include any public, private, or military organization engaging in this type of activity.

Fire Officer I.   The fire officer, at the supervisory level, who has met the job performance requirements specified in this standard for Level I.

Fire Officer II.   The fire officer, at the supervisory/managerial level, who has met the job performance requirements specified in this standard for Level II.

Fire Officer III.   The fire officer, at the managerial/administrative level, who has met the job performance requirements specified in this standard for Level III.

Fire Officer IV.   The fire officer, at the administrative level, who has met the job performance requirements specified in this standard for Level IV.

Incident Management System.   An organized system of roles, responsibilities, and standard operating procedures used to manage and direct emergency operations.

Job Performance Requirement.   A statement that describes a specific job task, lists the items necessary to complete the task, and defines measurable or observable outcomes and evaluation areas for the specific task.

Labeled.   Equipment or materials to which has been attached a label, symbol, or other identifying mark of an organization that is acceptable to the authority having jurisdiction and concerned with product evaluation, that maintains periodic inspection of production of labeled equipment or materials, and by whose labeling the manufacturer indicates compliance with appropriate standards or performance in a specified manner.

Listed.*   Equipment, materials, or services included in a list published by an organization that is acceptable to the authority having jurisdiction and concerned with evaluation of products or services, that maintains periodic inspection of production of listed equipment or materials or periodic evaluation of services, and whose listing states that either the equipment, material, or service meets identified standards or has been tested and found suitable for a specified purpose.

Member.   A person involved in performing the duties and responsibilities of a fire department under the auspices of the organization. A fire department member can be a full-time or part-time employee or a paid or unpaid volunteer, can occupy any position or rank within the fire department, and can engage in emergency operations.

Promotion.   The advancement of a member from one rank to a higher rank by a method such as election, appointment, merit, or examination.

Qualification.   Having satisfactorily completed the requirements of the objectives.

Shall.   Indicates a mandatory requirement.

Should.   Indicates a recommendation or that which is advised but not required.

Supervisor.   An individual responsible for overseeing the performance or activity of other members.

Unit.   An engine company, truck company, or other functional or administrative group.

1021–6

Chapter 2   Fire Officer I

2-1 General. For certification at Fire Officer Level I the candidate shall meet the requirements of Fire Fighter II as defined in NFPA 1001, *Standard on Fire Fighter Professional Qualifications*, and the job performance requirements defined in Sections 2-2 through 2-7 of this standard.

2-1.1 General Prerequisite Knowledge. The organizational structure of the department; departmental operating procedures for administration, emergency operations, and safety; departmental budget process; information management and record keeping; the fire prevention and building safety codes and ordinances applicable to the jurisdiction; incident management system; socioeconomic and political factors that impact the fire service; cultural diversity; methods used by supervisors to obtain cooperation within a group of subordinates; the rights of management and members; agreements in force between the organization and members; policies and procedures regarding the operation of the department as they involve supervisors and members.

2-1.2 General Prerequisite Skills. The ability to communicate verbally and in writing, to write reports, and to operate in the incident management system.

2-2 Human Resource Management. This duty involves utilizing human resources to accomplish assignments in a safe and efficient manner and supervising personnel during emergency and nonemergency work periods, according to the following job performance requirements.

2-2.1 Assign tasks or responsibilities to unit members, given an assignment at an emergency operation, so that the instructions are complete, clear, and concise; safety considerations are addressed; and the desired outcomes are conveyed.

(a) *Prerequisite Knowledge:* Verbal communications during emergency situations, techniques used to make assignments under stressful situations, methods of confirming understanding.

(b) *Prerequisite Skills:* The ability to condense instructions for frequently assigned unit tasks based upon training and standard operating procedures.

2-2.2 Assign tasks or responsibilities to unit members, given an assignment under nonemergency conditions at a station or other work location, so that the instructions are complete, clear, and concise; safety considerations are addressed; and the desired outcomes are conveyed.

(a) *Prerequisite Knowledge:* Verbal communications under nonemergency situations, techniques used to make assignments under routine situations, methods of confirming understanding.

(b) *Prerequisite Skills:* The ability to issue instructions for frequently assigned unit tasks based upon department policy.

2-2.3 Direct unit members during a training evolution, given a company training evolution and training policies and procedures, so that the evolution is performed safely, efficiently, and as directed.

(a) *Prerequisite Knowledge:* Verbal communication techniques to facilitate learning.

(b) *Prerequisite Skills:* The ability to distribute issue-guided directions to unit members during training evolutions.

2-2.4 Recommend action for member-related problems, given a member with a situation requiring assistance and the member assistance policies and procedures, so that the situation is identified and the actions taken are within the established policies and procedures.

(a)* *Prerequisite Knowledge:* The signs and symptoms of member-related problems, causes of stress in emergency services personnel, adverse effects of stress on the performance of emergency service personnel.

(b) *Prerequisite Skills:* The ability to recommend a course of action for a member in need of assistance.

2-2.5* Apply human resource policies and procedures, given an administrative situation requiring action, so that policies and procedures are followed.

(a) *Prerequisite Knowledge:* Human resource policies and procedures.

(b) *Prerequisite Skills:* The ability to communicate verbally and in writing and to relate interpersonally.

2-2.6 Coordinate the completion of assigned tasks and projects by members, given a list of projects and tasks and the job requirements of subordinates, so that the assignments are prioritized, a plan for the completion of each assignment is developed, and members are assigned to specific tasks and supervised during the completion of the assignments.

(a) *Prerequisite Knowledge:* Principles of supervision and basic human resource management.

(b) *Prerequisite Skills:* The ability to plan and to set priorities.

2-3 Community and Government Relations. This duty involves dealing with inquiries and concerns from members of the community and projecting the role of the department to the public, according to the following job performance requirements.

2-3.1 Initiate action to a citizen's concern, given policies and procedures, so that the concern is answered or referred to the appropriate individual for action and all policies and procedures are complied with.

(a) *Prerequisite Knowledge:* Interpersonal relationships and verbal and nonverbal communication.

(b) *Prerequisite Skills:* Familiarity with public relations and the ability to communicate verbally.

2-3.2 Respond to a public inquiry, given the policies and procedures, so that the inquiry is answered accurately, courteously, and in accordance with applicable policies and procedures.

(a) *Prerequisite Knowledge:* Written and verbal communication techniques.

(b) *Prerequisite Skills:* The ability to relate interpersonally and to respond to public inquiries.

2-4 Administration. This duty involves general administrative functions and the implementation of departmental policies and procedures at the unit level, according to the following job performance requirements.

2-4.1 Implement a new departmental policy at the unit level, given a new departmental policy, so that the policy is communicated to and understood by unit members.

(a) *Prerequisite Knowledge:* Written and verbal communication.

(b) *Prerequisite Skills:* The ability to relate interpersonally.

FIRE OFFICER II

1021–7

2-4.2 Execute routine unit-level administrative functions, given forms and record management systems, so that the reports and logs are complete and files are maintained in accordance with policies and procedures.

(a) *Prerequisite Knowledge:* Administrative policies and procedures and records management.

(b) *Prerequisite Skills:* The ability to communicate verbally and in writing.

2-5* Inspection and Investigation. This duty involves performing a fire investigation to determine preliminary cause, securing the incident scene, and preserving evidence, according to the following job performance requirements.

2-5.1 Evaluate available information, given a fire incident, observations, and interviews of first-arriving members and other individuals involved in the incident, so that a preliminary cause of the fire is determined, reports are completed, and, if required, the scene is secured and all pertinent information is turned over to an investigator.

(a) *Prerequisite Knowledge:* Common causes of fire, fire growth and development, and policies and procedures for calling for investigators.

(b) *Prerequisite Skills:* The ability to determine basic fire cause and the ability to conduct interviews and write reports.

2-5.2 Secure an incident scene, given rope or barrier tape, so that unauthorized persons can recognize the perimeters of the scene, are kept from restricted areas, and all evidence or potential evidence is protected from damage or destruction.

(a) *Prerequisite Knowledge:* Types of evidence, the importance of fire scene security, and evidence preservation.

(b) *Prerequisite Skill:* The ability to establish perimeters at an incident scene.

2-6* Emergency Service Delivery. This duty involves supervising emergency operations, conducting preincident planning, and deploying assigned resources, according to the following job performance requirements.

2-6.1 Develop a preincident plan, given an assigned facility and preplanning policies, procedures, and forms, so that all required elements are identified and the appropriate forms are completed and processed in accordance with policies and procedures.

(a) *Prerequisite Knowledge:* Elements of a preincident plan, basic building construction, basic fire protection systems and features, basic water supply, basic fuel loading, and fire growth and development.

(b) *Prerequisite Skills:* The ability to write reports, to communicate verbally, and to evaluate skills.

2-6.2 Develop an initial action plan, given size-up information for an incident and assigned emergency response resources, so that resources are deployed to control the emergency.

(a)* *Prerequisite Knowledge:* Elements of a size-up, standard operating procedures for emergency operations, and fire behavior.

(b) *Prerequisite Skills:* The ability to analyze emergency scene conditions, to allocate resources, and to communicate verbally.

2-6.3* Implement an action plan at an emergency operation, given assigned resources, type of incident, and a prelim-

inary plan, so that resources are deployed to mitigate the situation.

(a) *Prerequisite Knowledge:* Standard operating procedures, resources available, basic fire control and emergency operation procedures, an incident management system, and a personnel accountability system.

(b) *Prerequisite Skills:* The ability to implement an incident management system, to communicate verbally, and to supervise and account for assigned personnel under emergency conditions.

2-7* Safety. This duty involves integrating safety plans, policies, and procedures into the daily activities to ensure a safe work environment for all assigned members, according to the following job performance requirements.

2-7.1 Apply safety regulations at the unit level, given safety policies and procedures, so that required reports are completed, in-service training is conducted, and member responsibilities are conveyed.

(a) *Prerequisite Knowledge:* The most common causes of personal injury and accident to the member, safety policies and procedures, basic workplace safety, and the components of an infectious disease control program.

(b) *Prerequisite Skills:* The ability to identify safety hazards and to communicate verbally and in writing.

2-7.2 Conduct an initial accident investigation, given an incident and investigation forms, so that the incident is documented and reports are processed in accordance with policies and procedures.

(a) *Prerequisite Knowledge:* Procedures for conducting an accident investigation, and safety policies and procedures.

(b) *Prerequisite Skills:* The ability to communicate verbally and in writing and to conduct interviews.

Chapter 3 Fire Officer II

3-1 General. For certification at Level II, Fire Officer I shall meet the requirements of Fire Instructor I as defined in NFPA 1041, *Standard for Fire Service Instructor Professional Qualifications,* and the job performance requirements defined in Sections 3-2 through 3-7 of this standard.

3-1.1 General Prerequisite Knowledge. The organization of local government; the law-making process at the local, state/provincial, and federal level; functions of other bureaus, divisions, agencies and organizations; and their roles and responsibilities that relate to the fire service.

3-1.2 General Prerequisite Skills. Intergovernmental and interagency cooperation.

3-2 Human Resource Management. This duty involves evaluating member performance, according to the following job performance requirements.

3-2.1 Initiate actions to maximize member performance and/or to correct unacceptable performance, given human resource policies and procedures, so that member and/or unit performance improves or the issue is referred to the next level of supervision.

(a) Prerequisite Knowledge: Human resource policies and procedures, problem identification, organizational behavior, group dynamics, leadership styles, types of power, and interpersonal dynamics.

(b) Prerequisite Skills: The ability to communicate verbally and in writing, to solve problems, to increase team work, and to counsel members.

3-2.2 Evaluate the job performance of assigned members, given personnel records and evaluation forms, so each member's performance is evaluated accurately and reported according to human resource policies and procedures.

(a) Prerequisite Knowledge: Human resource policies and procedures, job descriptions, objectives of a member evaluation program, and common errors in evaluating.

(b) Prerequisite Skills: The ability to communicate verbally and in writing and to plan and conduct evaluations.

3-3 Community and Government Relations. This duty involves delivering life safety, injury, and fire prevention education programs, according to the following job performance requirements.

3-3.1 Deliver a public education program, given the target audience and topic, so that the intended message is conveyed clearly.

(a) Prerequisite Knowledge: Contents of the fire department's public education program as it relates to the target audience.

(b) Prerequisite Skill: The ability to communicate to the target audience.

3-4 Administration. This duty involves preparing budget requests, news releases, and policy changes, according to the following job performance requirements.

3-4.1 Prepare recommendations for changes to an existing policy or procedure, given a policy or procedure in need of change, so that the recommendations identify the problem and propose a solution.

(a) Prerequisite Knowledge: Policies and procedures and problem identification.

(b) Prerequisite Skills: The ability to communicate in writing and to solve problems.

3-4.2 Prepare a budget request, given a need and budget forms, so that the request is in the proper format and is supported with data.

(a) Prerequisite Knowledge: Policies and procedures and the revenue sources and budget process.

(b) Prerequisite Skill: The ability to communicate in writing.

3-4.3 Prepare a news release, given an event or topic, so that the information is accurate and formatted correctly.

(a) Prerequisite Knowledge: Policies and procedures and the format used for news releases.

(b) Prerequisite Skills: The ability to communicate verbally and in writing.

3-4.4 Prepare a concise report for transmittal to a supervisor, given fire department record(s) and a specific request for details such as trends, variances, or other related topics.

(a) Prerequisite Knowledge: The data processing system.

(b) Prerequisite Skills: The ability to communicate in writing and to interpret data.

3-5 Inspection and Investigation. This duty involves conducting inspections to identify hazards and address violations and conducting fire investigations to determine origin and preliminary cause, according to the following job performance requirements.

3-5.1 Describe the procedures for conducting fire inspections, given any of the following occupancies:

(a) Assembly

(b) Educational

(c) Health care

(d) Detention and correctional

(e) Residential

(f) Mercantile

(g) Business

(h) Industrial

(i) Storage

(j) Unusual structures

(k) Mixed occupancies

so that all hazards, including hazardous materials, are identified, appropriate forms are completed, and appropriate action is initiated.

(a) Prerequisite Knowledge: Inspection procedures; fire detection, alarm, and protection systems; identification of fire and life safety hazards; and marking and identification systems for hazardous materials.

(b) Prerequisite Skills: The ability to communicate in writing and to apply the appropriate codes.

3-5.2 Determine the point of origin and preliminary cause of a fire, given a fire scene, photographs, diagrams, pertinent data and/or sketches, to determine if arson is suspected.

(a) Prerequisite Knowledge: Methods used by arsonists, common causes of fire, basic cause and origin determination, fire growth and development, and documentation of preliminary fire investigative procedures.

(b) Prerequisite Skills: The ability to communicate verbally and in writing and to apply knowledge using deductive skills.

3-6 Emergency Service Delivery. This duty involves supervising multi-unit emergency operations, conducting preincident planning, and deploying assigned resources, according to the following job requirements.

3-6.1 Produce operational plans, given a hazardous materials incident and another emergency requiring multi-unit operations, so that required resources, their assignments, and safety considerations for successful control of the incident are identified.

(a) Prerequisite Knowledge: Standard operating procedures; national, state/provincial, and local information resources available for the handling of hazardous materials under emergency situations; basic fire control and emergency operation procedures; an incident management system; and a personnel accountability system.

(b) Prerequisite Skills: The ability to implement an incident management system, to communicate verbally, and to

supervise and account for assigned personnel under emergency conditions.

3-7 Safety. This duty involves reviewing injury, accident, and health exposure reports, identifying unsafe work environments or behaviors, and taking appropriate action to prevent reoccurrence, according to the following job requirements.

3-7.1 Analyze a member's accident, injury, or health exposure history, given the case study, so that a report is prepared for a supervisor and includes action taken and recommendations given.

(a) Prerequisite Knowledge: The causes of unsafe acts, health exposures, or conditions that result in accidents, injuries, occupational illnesses, or deaths.

(b) Prerequisite Skills: The ability to communicate in writing and to interpret accidents, injuries, occupational illnesses, or death reports.

Chapter 4   Fire Officer III

4-1 General. For certification at Level III, the Fire Officer II shall meet the requirements of Fire Instructor II as defined in NFPA 1041, *Standard for Fire Service Instructor Professional Qualifications,* and the job performance requirements defined in Sections 4-2 through 4-7 of this standard.

4-1.1 General Prerequisite Knowledge. Current national and international trends and developments related to fire service organization, management, and administrative principles; public and private organizations that support the fire and emergency services and the functions of each.

4-1.2 General Prerequisite Skills. The ability to use evaluative methods, to analyze data, to communicate verbally and in writing, and to motivate members.

4-2 Human Resource Management. This duty involves establishing procedures for hiring, assigning, promoting, and encouraging professional development of members, according to the following job performance requirements.

4-2.1 Establish personnel assignments to maximize efficiency, given knowledge, training, and experience of the members available in accordance with policies and procedures.

(a) Prerequisite Knowledge: Minimum staffing requirements, available human resources, and policies and procedures.

(b) Prerequisite Skills: The ability to relate interpersonally and to communicate verbally and in writing.

4-2.2 Develop procedures for hiring members, given applicable policies and legal requirements, so that the process is valid and reliable.

(a) Prerequisite Knowledge: Applicable federal, state/provincial, and local laws, regulations and standards, and policies and procedures.

(b) Prerequisite Skills: The ability to communicate verbally and in writing.

4-2.3 Develop procedures for promoting members, given applicable policies and legal requirements, so that the process is valid and reliable.

(a) Prerequisite Knowledge: Applicable federal, state/provincial, and local laws; regulations and standards; and policies and procedures.

(b) Prerequisite Skills: The ability to communicate verbally and in writing, to encourage professional development, and to mentor members.

4-2.4 Describe methods to facilitate and encourage members to participate in professional development to achieve their full potential.

(a) Prerequisite Knowledge: Interpersonal and motivational techniques.

(b) Prerequisite Skills: The ability to evaluate potential, to communicate verbally, and to counsel members.

4-3 Community and Government Relations. This duty involves developing programs that improve and expand service and that build partnerships with the public, according to the following job performance requirements.

4-3.1 Prepare community awareness programs to enhance the quality of life by developing nontraditional services that provide for increased safety, injury prevention, and convenient public services.

(a) Prerequisite Knowledge: Community demographics, resource availability, community needs, and customer service principles.

(b) Prerequisite Skills: The ability to relate interpersonally and to communicate verbally and in writing.

4-4 Administration. This duty involves preparing a budget, developing a budget management system, soliciting bids, planning for resource allocation, and working with information management systems, according to the following job performance requirements.

4-4.1 Develop a budget, given schedules and guidelines concerning its preparation, so that capital, operating, and personnel costs are determined and justified.

(a) Prerequisite Knowledge: The supplies and equipment necessary for existing and new programs; repairs to existing facilities; new equipment, apparatus maintenance, and personnel costs; and appropriate budgeting system.

(b) Prerequisite Skills: The ability to allocate finances, to relate interpersonally, and to communicate verbally and in writing.

4-4.2* Develop a budget management system, given fiscal and financial policies, in order to stay within the budgetary authority.

(a) Prerequisite Knowledge: Revenue to date, anticipated revenue, expenditures to date, encumbered amounts, and anticipated expenditures.

(b) Prerequisite Skills: The ability to interpret financial data and to communicate verbally and in writing.

4-4.3 Describe the process of soliciting and awarding bids, given established specifications, in order to assure competitive bidding.

(a) Prerequisite Knowledge: Purchasing laws, policies, and procedures.

(b) Prerequisite Skills: The ability to use evaluative methods and to communicate verbally and in writing.

4-4.4   Direct the development, maintenance, and evaluation of a department record-keeping system, given policies and procedures, so as to attain completeness and accuracy.

(a) Prerequisite Knowledge: The principles involved in the acquisition, implementation, and retrieval of information by data processing as it applies to the record and budgetary processes, capabilities, and limitations of information management systems.

(b) Prerequisite Skills: The ability to use evaluative methods, to communicate verbally and in writing, and to organize.

4-4.5   Analyze and interpret records and data, given fire department records system, to determine validity and recommend improvements.

(a) Prerequisite Knowledge: The principles involved in the acquisition, implementation, and retrieval of information and data.

(b) Prerequisite Skills: The ability to use evaluative methods, to communicate verbally and in writing, to organize, and to analyze.

4-4.6   Develop a model plan, given a prescribed quantity of personnel and equipment for a given area to be protected, for the maximum utilization of resources.

(a) Prerequisite Knowledge: Demographics of the area, hazards, geographic area, and established maximum response times.

(b) Prerequisite Skills: The ability to use evaluative methods, to communicate verbally and in writing, and to organize.

4-5 Inspection and Investigation.   This duty involves evaluating inspection programs to determine effectiveness and developing public safety plans, according to the following job performance requirements.

4-5.1   Evaluate and identify construction, alarm, detection, and suppression features that contribute to or prevent the spread of fire, heat, and smoke throughout the building or from one building to another, given an occupancy, to evaluate the development of a preincident plan for any of the following occupancies:

(a) Public assembly

(b) Educational

(c) Institutional

(d) Residential

(e) Business

(f) Industrial

(g) Manufacturing

(h) Storage

(i) Mercantile

(j) Special properties

(a) Prerequisite Knowledge: Fire behavior, program evaluation, building construction, inspection and incident reports, detection, alarm and suppression systems, and applicable codes, ordinances, and standards.

(b) Prerequisite Skills: The ability to use evaluative methods, to communicate verbally and in writing, and to organize.

4-5.2   Develop a plan, given an identified fire safety problem, to facilitate the approval for a new program, legislation, public education, or fire safety code.

(a) Prerequisite Knowledge: Policies and procedures; and applicable codes, ordinances, and standards and their development process.

(b) Prerequisite Skills: The ability to use evaluative methods, to use consensus-building techniques, to communicate verbally and in writing, and to organize.

4-6 Emergency Service Delivery.   This duty involves managing multi-agency planning, deployment, and operations, according to the following job performance requirements.

4-6.1   Prepare an action plan, given an emergency incident requiring multiple agency operations, so that the required resources are determined and the assignment and placement of the resources are designated in order to mitigate the incident.

(a) Prerequisite Knowledge: Policies and procedures, resources, capabilities, roles, responsibilities, and authority of support agencies.

(b) Prerequisite Skills: The ability to use evaluative methods, to delegate authority, to communicate verbally and in writing, and to organize.

4-7 Safety.   This duty involves developing, managing, and evaluating a departmental safety program, according to the following job performance requirements.

4-7.1   Develop a measurable accident and injury prevention program, given specific data, so that the results are evaluated to determine effectiveness.

(a) Prerequisite Knowledge: Policies and procedures; accepted safety practices; and applicable codes, standards, and laws.

(b) Prerequisite Skills: The ability to use evaluative methods, to analyze data, and to communicate verbally and in writing.

Chapter 5   Fire Officer IV

5-1 General.   For certification at Level IV the Fire Officer III shall meet the job performance requirements defined in Sections 5-2 through 5-7 of this standard.

5-1.1 General Prerequisite Knowledge.   Advanced administrative, financial, communications, political, legal, managerial, analytical, and information management.

5-1.2 General Prerequisite Skill.   The ability to effectively apply prerequisite knowledge.

5-2 Human Resource Management.   This duty involves administrating job performance requirements and evaluating and improving the department, according to the following job performance requirements.

5-2.1   Appraise a grievance program, given appropriate data, to determine if the program is effective, consistent, and produces resolution at the appropriate level.

(a) Prerequisite Knowledge: Policies and procedures; contractual agreements; and local, state/provincial, and federal regulations.

(b) Prerequisite Skills: The ability to communicate verbally and in writing, to negotiate, to conduct interviews, to relate interpersonally, and to detect and analyze the cause of grievances.

5-2.2   Establish and evaluate a list of education and in-service training goals, given a summary of the job requirements for all positions within the department, so that all members can achieve and maintain required proficiencies.

(a) *Prerequisite Knowledge:* Training resources; community needs; internal and external customers; policies and procedures; contractual agreements; and local, state/provincial, and federal regulations.

(b) *Prerequisite Skills:* The ability to communicate verbally and in writing and to organize.

5-2.3 Appraise a member-assistance program, given appropriate data, to determine if the program, when used, produces the desired results and benefits.

(a) *Prerequisite Knowledge:* Policies and procedures; available assistance programs; contractual agreements; and local, state/provincial, and federal regulations.

(b) *Prerequisite Skills:* The ability to communicate verbally and in writing, to relate interpersonally, and to analyze.

5-2.4 Evaluate an incentive program, given appropriate data, so that a determination is made regarding achievement of the desired results.

(a) *Prerequisite Knowledge:* Policies and procedures; available incentive programs; contractual agreements; and local, state/provincial, and federal regulations.

(b) *Prerequisite Skills:* The ability to communicate verbally and in writing, to relate interpersonally, and to analyze.

5-3 Community and Government Relations. This duty involves projecting a positive image of the fire department to the community, according to the following job performance requirements.

5-3.1 Attend, participate, and play a leadership role in given community events in order to enhance the image of the fire department.

(a) *Prerequisite Knowledge:* Community demographics, community issues, and formal and informal community leaders.

(b) *Prerequisite Skills:* The ability to communicate verbally and familiarity with public relations

5-4 Administration. This duty involves long-range planning and fiscal projections, according to the following job performance requirements.

5-4.1 Develop a comprehensive long-range plan, given community requirements, current department status, and resources, so that the projected needs of the community are met.

(a) *Prerequisite Knowledge:* Policies and procedures; physical and geographical characteristics; demographics; community plan; staffing requirements; response time benchmarks; contractual agreements; and local, state/provincial, and federal regulations.

(b) *Prerequisite Skills:* The ability to communicate verbally and in writing; and familiarity with fiscal analysis, public policy processes, forecasting, and analyzing.

5-4.2 Evaluate and project training requirements, facilities, and buildings, given appropriate data that reflect community needs and resources, to meet departmental training goals.

(a) *Prerequisite Knowledge:* Policies and procedures; physical and geographical characteristics; building and fire codes; departmental plan; staffing requirements; training standards; needs assessment; contractual agreements; and local, state/provincial, and federal regulations.

(b) *Prerequisite Skills:* The ability to communicate verbally and in writing; and familiarity with fiscal analysis, forecasting, and analyzing.

5-5 Inspection and Investigation.

5-5.1 Definition of Duty. No additional job performance requirements at this level.

5-6 Emergency Services Delivery. This duty involves developing plans for major disasters, according to the following job performance requirements.

5-6.1 Develop a comprehensive disaster plan that integrates other agencies' resources, given appropriate data, in order to rapidly and effectively mitigate the impact on a community.

(a) *Prerequisite Knowledge:* Major incident policies and procedures; physical and geographical characteristics; demographics; target hazards; incident management system; communications systems; contractual and mutual-aid agreements; and local, state/provincial, and federal regulations and resources.

(b) *Prerequisite Skills:* The ability to communicate verbally and in writing and to organize; and familiarity with interagency planning and coordination.

5-6.2 Develop a comprehensive plan, given appropriate data (including agency data), so that the agency operates at a civil disturbance, integrates with other agencies' actions, and provides for the safety and protection of members.

(a) *Prerequisite Knowledge:* Major incident policies and procedures; physical and geographical characteristics; demographics; incident management system; communications systems; contractual and mutual-aid agreements; and local, state/provincial, and federal regulations and resources.

(b) *Prerequisite Skills:* The ability to communicate verbally and in writing and to organize; and familiarity with interagency planning and coordination.

5-7 Safety. This duty involves administering a comprehensive risk management program, according to the following job performance requirements.

5-7.1 Maintain, develop, and provide leadership for a risk management program, given specific data, so that injuries and property damage accidents are reduced.

(a) *Prerequisite Knowledge:* Risk management concepts; retirement qualifications; occupational hazards analysis; and disability procedures, regulations, and laws.

(b) *Prerequisite Skills:* The ability to communicate verbally and in writing, to analyze, and to use evaluative methods.

Chapter 6   Referenced Publications

6-1 The following documents or portions thereof are referenced within this standard as mandatory requirements and shall be considered part of the requirements of this standard. The edition indicated for each referenced mandatory document is the current edition as of the date of the NFPA issuance of this standard. Some of these mandatory documents might also be referenced in this standard for specific informational purposes and, therefore, are also listed in Appendix B.

1021–12

6-1.1 NFPA Publications. National Fire Protection Association, 1 Batterymarch Park, P.O. Box 9101, Quincy, MA 02269-9101.

NFPA 1001, *Standard for Fire Fighter Professional Qualifications,* 1997 edition.

NFPA 1041, *Standard for Fire Service Instructor Professional Qualifications,* 1996 edition.

Appendix A   Explanatory Material

This appendix is not a part of the requirements of this NFPA document but is included for informational purposes only.

A-1-1 It is envisioned that in addition and supplemental to the requirements of NFPA 1021 appropriate educational credentials are necessary. These can include fire degree programs and general education in business, management, science, and associated degree curriculums.

A-1-3 Fire officers are expected to be ethical in their conduct. Ethics implies honesty, doing what's right, and performing to the best of one's ability. For public safety personnel, ethical responsibility extends beyond one's individual performance. In serving the citizens, public safety personnel are charged with the responsibility of ensuring the provision of the best possible safety and service.

Ethical conduct requires honesty on the part of all public safety personnel. Choices must be made on the basis of maximum benefit to the citizens and the community. The process of making these decisions must also be open to the public. The means of providing service, as well as the quality of the service provided, must be above question and must maximize the principles of fairness and equity as well as those of efficiency and effectiveness.

A-1-4 Definitions of action verbs used within this document are based upon the first definition of the word found in *Webster's Dictionary.* (*Webster's Third New International Dictionary of the English Language,* Unabridged, G. & C. Merriam Company.)

A-1-4 Approved. The National Fire Protection Association does not approve, inspect, or certify any installations, procedures, equipment, or materials; nor does it approve or evaluate testing laboratories. In determining the acceptability of installations, procedures, equipment, or materials, the authority having jurisdiction may base acceptance on compliance with NFPA or other appropriate standards. In the absence of such standards, said authority may require evidence of proper installation, procedure, or use. The authority having jurisdiction may also refer to the listings or labeling practices of an organization that is concerned with product evaluations and is thus in a position to determine compliance with appropriate standards for the current production of listed items.

A-1-4 Authority Having Jurisdiction. The phrase "authority having jurisdiction" is used in NFPA documents in a broad manner, since jurisdictions and approval agencies vary, as do their responsibilities. Where public safety is primary, the authority having jurisdiction may be a federal, state, local, or other regional department or individual such as a fire chief; fire marshal; chief of a fire prevention bureau, labor department, or health department; building official; electrical inspector; or others having statutory authority. For insurance purposes, an insurance inspection department, rating bureau, or other insurance company representative may be the authority having jurisdiction. In many circumstances, the property owner or his or her designated agent assumes the role of the authority having jurisdiction; at government installations, the commanding officer or departmental official may be the authority having jurisdiction.

A-1-4 Listed. The means for identifying listed equipment may vary for each organization concerned with product evaluation; some organizations do not recognize equipment as listed unless it is also labeled. The authority having jurisdiction should utilize the system employed by the listing organization to identify a listed product.

A-2-2.4(a) Member-related problems could include substance abuse; acute, chronic, and delayed stress; and health, financial, personal, family, and other situations that adversely affect the member's job performance.

A-2-2.5 The Fire Officer I should be able to deal with administrative procedures that might include the following: transfers, promotions, compensation/member benefits, sick leave, vacation, requests for pay or benefits while acting in a temporary position, change in member benefits, commendations, disciplinary actions, and grievances.

A-2-5 The committee's intent is to instill an awareness of those areas that officers might address in the performance of their duties. Organizations that desire higher levels of competency in these areas should refer to the applicable NFPA professional qualifications standards: NFPA 1031, *Standard for Professional Qualifications for Fire Inspector,* and NFPA 1033, *Standard for Professional Qualifications for Fire Investigator.*

A-2-6 Emergency service delivery is the component of fire department organization providing responses to emergency incidents.

A-2-6.2(a) Size-up includes the many variables that the officer collects from the time of the alarm, during response, and upon arrival in order to develop an initial action plan to control an emergency incident. These observations can include building type and occupancy, fire involvement, number of occupants, mechanism of injury, materials spilled or involved in fire, wind direction, topography, demographics, and other observations relevant to the incident.

A-2-6.3 This requirement takes into consideration the officer's ability to give orders, direct personnel, evaluate information, and allocate resources to respond to the wide variety of emergency situations the fire service encounters.

A-2-7 One of the fire officer's primary responsibilities is safety both on the fire ground and during normal operations. This standard defines the minimum requirements for the fire officer. NFPA 1521, *Standard for Fire Department Safety Officer;* NFPA 1500, *Standard on Fire Department Occupational Safety and Health Program;* and applicable OSHA regulations define additional requirements for the officer who might be assigned those duties.

A-4.4.2 The following are some of the budgeting systems commonly used:

(a) Planning programming budgeting system (PPBS)

(b) Line item budgets

(c) Zero-based budgeting (ZBB)

(d) Program budgeting

(e) Performance budgeting

(f) Matrix budgets

Appendix B   Referenced Publications

B-1 The following documents or portions thereof are referenced within this standard for informational purposes only and are thus not considered part of the requirements of this standard unless also listed in Chapter 6. The edition indicated here for each reference is the current edition as of the date of the NFPA issuance of this standard.

B-1.1 NFPA Publications. National Fire Protection Association, 1 Batterymarch Park, P.O. Box 9101, Quincy, MA 02269-9101.

NFPA 1031, *Standard for Professional Qualifications for Fire Inspector,* 1993 edition.

NFPA 1033, *Standard for Professional Qualifications for Fire Investigator,* 1993 edition.

NFPA 1500, *Standard on Fire Department Occupational Safety and Health Program,* 1997 edition.

NFPA 1521, *Standard for Fire Department Safety Officer,* 1997 edition.

B-1.2 Other Publication.

Webster's Third New International Dictionary of the English Language, Unabridged, G. & C. Merriam Company.

Appendix C

This appendix is not a part of the requirements of this NFPA document but is included for informational purposes only.

C-1 Explanation of the Standard and Concepts of JPRs.

The primary benefit of establishing national professional qualification standards is to provide both public and private sectors with a framework of the job requirements for the fire service. Other benefits include enhancement of the profession, individual as well as organizational growth and development, and standardization of practices.

NFPA professional qualification standards identify the minimum job performance requirements for specific fire service positions. The standards can be used for training design and evaluation, certification, measuring and critiquing on-the-job performance, defining hiring practices, and setting organizational policies, procedures, and goals. (Other applications are encouraged.)

Professional qualification standards for a specific job are organized by major areas of responsibility defined as duties. For example, the fire fighter's duties might include fire suppression, rescue, and water supply; and the public fire educator's duties might include education, planning and development, and administration. Duties are major functional areas of responsibility within a job.

The professional qualification standards are written as JPRs. JPRs describe the performance required for a specific job. JPRs are grouped according to the duties of a job. The complete list of JPRs for each duty defines what an individual must be able to do in order to successfully perform that duty.

Together, the duties and their JPRs define the job parameters; that is, the standard as a whole is a description of a job.

C-2 Breaking Down the Components of a JPR.

The JPR is the assembly of three critical components. *(See Table C-2.)* These components are as follows:

(a) Task that is to be performed

(b) Tools, equipment, or materials that must be provided to successfully complete the task

(c) Evaluation parameters and/or performance outcomes

The task to be performed. The first component is a concise, brief statement of what the person is supposed to do.

Tools, equipment, or materials that must be provided to successfully complete the task. This component ensures that all individuals completing the task are given the same minimal tools, equipment, or materials when being evaluated. By listing these items, the performer and evaluator know what must be provided in order to complete the task.

Evaluation parameters and/or performance outcomes. This component defines how well one must perform each task—for both the performer and the evaluator. The job performance requirements guide performance towards successful completion by identifying evaluation parameters and/or performance outcomes. This portion of the job performance requirements promotes consistency in evaluation by reducing the variables used to gauge performance.

Table C-2   Example of a JPR

(a) Task	(a) Ventilate a pitched roof
(b) Tools, equipment, or materials	(b) Given an ax, a pike pole, an extension ladder, and a roof ladder
(c) Evaluation parameters and performance outcomes	(c) So that a 4-ft ∞ 4-ft hole is created; all ventilation barriers are removed; ladders are properly positioned for ventilation; ventilation holes are correctly placed; and smoke, heat, and combustion by-products are released from the structure

In addition to these three components, the job performance requirements contain prerequisite knowledge and skills. Just as the term prerequisite suggests, these are the necessary knowledge and skills one must have prior to being able to perform the task. Prerequisite knowledge and skills are the foundation for task performance.

Once the components and prerequisites are put together, the job performance requirements might read as follows.

Example 1. The **Fire Fighter I** shall ventilate a pitched roof, given an ax, a pike pole, an extension ladder, and a roof ladder, so that a 4-ft ∞ 4-ft hole is created, all ventilation barriers are removed, ladders are properly positioned for ventilation, and ventilation holes are correctly placed.

(a) *Prerequisite Knowledge:* Pitched roof construction, safety considerations with roof ventilation, the dangers associated with improper ventilation, knowledge of ventilation tools, the effects of ventilation on fire growth, smoke movement in structures, signs of backdraft, and the knowledge of vertical and forced ventilation.

(b) *Prerequisite Skills:* The ability to remove roof covering; properly initiate roof cuts; use the pike pole to clear ventilation barriers; use ax properly for sounding, cutting, and stripping; position ladders; and climb and position self on ladder.

Example 2. The **Fire Investigator** shall interpret burn patterns, given standard equipment and tools and some structural/content remains, so that each individual pattern is evaluated with respect to the burning characteristics of the material involved.

(a) *Prerequisite Knowledge:* Knowledge of fire development and the interrelationship of heat release rate, form, and ignitability of materials.

(b) *Prerequisite Skill:* The ability to interpret the effects of burning characteristics on different types of materials.

C-3 Examples of Potential Uses.

Certification. JPRs can be used to establish the evaluation criteria for certification at a specific job level. When used for certification, evaluation must be based on the successful completion of JPRs.

First, the evaluator would verify the attainment of prerequisite knowledge and skills prior to job performance requirements evaluation. This might be through documentation review or testing.

Next, the candidate would then be evaluated on completing the JPRs. The candidate would perform the task and be *evaluated* based on the evaluation parameters and/or performance outcomes. This performance-based evaluation can be either practical (for psychomotor skills such as "ventilate a roof") or written (for cognitive skills such as "interpret burn patterns").

> **NOTE:** Psychomotor skills are those physical skills that can be demonstrated or observed. Cognitive skills (or mental skills) cannot be observed, but are rather evaluated on how one completes the task (process oriented) or the task outcome (product oriented).

Using **Example 1**, a practical performance-based evaluation would measure one's ability to "ventilate a pitched roof." The candidate passes this particular evaluation if the standard was met—that is, a 4-ft ∞ 4-ft hole was created; all ventilation barriers were removed; ladders were properly positioned for ventilation; ventilation holes were correctly placed; and smoke, heat, and combustion by-products were released from the structure.

For **Example 2**, when evaluating the task "interpret burn patterns," the candidate could be given a written assessment in the form of a scenario, photographs, and drawings and then be asked to respond to specific written questions related to the JPR's evaluation parameters.

Remember, when evaluating performance, you must give the person the tools, equipment, or materials listed in the job performance requirements—for example, an ax, a pike pole, an extension ladder, and a roof ladder—before he or she can be properly evaluated.

C-4 Curriculum Development/Training Design and Evaluation.

The statements contained in this document that refer to job performance were designed and written as JPRs. While a resemblance to instructional objectives might be present, these statements should not be used in a teaching situation until after they have been modified for instructional use.

JPRs state the behaviors required to perform specific skill(s) on the job as opposed to a learning situation. These statements should be converted into instructional objectives with behaviors, conditions, and standards that can be measured within the teaching/learning environment. A JPR that requires a fire fighter to "ventilate a pitched roof" should be converted into a measurable instructional objective for use when teaching the skill. *[See Figure C-4(a).]*

Using **Example 1**, a terminal instructional objective might read as follows:

The learner will ventilate a pitched roof, given a simulated roof, an ax, a pike pole, an extension ladder, and a roof ladder, so that 100-percent accuracy is attained on a skills checklist. (At a minimum, the skills checklist should include each of the measurement criteria from the job performance requirements).

Figure C-4(b) is a sample checklist for use in evaluating this objective.

While the differences between job performance requirements and instructional objectives are subtle in appearance, the purpose of each statement differs greatly. JPRs state what is necessary to perform the job in the "real world." Instructional objectives, however, are used to identify what students must do at the end of a training session and are stated in behavioral terms that are measurable in the training environment.

By converting JPRs into instructional objectives, instructors will be able to clarify performance expectations and avoid confusion related to using statements designed for purposes other than teaching. Additionally, instructors will be able to add local/state/regional elements of performance into the standards as intended by the developers.

Prerequisite skills and knowledge should be converted into enabling objectives. These help to define the course content. The course content would include each of the prerequisite knowledge and skills. Using the above example, the enabling objectives would be pitched roof construction, safety considerations with roof ventilation, removal of roof covering, properly initiated roof cuts, and so on. This ensures that the course content supports the terminal objective.

> **NOTE:** It is assumed that the reader is familiar with curriculum development or training design and evaluation.

C-5 Other Uses.

While the professional qualifications standards are principally used to guide the development of training and certification programs, there are a number of other potential uses for the documents. Because the documents are written using terms specific to job performance requirements, they lend themselves well to any area of the profession where a level of performance or expertise must be determined.

These areas might include the following.

Employee Evaluation/Performance Critiquing. The JPRs can be used as a guide by both the supervisor and the employee during an evaluation. The JPRs for a specific job define tasks that are essential to perform on the job as well as the evaluation criteria to measure when those tasks are completed.

Establishing Hiring Criteria. The professional qualifications standards can be used in a number of ways to further the establishment of hiring criteria. The authority having jurisdiction could simply require certification at a specific job level—for example, Fire Fighter I. The JPRs could also be used as the basis for pre-employment screening by establishing essential

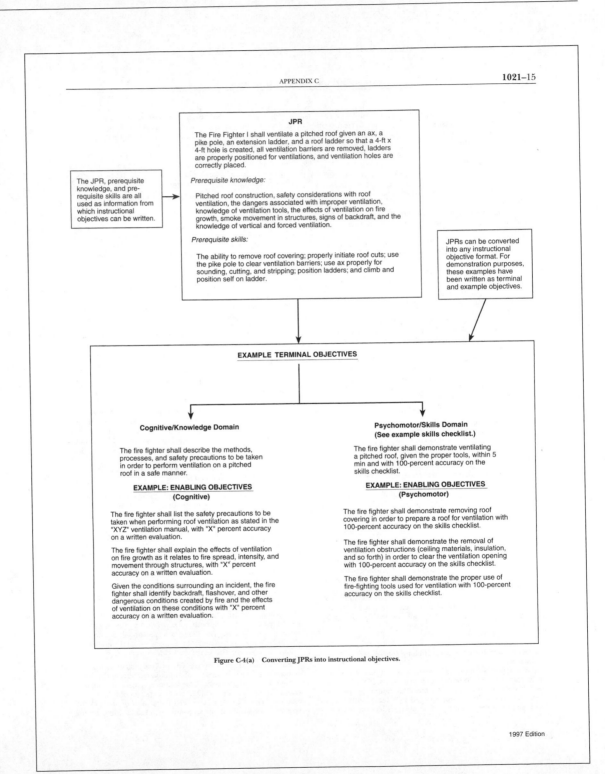

APPENDIX C   1021–15

JPR

The Fire Fighter I shall ventilate a pitched roof given an ax, a pike pole, an extension ladder, and a roof ladder so that a 4-ft x 4-ft hole is created, all ventilation barriers are removed, ladders are properly positioned for ventilations, and ventilation holes are correctly placed.

Prerequisite knowledge:

Pitched roof construction, safety considerations with roof ventilation, the dangers associated with improper ventilation, knowledge of ventilation tools, the effects of ventilation on fire growth, smoke movement in structures, signs of backdraft, and the knowledge of vertical and forced ventilation.

Prerequisite skills:

The ability to remove roof covering; properly initiate roof cuts; use the pike pole to clear ventilation barriers; use ax properly for sounding, cutting, and stripping; position ladders; and climb and position self on ladder.

The JPR, prerequisite knowledge, and pre-requisite skills are all used as information from which instructional objectives can be written.

JPRs can be converted into any instructional objective format. For demonstration purposes, these examples have been written as terminal and example objectives.

EXAMPLE TERMINAL OBJECTIVES

Cognitive/Knowledge Domain

The fire fighter shall describe the methods, processes, and safety precautions to be taken in order to perform ventilation on a pitched roof in a safe manner.

EXAMPLE: ENABLING OBJECTIVES
(Cognitive)

The fire fighter shall list the safety precautions to be taken when performing roof ventilation as stated in the "XYZ" ventilation manual, with "X" percent accuracy on a written evaluation.

The fire fighter shall explain the effects of ventilation on fire growth as it relates to fire spread, intensity, and movement through structures, with "X" percent accuracy on a written evaluation.

Given the conditions surrounding an incident, the fire fighter shall identify backdraft, flashover, and other dangerous conditions created by fire and the effects of ventilation on these conditions with "X" percent accuracy on a written evaluation.

Psychomotor/Skills Domain
(See example skills checklist.)

The fire fighter shall demonstrate ventilating a pitched roof, given the proper tools, within 5 min and with 100-percent accuracy on the skills checklist.

EXAMPLE: ENABLING OBJECTIVES
(Psychomotor)

The fire fighter shall demonstrate removing roof covering in order to prepare a roof for ventilation with 100-percent accuracy on the skills checklist.

The fire fighter shall demonstrate the removal of ventilation obstructions (ceiling materials, insulation, and so forth) in order to clear the ventilation opening with 100-percent accuracy on the skills checklist.

The fire fighter shall demonstrate the proper use of fire-fighting tools used for ventilation with 100-percent accuracy on the skills checklist.

Figure C-4(a)   Converting JPRs into instructional objectives.

1997 Edition

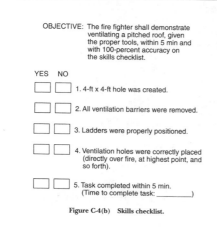

OBJECTIVE: The fire fighter shall demonstrate ventilating a pitched roof, given the proper tools, within 5 min and with 100-percent accuracy on the skills checklist.

YES   NO

☐   ☐   1. 4-ft x 4-ft hole was created.

☐   ☐   2. All ventilation barriers were removed.

☐   ☐   3. Ladders were properly positioned.

☐   ☐   4. Ventilation holes were correctly placed (directly over fire, at highest point, and so forth).

☐   ☐   5. Task completed within 5 min. (Time to complete task: _____)

Figure C-4(b)   Skills checklist.

minimal tasks and the related evaluation criteria. An added benefit is that individuals interested in employment can work towards the minimal hiring criteria at local colleges.

Employee Development. The professional qualifications standards can be useful to both the employee and the employer in developing a plan for the individual's growth within the organization. The JPRs and the associated prerequisite knowledge and skills can be used as a guide to determine additional training and education required for the employee to master his job or profession.

Succession Planning. Succession planning or career pathing addresses the efficient placement of people into jobs in response to current needs and anticipated future needs. A career development path can be established for targeted individuals to prepare them for growth within the organization. The JPRs and prerequisite knowledge and skills could then be used to develop an educational path to aid in the individual's advancement within the organization or profession.

Establishing Organizational Policies, Procedures, and Goals. The JPRs can be incorporated into organizational policies, procedures, and goals where employee performance is addressed.

C-6 Bibliography.

Boyatzis, *R. E. The Competent Manager: A Model For Effective Performance.* New York: John Wiley & Sons, 1982.

Castle, D. K. "Management Design: A Competency Approach to Create Exemplar Performers." *Performance and Instruction* 28: 1989; 42–48.

Cetron, M., and O'Toole, T. *Encounters with the Future: A Forecast into the 21st Century.* New York: McGraw Hill, 1983.

Elkin, G. "Competency-Based Human Resource Development: Making Sense of the Ideas." *Industrial & Commercial Training* 22: 1990; 20–25.

Furnham, A. "The Question of Competency." *Personnel Management* 22: 1990; 37.

Gilley, J. W., and Eggland, S. A. *Principles of Human Resource Development.* Reading, MA: Addison-Wesley, 1989.

Hooton, J. Job *Performance = Tasks + Competency ∞ Future Forces.* Unpublished manuscript, Vanderbilt University, Peabody College, Nashville, TN, 1990.

McLagan, P. A. "Models for HRD Practice." *Training & Development Journal.* Reprinted, 1989.

McLagan, P. A., and Suhadolnik, D. *The Research Report.* Alexandria, VA. American Society for Training and Development, 1989.

Nadler, L. "HRD on the Spaceship Earth." *Training and Development Journal,* October 1983; 19–22.

Nadler, L. *The Handbook of Human Resource Development.* New York: Wiley-Interscience, 1984.

Naisbitt, J., *Megatrends,* Chicago: Nightingale-Conant, 1984.

Spellman, B. P. "Future Competencies of the Educational Public Relations Specialist" (Doctoral dissertation, University of Houston, 1987.) *Dissertation Abstracts International* 49: 1987; 02A.

Springer, J. *Job Performance Standards and Measures.* A Series of Research Presentation and Discussions for the ASTD Second Annual Invitational Research Seminar, Savannah, GA (November 5–8, 1979). Madison, WI: American Society for Training and Development, 1980.

Tracey, W. R. *Designing Training and Development Systems.* New York: AMACOM, 1984.

Appendix

B

Suggested Training Evolutions

Although firefighters may not be confronted with working fire and rescue situations on a daily basis, it is important that they maintain basic skills through effective training and evolution programs. These standard evolutions can serve as a guide for conducting practical instruction and drills or to measure the performance capabilities of individuals and companies.

Individual Skills

1. Donning the self-contained breathing apparatus (SCBA)
2. Carrying and raising a 24-foot extension ladder
3. Tying five fireground knots
4. Tying knots for hoisting tools and equipment
5. Tying the rescue knot

Single Engine Company Evolutions

1. Conducting a primary search
2. Laying a supply line and putting an attack line into operation
3. Laying a reverse-lay supply line and putting an attack line into operation
4. Advancing an attack line over ground ladders
5. Placing a deck gun into operation

Multiple Engine Company Evolutions

1. Establishing a water supply and advancing two preconnected attack lines
2. Placing a deck gun and two attack lines into operation
3. Placing a 2 1/2-inch handline into operation
4. Supplying the fireground from a draft
5. Supplying a standpipe

Squad or Truck Company Evolutions

1. Laddering a building with ground ladders
2. Rescuing a victim from an elevator
3. Rescuing a victim from an elevated position
4. Deploying fans and smoke ejectors
5. Hoisting tools for roof ventilation

The following sections provide an objective, equipment to be used, and a suggested time allowed for each of the standard evolutions. The criteria for conducting these events should be clearly defined. All of these events can be modified as needed to meet local conditions and to conform to existing department policies.

INDIVIDUAL SKILLS

Evolution Number 1: Donning the SCBA

Objective:	To demonstrate the ability to correctly don a SCBA.
Given:	One SCBA mounted in the jump seat of a fire apparatus. The firefighter shall be dressed in full protective clothing.
Procedure:	The firefighter shall:

1. Sit in the jump seat and don the SCBA backpack.
2. Step to the ground, move at least 10 feet, don the facepiece, make all adjustments/connections necessary, and begin breathing from the SCBA.
3. Don all protective clothing including hood, helmet, and gloves.

Time:	1 minute
Note:	Time to start when the firefighter is seated. Time to stop when the firefighter is breathing from the SCBA and all protective clothing is in place.

Evolution Number 2: Carrying and Raising a 24-Foot Extension Ladder

Objective: To demonstrate the ability to properly carry, raise, and extend a 24-foot extension ladder.

Given: 24-foot fire service extension ladder

Procedure: The firefighter shall remove the ladder from the apparatus and advance the ladder to the building (25 to 50 foot distance), utilizing a proper shoulder carry. After observing for proper placement and freedom from overhead obstructions, heel the ladder flat against the base of the building with the fly section in (closest to the building), raise the ladder, and while steadying the ladder with one leg, extend the ladder to the desired height using a hand-over-hand pulling motion on the halyard. When the ladder is at the proper height, the firefighter shall set the ladder to the proper climbing angle by lifting the ladder and placing the hands in a one high and one low position and moving the heel of the ladder away from the building, roll the ladder over so that the fly section is on the outside, place the ladder against the building, and tie off the halyard.

Time: 2 minutes

Note: Additional firefighter(s) should be available to assist to avoid injury or damage while raising the ladder and to assist in taking the ladder down and replacing it on the apparatus.

Evolution Number 3: Tying Five Fireground Knots

Objective: To demonstrate the ability to tie common knots.

Given: A length of 1/2-inch utility rope

Procedure: The firefighter shall tie the following knots: a clove hitch in 20 seconds, a becket bend in 25 seconds, a bowline around an object in 30 seconds, a chimney hitch in 30 seconds, and a figure-eight knot on the bight in 30 seconds.

Time: As indicated

Notes: Time starts when the firefighter picks up the rope and ends when the firefighter indicates that the knot is tied. The rope should be flexible, free of kinks and knots, and uncoiled before the evolution starts. No protective clothing is required, but to enhance the realism of this event, the firefighter should be asked to demonstrate this skill while wearing the gloves normally used on the fireground.

Evolution Number 4: Tying Knots for Hoisting Tools and Equipment

Objective:	To demonstrate the ability of the firefighter to tie off given tools and equipment to be hoisted.
Given:	A length of 1/2-inch utility rope, a pick-head ax, a pike pole, a straight ladder, and a section of hose with the nozzle attached
Procedure:	The firefighter shall tie the rope: to the pick-head axe in 30 seconds, to the pike pole in 30 seconds, to the hose and nozzle in 50 seconds, and to the ladder in 50 seconds.
Note:	The rope should be flexible, free of kinks and knots, and uncoiled before the evolution starts. No protective clothing is required.

Evolution Number 5: Tying the Rescue Knot

Objective:	To demonstrate the ability to tie the rescue knot.
Given:	A length of Kernmantle (or similar) rope
Procedure:	The firefighter shall tie the rescue knot:
	On him/herself
	On another firefighter
Time:	2 minutes
Note:	The rope should be flexible, free of kinks and knots, and uncoiled before the evolution starts. The evaluator may elect to have the firefighter tie the knot on him- or herself or another person. No protective clothing is required.

SINGLE ENGINE COMPANY EVOLUTIONS

Unless otherwise noted, in the following evolutions all personnel except the driver are to be in full protective clothing including SCBA. When required, the hydrant or other water supply should be 100 to 200 feet from the simulated fire building.

Evolution Number 1: Conducting a Primary Search

Objective:	To demonstrate the ability of the company to search a floor of a building.
Given:	A space similar to a private residence
Procedure:	The company shall enter the building and search the space indicated.

Time:	5 minutes. Time is started when the first member of the team enters the building and ends when the officer reports to the evaluator that the search is complete.
Notes:	Visibility should be reduced by the use of darkness, obscuring the facepiece, or by the use of nontoxic chemical smoke. A manikin or similar object may be used to ensure a complete search. If used, the manikin must be "rescued" from the hazardous environment.

Evolution Number 2: Laying a Supply Line and Putting an Attack Line into Operation

Objective:	To demonstrate the ability of a single engine company to put a line in operation in a simulated fire building.
Given:	A single engine company with apparatus, and a building
Procedure:	Engine to start at a designated point, proceed to a hydrant, drop a supply line, and continue to an appropriate position on side 1 of the fire building. Personnel are to advance a 1 3/4-inch attack line to the second floor via the front door and interior stairway, and discharge water through a designated window.
Time:	4 minutes. Time starts when the pumper stops at the hydrant and is recorded when water is flowing at the proper pressure.

Evolution Number 3: Laying a Reverse-Lay Supply Line and Putting an Attack Line into Operation

Objective:	To demonstrate the ability of a single engine company to put an attack line in operation using a reverse lay for water supply.
Given:	A single engine company with apparatus, and a building
Procedure:	Engine to start at a designated point and proceed to side 1 of the fire building. Personnel shall pull an attack line (typically a 1 3/4-inch line), extension ladder, and forcible entry tools. The ladder is be placed at a second-story window. A supply line is pulled and connected to attack line. Driver proceeds with engine to the hydrant and establishes a water supply.
Time:	5 minutes. Time starts when the pumper stops at the fire building and is recorded when water is flowing at the proper pressure.
Note:	The size and length of the attack line and size of the supply line should follow the department's standard procedure. The ladder is for rescue. The line would be advanced in the door. The next event combines these activities.

Evolution Number 4: Advancing an Attack Line over Ground Ladders

Objective: To demonstrate the ability of a single engine company to put an attack line in service on an upper floor using a ground ladder.

Given: A single engine company with apparatus, and a building

Procedure: Engine to start at a designated point and proceed to side 1 of the fire building. Personnel shall pull a preconnected 1 3/4-inch attack line and remove a 24-foot extension ladder, raise the ladder to a second story window, advance a dry 1 3/4-inch attack line up the ladder and into the building. Line is charged upon command and water discharged through a designated window.

Time: 4 minutes. Time starts when the pumper stops at the fire building and is recorded when water is flowing at the proper pressure.

Evolution Number 5: Placing a Deck Gun into Operation

Objective: To demonstrate the ability of a single engine company to place a master stream into service.

Given: A single engine company, with apparatus, and a building

Procedure: Engine to start at a designated point and proceed to side 1 of the simulated fire building. Personnel shall remove the portable monitor, place it on the ground, and connect two supply lines. The driver will proceed to a hydrant making a duel reverse lay en route, establish a water supply, and charge both lines upon command.

Time: 4 minutes. Time is recorded when proper pressure is shown.

MULTIPLE ENGINE COMPANY EVOLUTIONS

Unless otherwise noted, in the following evolutions all personnel except the driver are to be in full protective clothing including SCBA. When required, the hydrant or other water supply should be 100 to 200 feet from the simulated fire building.

Evolution Number 1: Establishing a Water Supply and Advancing Two Preconnected Attack Lines into Operation

Objective: To demonstrate the ability of two companies, working together, to advance two preconnected handlines into a structure.

Given: Two engine companies with apparatus, and a building

Procedure: Both pumpers to start at the same designated point. Engine 1 will proceed to a hydrant, drop a supply line, and continue to an appropriate position on side 1 (front) of the fire building.

Personnel of engine 1 will advance a 1 3/4-inch handline into a simulated fire building via the front door and flow water from a designated window.

Engine 2, after a 30-second delay, will proceed to the hydrant and establish a water supply for engine 1. Personnel from engine 2 will proceed to engine 1, pull a second 1 3/4-inch preconnected attack line, advance the line via the front door and interior stairs to the second floor, and flow water from a designated window.

Time: For engine 1, 4 minutes. For engine 2 to charge the supply line, 4 minutes. To flow water from a window on the second floor, 6 minutes. Time is recorded when proper pressure is obtained.

Evolution Number 2: Placing a Deck Gun and Two Attack Lines into Operation

Objective: To place a deck gun into operation to control a large volume of fire.

Given: Two engine companies, with apparatus, and a building

Procedure: Both pumpers to start at the same designated point. Engine 1 will proceed to a hydrant, lay dual supply lines, and proceed to a proper position on side 1 of the simulated fire building. Engine 1 will operate the deck gun as a master stream (1 1/4-inch tip) into a designated window for a period of 30 to 60 seconds. Engine 1 crew will pull a 1 3/4-inch preconnected attack line and advance it to the front door. When the master stream is shut down, the crew will advance the 1 3/4-inch line into the structure and flow water from a designated window.

Engine 2, after a 30-second delay, will proceed to the hydrant and establish a water supply for engine 1. The crew from engine 2 will proceed to engine 1, pull a second 1 3/4-inch preconnected attack line, advance the line via the front door and interior stairs to the second floor, and flow water from a designated window.

Time: For engine 1, 5 minutes. For engine 2 to charge the supply line, 4 minutes and to flow water from a window on the second floor, a total of 6 minutes. Time is started when the pumper stops at the hydrant and is recorded when proper pressure is obtained.

Evolution Number 3: Placing a 2 1/2-inch Handline into Operation

Objective: To demonstrate the ability to place a 2 1/2-inch handline into operation.

Given: Two engine companies with apparatus, and a building

Procedure: Both pumpers to start at the same designated point. Engine 1 will proceed to side 1 of the simulated fire building. Personnel from engine 1 will begin advancing a 2 1/2-inch handline to the building.

Engine 2 will proceed to a position near engine 1, drop a single 3-inch supply line, and proceed to the hydrant. The driver of engine 2 will make necessary connections and charge the supply line. Two members of the crew of engine 2 will assist engine 1 personnel in advancing the attack line to the door of the structure.

Time: For engine 1 from stopping in front of the building until flowing an effective stream, 3 minutes. For engine 2 from the time from stopping in front of the building until the supply is charged, 4 minutes. Time is recorded when proper pressure is shown.

Evolution Number 4: Supplying the Fireground from a Draft

Objective: To demonstrate the ability to support fireground operations from an alternate water supply.

Given: Two engine companies with apparatus, and a building

Procedure: Both pumpers to start at the same designated point. Pumper 1 will proceed to a position on side 1 of the simulated fire building. Personnel of engine 1 will advance a 1 3/4-inch handline into a simulated fire building via the front door and flow water from a designated window.

Engine 2, after a 30-second delay, will proceed to a position in the vicinity of engine #1, drop a single 3-inch supply line, proceed to a water source, establish a draft, and charge the supply line.

Time: For engine 1, 3 minutes. For engine 2, 5 minutes. Time is started when the first engine stops at the fire building and is recorded when proper pressure is shown.

Evolution Number 5: Supplying a Standpipe

Objective:	To demonstrate the ability to use a standpipe.
Given:	Two engine companies with apparatus, and a building with a standpipe
Procedure:	Both engine companies are to start at the same designated point. Engine 1 will proceed to a position on side 1 of a simulated fire building and make connection with the standpipe connection. Personnel will remove the high-rise pack, enter the building and proceed to the designated floor. Upon arrival, the crew shall connect to the standpipe, advance their attack line to the designated room, and flow water from a designated window.
	Engine 2, after a 30-second delay, will proceed to a position in the vicinity of engine 1, drop two 3-inch supply lines, proceed to a hydrant, make connections, and charge the supply lines.
Time:	For engine 1, 5 minutes. For engine 2, 5 minutes. Time is started when the first engine stops at the fire building and is recorded when proper pressure is shown.

SQUAD OR TRUCK COMPANY EVOLUTIONS

Unless otherwise noted, in the following evolutions all personnel except the driver are to be in full protective clothing including SCBA. When required, the hydrant or other water supply should be 100 to 200 feet from the simulated fire building.

Evolution Number 1: Laddering a Building with Ground Ladders

Objective:	To demonstrate the ability to raise ground ladders to support firegound operations.
Given:	One squad or truck company, with apparatus, and a building. Additional apparatus may be required to provide an adequate supply of ladders. All personnel to be in full protective clothing. SCBA is not required.
Procedure:	With the apparatus parked on side 1 of the simulated fire building, the crew of the squad or truck company will place and raise one ladder on each of four sides of the building. Ladders shall be raised to either the second or third floors of the building as instructed by the evaluator. The ladders shall be placed at windows in a position to facilitate rescue and/or escape.
Time:	5 minutes. Time will start upon signal from the evaluator and be recorded when the last ladder is properly in place.

Evolution Number 2: Rescuing a Victim from an Elevator

Objective:
To demonstrate the ability to rescue a victim from a stalled elevator.

Given:
A single hoist elevator located between the first and second floor, proper tools, and personnel from a squad or truck company. Full protective clothing is not needed, but helmets, gloves, and safety shoes are required.

Procedure:
The officer shall determine (from the evaluator) the nature of the situation, the number of people stranded, and their condition. In addition, the officer will simulate taking action to obtain the services of a certified elevator repair person.

Personnel shall proceed as follows: Secure the car by disconnecting the main power source and instructing the victim to activate the Emergency Stop switch in the car; open the hoist way door on floor above the car, and secure the doors in an open position with door wedges. One or more personnel, properly secured with a life belt and lying on the floor, should reach down with a pole and unlock the hoist way doors on the second floor.

Personnel on the second floor shall secure the doors in an open position with door wedges, make contact with victims, extend a ladder into elevator car, and one rescuer should enter the car to assist victims and direct their exit from the car. When all of the victims have been rescued, remove the ladder and close the hoist way door.

The officer should simulate advising the building official that the rescue is complete, that the power is off, and that the car is out of service until checked by a certified elevator mechanic.

Time:
10 minutes. Time is started when the crew arrives at the building and is recorded when the hoist way doors are closed after the rescue is complete.

Evolution Number 3: Rescuing a Victim from an Elevated Position

Objective:
To demonstrate the ability to rescue a victim from an elevated position.

Given:
A squad or ladder company. Full protective clothing is not needed, but helmets, gloves, and safety shoes are required.

Procedure:
Personnel will proceed to the victim, properly stabilize the victim in a Stokes basket or similar device, and safely lower the victim to the ground.

Time: 10 minutes. Time starts when crew arrives on the scene and is recorded when the victim is on the ground.

Notes: For purposes of this event, the victim may be located on a porch, balcony, flat roof or other accessible site. Several lengths of hose or other objects should be placed in the basket to represent the weight of a victim.

Evolution Number 4: Deploying Fans and Smoke Ejectors

Objective: To demonstrate the ability of personnel to place fans or smoke ejectors in service in an effective manner.

Given: A squad or truck company with two fans or smoke ejectors. All personnel to be in full protective clothing. SCBA is not required.

Procedure: One fan or smoke ejector shall be placed in the doorway on side 1; if fans are used, a second fan should used to augment the fan already in place. If smoke ejectors are used, the second unit should be placed in a separate doorway.

Time: 4 minutes. Time will start upon direction from the evaluator and be recorded when both fans or smoke ejector are operating at full power.

Evolution Number 5: Hoisting Tools for Roof Ventilation

Objective: To demonstrate the ability to ladder the roof and hoist the tools needed for ventilation.

Given: One truck or squad company, suitable ladders to reach the roof, and a roof ladder. In addition, one line for hoisting equipment, one power-driven chain saw or rotary saw, one 6-foot pike pole or similar, and one pick-head axe. All personnel shall be in full protective clothing. Personnel on the roof shall, in addition, wear SCBA.

Procedure: Remove ladders from apparatus and ladder the roof. Carry roof ladder to roof and place in position. Drop the hoisting line. A member on the ground will attach the saw, pole, and ax, one at a time, to be hoisted to the roof.

Time: 10 minutes. Time starts upon command from the evaluator and is recorded when all of the tools are on the roof.

Appendix

C

A Brief Introduction to the Incident Management System

Prior to the 1970s, the concept of incident command was pretty much a local function. Since then, several significant events have helped the nation's fire service move to a standardized program. During the 1970s, the FIRESCOPE incident command system, and the Phoenix Fireground Command system became popular.

FIRESCOPE stands for FIre REsources of Southern California Organized for Potential Emergencies. Serious wildland fires in southern California in those years challenged the fire fighting resources of southern California. As a result federal, state, and local agencies combined their talents to form a more organized response to the disastrous fires that seemed to be occurring every year. FIRESCOPE focused on command procedures, resource management, terminology, and communications.

Chief Alan Brunacini of the Phoenix Fire Department realized that the organization and management concerns that faced firefighters in wildland firefighting operations were also faced by firefighters dealing with structural firefighting back home. He took some of the ideas from FIRESCOPE and applied them to his department. Others followed suit. This led to the **incident command system** (ICS). While originally intended to serve as a command system for large structural fires, ICS became a tool for dealing with emergencies of all kinds. In 1986, the

incident command system
a tool for dealing with emergencies of all kinds

373

National Fire Academy adopted the ICS and has been teaching it ever since. ICS is a tool for dealing with emergencies of all kinds.

While this activity was taking place, another part of the federal government, the Occupational Safety and Health Administration (OSHA) was working on developing incident management system requirements for agencies responding to hazardous materials incidents. (Later they would look at high-risk rescue operations as well.) Now federal law requires the use of an incident command system at these events.

During this same time, the National Fire Protection Association started working on a new document that was to become known as NFPA 1561, *Standard for Fire Department Incident Management System.*

NFPA 1561 was a logical follow-on for the then brand-new NFPA 1500, *Standard for Fire Department Occupational Safety and Health.* One of the areas addressed in NFPA 1500 was a need for efficient incident command. There is an obvious correlation in effective command and firefighter safety. It should not come as a surprise to you that the chairman of the original committee that put these documents together was Alan Brunacini. His contributions to the fire service and to the public we serve will be with us for many years to come.

Incident management system (IMS) implies something more than just incident command; it implies that *every* aspect of controlling an incident is accomplished in an orderly and logical way. IMS looks upon emergency management as an extension of the type of management you should be doing on a daily basis. IMS is a management tool—it allows a department and its personnel to plan, organize, command, coordinate, and control emergency incidents (see Figure C-1). The focus of IMS is on managing people and resources. You cannot do much to manage fire and floods: Those events have already happened before you arrive. However, IMS allows you to manage your response to these events.

The command function must be clearly established with the first-arriving unit. This implies that the first arriving unit is qualified and has the authority (and the ability) to take command, and that it is clear to all responding personnel who is in charge. IMS also incorporates the rules of organization and management discussed in Chapters 3 and 4. When applied to the management of resources at a **emergency incident**, they improve the operation while reducing confusion.

While IMS defines how the management of an incident starts and progresses during the incident, the success of the operation is usually directly related to how well the initial command function is established and how effective the communications are handled during the event. Effective communications are vital to incident management.

What Makes IMS Work

Several characteristics make IMS work. Two of these have to do with communications. The first is the use of plain language, the second is the use of common terminology. To be effective, the terms and definitions used in IMS must be understood

incident management system
an organized system of roles, responsibilities, and standard operating procedures used to manage an emergency operation

emergency incident
any situation to which a fire department or other emergency response organization responds to deliver emergency services

Figure C-1 *The incident commander is responsible for the overall management of the event.*

DESIGN REQUIREMENTS FOR AN INCIDENT MANAGEMENT SYSTEM

- Can be used with a single jurisdiction or agency or can include multiple jurisdictions and/or agencies.
- Structure can be adapted to fit any situation. The system is easy to use and can be used at all types of events.
- Must have common elements of organization, terminology, and procedures.
- Allows for expansion, transfer, and termination of operations.

---

and accepted and the words must have a single definition over a period of time and the space of distances. If we can agree on the use of terms and their definitions, we can better communicate our thoughts.

Another characteristic of IMS is that it has a modular organization, meaning that you can put it together as you see fit. In simple incidents, the IMS is simple. There are few problems, the event is resolved quickly, the resources are

COMMUNICATIONS MANAGEMENT = GOOD INCIDENT MANAGEMENT

Communications management is integral to effective incident management. Communications management involves making effective use of communications systems to provide for the timely and useful flow of information. We should have communications procedures that are easily used and widely understood. This implies the use of plain language and common terms that mean the same thing to all personnel.

Fire officers should also establish a procedure for providing initial reports and progress reports on a regular basis. Initial reports provide other responding personnel with necessary information so that they can mentally and physically prepare for action. Initial reports also provide information for fire department managers regarding the commitment of resources and the need to expand the incident command system to meet the particular needs of the situation. By the same token, progress reports provide vital information regarding the progress, prognoses, and expected duration of the event.

Whether you are speaking face-to-face or over a radio or telephone, good communications are enhanced by remembering to use the six C's of communications: conciseness, clarity, confidence, control, capability, and confirmation. Remember: A moment of thought before speaking usually makes the message shorter and clearer.

Conciseness is important during emergency operations Usually there is much to think about and to do, and there is usually a lot of radio traffic.

Clearly worded and concise messages speed the entire evolution. By the same token, clear and understandable messages increase the likelihood of comprehension and the probability that the requested action will be accomplished.

Confidence is exhibited with a calm voice and a normal speaking rate. Confidence on the part of the incident commander is communicated with the message, and shared by all those involved in the incident.

Control of communications is also important to the success of an operation. Communications control is established by the first arriving unit and maintained by the incident commander and the dispatch operator throughout the incident. Communications control requires that certain discipline be maintained and that certain rules be followed. Units must identify themselves when calling, and units being called must acknowledge the message. This ensures the message was received and understood.

Capability deals with ability and the willingness to communicate at every level. Communications is a two-way process. Capability implies that the incident commander is willing to listen to others as well as to transmit orders. Finally, *confirmation* means that you get an acknowledgement of orders. It is the feedback in the communications loop: "I got your message."

limited and adequate, and the incident commander (IC) can personally deal with all of the management functions. As the incident becomes more complicated, so does the IMS. You will need additional resources, and the event is likely to last longer. The IC may not be able to personally manage all of the activities. In such cases, the IMS must be able to expand to meet the needs of the situation.

When the IMS expands to fit the needs of a larger situation, the IC has to delegate some of the management functions to others. The IMS allows for an orderly and standardized way of doing this, and allows for delegation of specific responsibilities (see Figure C-2). These responsibilities can be along functional lines (fire attack) or at specific locations (third floor.)

IMS is a performance-oriented system rather than a rank-oriented system. The function is more important than the rank of the person filling the position. The qualifications of the individual assigned should be the only concern. Let us outline the nine basic components of the incident management system.

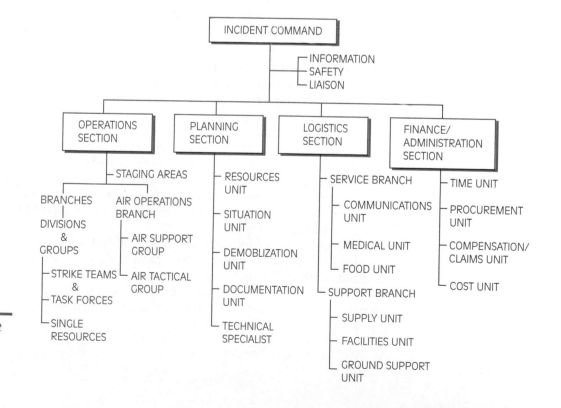

Figure C-2 *Incident Command System organizational flowchart.*

The Key Positions

The incident commander (IC) The IC is the one position of the IMS that must be staffed. Someone has to be in charge, responsible and accountable for the safe and effective mitigation of the problem. The IC must understand the roles of the various positions with the IMS structure. The IC is responsible for determining the strategy, selecting tactics, setting the action plan in motion, and developing the IMS organization.

Depending upon the size of the incident, the IC may also be responsible for managing resources, coordinating the activities of these resources, providing for scene safety, releasing information about the incident, and coordinating efforts with other agencies. At smaller incidents, the IC will likely personally retain these tasks. At larger incidents, these functions will likely be delegated. In all cases however, the IC is responsible for the overall outcome of the event. There is no requirement that the fire chief automatically becomes the IC.

We could add pages about the roles and responsibilities of the IC, but we will add just one comment here. The IC has to run the show, look at things from a strategic perspective, and focus on the *overall* situation. This means the IC has to operate from a fixed vantage point where he or she can see as much of the entire event as possible, a spot that will become known as the command post. The command post may be the hood of someone's car, or the inside of a custom vehicle designed expressly as a command vehicle. It must be visible, fixed, and capable of supporting the mission.

Operations The first position that is likely to be delegated is the operations sector or section. The terms **sectors**, **sections**, **divisions** and **groups** are tactical-level management groups that command companies. The term *division* indicates a geographical assignment whereas the term *group* indicates a functional assignment. Not all terms are used in all areas of the country. The **operations section** is responsible for management of all operational activity associated with the primary mission of the IMS. Operations assigns resources to control of the incident and directs the organization's tactical operations to meet the strategic goals developed by command. The operations section may have just one officer, usually one of the first officers to arrive on the scene. Operations reports to the incident commander (see Figure C-3).

The remaining functions are important but we are less likely to see these positions filled. Larger departments have the luxury of being able to fill these positions with staff officers. However, as a company officer, you should be aware of the functions of each of these positions, appreciate their importance, and be prepared to occupy any one of these positions if necessary. All of these positions report to the incident commander.

---

incident commander
the person in overall command of an incident

sector, section, division, group
• the above terms are tactical-level management groups that command companies

operations section
division in the incident management system that oversees the functions directly involved in rescue, fire suppression, or other activities within the mission of an organization

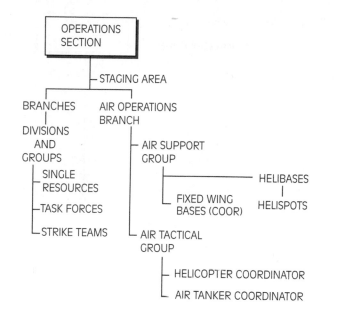

Figure C-3
Operations section.

planning section
that part of the
incident management
system that focuses
on the collection,
evaluation,
dissemination, and use
of information
(information
management) to
support the incident
command structure

Planning The **planning section** is responsible for collecting, evaluating, and disseminating appropriate information. The planning sector keeps track of the current situation, predicts the probable course of events, and prepares optional strategies and tactics. The planning section may also be tasked with keeping track of resources. The planning section may be one or two persons, one of whom may be a senior officer with considerable command experience (see Figure C-4).

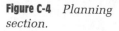

Figure C-4 *Planning section.*

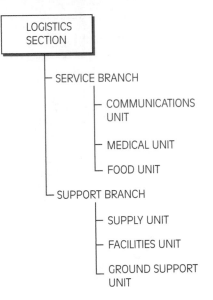

Figure C-5 *Logistics section.*

logistics section
that part of the incident management system that provides equipment, services, material, and other resources to support the response to an incident

finance section
that part of the incident management system that is responsible for facilitating the procurement of resources and for tracking the costs associated with such procurement

Logistics   The **logistics section** is responsible for obtaining the resources needed at the event. Logistics comes into play at long duration events (see Figure C-5).

Finance   Finance is the last of the major components of IMS. The **finance section** is responsible for accounting for the costs associated with the event. This is not usually a concern, but for long, drawn out operations, especially where multiple agencies are involved, the finance section becomes important. Finance and logistics are sometimes combined (see Figure C-6).

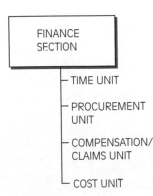

Figure C-6 *Finance section.*

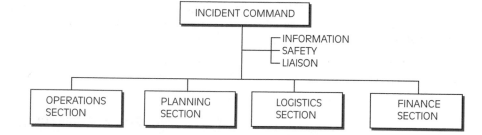

Figure C-7 *Incident Command System and general staff.*

staff assistants
those designated to aid the IC in fulfilling his/her responsibilities, typically, an incident safety officer, an information officer, and a liaison officer

incident safety officer
as part of the IMS, that person responsible for monitoring and assessing safety hazards and ensuring personnel safety

information officer
as part of the IMS, that person who acts as the contact between the IC and the news media, responsible for gathering and releasing incident information

Staff Assistants Depending upon the nature of the incident the incident commander may have three additional **staff assistants**. These include a safety officer, an information officer, and a liaison officer (see Figure C-7). The **incident safety officer** is one of the most important positions in the IMS organization. This position should be filled as soon as possible. The incident safety officer is responsible for monitoring working conditions and recommending appropriate action. Most organizations have policies that allow the safety officer to take direct action when warranted to protect firefighters from immediate danger. In such cases the safety officer should immediately advise the IC of his actions. In addition to spotting critical situations, the safety officer acts as an advisor to the IC. The safety officer has the advantage of being able to move about and see all sides of the action. The ideal safety officer is an experienced company officer.

The **information officer** is responsible for the development and release of timely information regarding the event and for serving as a point of contact for the media. The information officer should have access to the incident commander so that they have timely and correct information about the present and projected status of the event. However, their job is to feed that information to the media, thus allowing the incident commander to focus on the job at hand. Anyone can serve as information officer, but she should have good communications skills and enough understanding of the situation and the department's operations so that she can answer questions.

We often overlook the value of the media at events, thinking of them as a liability rather than an asset. The media can provide valuable information and services. If an evacuation is needed, the media are an excellent away to reach the public with timely information. Likewise when people are missing, getting the word to radio and television stations will bring many helpers. The media can also be used for effective life safety public education messages. While you are on television discussing the fire at Ms. Doe's house, you have a good opportunity to point out that Ms. Doe's first notice of her problem was her smoke detector. (Obviously it was working.) Remind viewers about the value of smoke detectors and how important it is to check them once a month.

liaison officer
as part of the incident management system, the contact between the incident commander and agencies not represented in the incident command structure

occupant services sector
that part of the incident management system that focuses on the needs of the citizens who are directly of indirectly affected by an inciden

Finally we come to the **liaison officer**, who provides liaison to other agencies. In some cases, these supporting agencies are not represented elsewhere in the command structure, so to ensure their effective use and to prevent conflicts, you need to have the ability to communicate with them. The liaison officer works with the representatives of the other agencies.

TAKING CARE OF MS. DOE, OUR CUSTOMER

Some departments have introduced a new position called the **occupant services sector**. The basic function of this sector is to treat the uninjured survivors of the event. Leaving Ms. Doe standing on the sidewalk while you attack her house and its fire will not be a good experience for her. Watching a dozen firefighters drag hoses and tools through her superneat house will not make her feel any better. Maybe you should find a nice place for Ms. Doe to sit down. Someone should check her over; make sure she is okay. An ambulance would be a good choice.

But lacking an ambulance, a staff car, the cab of a fire truck, almost anything will do. You want to move her out of your way and get her to a safe and comfortable place, protect her from the weather, and keep track of her whereabouts. Someone should stay with her, treat her injuries, explain what is going on, and calm her. You might even let her use your fancy cellphone to call her husband.

Acronyms

ADA	Americans with Disabilities Act	JPR	Job performance requirement
BOCA	Building Officials and Code Administrators	MBE	Management by exception
		MBO	Management by objectives
BTU	British thermal unit	NFA	National Fire Academy
CIS	Critical incident stress	NFIRS	National Fire Incident Reporting System
CPSC	Consumer Product Safety Commission		
DOE	Department of Energy	NFPA	National Fire Protection Association
EEOC	Equal Employment Opportunity Commission	NIH	National Institutes of Health
		OSHA	Occupational Safety and Health Administration
EMT	Emergency medical technician		
EMS	Emergency medical services	PASS	Personal alert safety system
HSO	Health and safety officer	PCFD	Plain City Fire Department
IAFF	International Association of Fire Fighters	PPE	Personal protective equipment
		PPV	Positive pressure ventilation
IC	Incident commander	SCBA	Self-contained breathing apparatus
ICS	Incident command system	SOP	Standard operating procedure
IDLH	Immediately dangerous to life and health	TQM	Total quality management
		USFA	United States Fire Administration
IMS	Incident management system		

Glossary

Access factors   An assessment of the department's access to and into a building.

Accidental causes   As used as a cause of fire, it refers to those fires that are the result of unplanned or unintentional events. Accidental fires include "acts of God" as well as the actions or inactions of individuals.

Accountability   To be responsible for one's personal activities. In the organizational context, accountability includes being responsible for the actions of one's subordinates.

Action plan   An organized course of action that addresses all phases of incident control within a specified period of time. Action plans should address overall incident strategy, tactics, risk management, and member safety.

Active listening   The deliberate and apparent process by which one focuses his or her attention on the communications of another.

Administration   Management. In an organization, those charged with managing the affairs of the organization.

Administrative process   A process derived from the powers assigned by law to the executive branch of government that deals with issues such as police powers and public safety.

Administrative warrant   (See administrative search warrant.)

Administrative search warrant   A written order issued by a court specifying the place to be searched and the reason for the search.

Agency shop   In labor relations, an arrangement whereby employees are not required to join a union, but are required to pay a service charge for representation.

Arbitration   In labor relations, a term that describes the process by which contract disputes are resolved. Arbitration differs from mediation in that a third party decides the issue.

Area of refuge   A portion of a structure that is relatively safe from fire and the products of combustion.

Arson   A deliberate and unlawful burning of property. Two elements are required for arson: (1) There must have been a burning and (2) the burning must have been the result of a willful criminal act. The word *arson* is a legal term.

Attack mode   A situation in which the first arriving officer elects to take immediate action (rescue, fire attack, etc.) and pass command to another officer.

Authority   The right and power to command.

Automatic fire protection sprinkler   A self-operating thermosensitive device that releases a spray of water over a designed area to control or extinguish a fire.

Backdraft   A type of explosion caused by sudden influx of air into a mixture of burning a gases, which have been heated to the ignition temperature of at least one of them.

Barriers   An obstacle. In communications, a barrier prevents the message from being understood by the receiver.

Benchmarks   Significant points in the emergency event usually marking the accomplishment of one of the three incident priorities: life safety, incident stabilization, or property conservation.

British thermal unit (BTU)   The amount of heat required to raise the temperature of one pound of water one degree Fahrenheit.

Budget   A financial plan for an individual or organization.

Building code   A law or regulation that establishes minimum requirements for the design and construction of buildings.

Capital budget   A financial plan to purchase high-dollar items that have a life expectancy in excess of one year. For the fire service, apparatus and fire stations are good examples of capital budget items.

Certificate   A document serving as evidence of the completion of an educational or training program. The term certificate also describes a document issued to an individual or company as a fire prevention tool.

Certification   The issuing of a document that attests that one has demonstrated the knowledge and skills necessary to function in a particular craft or trade.

Certified   A guarantee that something is authentic.

Closed shop   A term used in labor relations denoting that union membership is a condition of employment.

Coach   A person who helps another develop a skill.

Codes   A systematic arrangement of a body of rules.

Commanding   The third step in the management process. Commanding involves using the talents of others, bringing others into the task, giving them direction, and setting them to work. Commanding involves having authority and control and giving direction to others.

Command role   A situation in which the first-arriving officer elects to take command until relieved by a senior officer.

Community consequences   An assessment of the consequences on the community. The community includes the people, their property, and the environment.

Complaint   An expression of discontent.

Confinement   An activity required to prevent fire from extending to an uninvolved area or another structure.

Conflict   A disagreement, quarrel, or struggle between two individuals or groups.

Consensus document   The result of general agreement among members who contribute to the document.

Consulting   Seeking advice or getting information from another. As a leadership style, consulting implies that the leader seeks ideas and allows contributions to the decision-making process.

Controlling   The fifth function of management. Controlling involves monitoring the process to ensure that the work is accomplishing the intended goals and objectives, and taking corrective action when it is not. Controlling involves using feedback, measuring accomplishments against expectations, and making adjustments to improve performance.

Cooking equipment   A leading cause of fires. In this context, cooking equipment includes the heating equipment used in cooking. Cooking equipment itself is seldom the cause; it is the improper installation, use, and maintenance that causes fires.

Coordinating   Coordinating is the fourth management function. Once the management process is implemented, managers must control the efforts of others.

Counseling   Counseling is one of several leadership tools that focuses on improving employee performance. As a part of the critical incident stress recovery process, counseling usually involves having the employee meet with a well-trained professional for the purpose of mitigating the effects of stress.

Criminal process   Pertaining to the apprehension, trial or prosecution, and the fixing of punishment of a person or persons who are supposed to have broken the law.

Criminal search warrant   (See search warrant.)

Customer service   Service to members of the community.

Defensive mode   Actions intended to control a fire by limiting its spread to a defined area. Defensive operations are generally performed from the exterior of buildings and are based on the premise that the value of the property to be saved does not warrant the risks associated with interior firefighting.

Delegate   To grant to another a part of one's authority or power. As a leadership style, delegating implies that the supervisor turns much of the management of the project over to subordinates.

Delegation   The act of delegating; see *delegate*.

Demotion   To reduce someone to a lower grade.

Dillon's rule   A legal ruling issued by Judge John Forrest Dillon declared that local governments possess only those powers expressly granted by charter or statute.

Directing   Controlling the course of action. As a leadership style, it is characterized by a rather authoritarian style.

Disciplinary action   An administrative process whereby an employee is punished for not conforming to the organizational rules or regulations.

Discipline   A system of rules and regulations. Acting in accordance with a system of rules and regulations. Maintaining order within an organization by requiring conformance with the rules and regulations. Imposing punishment as a method of maintaining conformance with the system of rules and regulations.

Diversity   A quality of being diverse, different, or not all alike.

Division of labor   The dividing of an organization's overall activity into smaller tasks.

Electrical equipment   A leading cause of fires. In this context, electrical equipment includes electrical distribution equipment, including installed and temporary wiring.

Emergency incident   Any situation to which a fire department or other emergency response organization responds to deliver emergency services including rescue, fire suppression, medical treatment, and so on.

Empowerment   To give power or authority to another.

Energy   The capacity to do work; the property diminishes when the work is done. In the fire triangle, heat represents energy.

Environmental factors   Factors like weather that impact on firefighting operations.

Equal Employment Opportunity (EEO)   A condition in which all applicants for a position are treated fairly and the employer does not discriminate against any group or individual on the basis of race, color, religion, gender, national origin, age, or disability.

Equal Employment Opportunity Commission (EEOC)   The federal government agency charged with administering laws related to nondiscrimination on the basis of race, color, religion, gender, national origin, and age.

Ethics   A system of values. A standard of conduct. Principles of honor and morality, or guidelines for human action.

Evacuation   The removal of persons or things from an endangered area.

Expense budget   A financial plan to facilitate the orderly acquisition of the goods and services needed to run an organization for a specific period of time, usually one year.

Expert power   A recognition of authority by virtue of an individual's skill or knowledge.

Exposures   Any property that may be exposed to a fire that has started in another area.

Fact finding   A collective bargaining process.

Fayol's bridge   An organizational principle that recognized the practical necessity for horizontal as well as vertical communications within an organization.

Feedback   A reaction or evaluation, which may serve to alter or reinforce that condition.

Finance section   As a component of the incident management system, the finance section is responsible for facilitating the procurement of resources and for tracking the costs associated with such procurement.

Fire behavior   The phenomena of the science and consequences of fire.

Fire confinement   Activities associated with keeping a fire from making any further advance.

Fire control   Activities associated with confining and extinguishing a fire.

Fire extension   The movement of fire from one area to another by conduction, convection, or radiation.

Fire extinguishment   Activities associated with putting out all visible fire.

Fire flow   The water flow requirements expressed in gallons per minute needed to control a fire in a given area.

Fire load   Stuff that will burn.

Fire prevention   Action taken to prevent a fire from occurring, or if one does occur, to minimize the loss.

Fire prevention code   A legal document that sets forth the requirements for life safety and property protection in the event of fire, explosion, or similar emergency.

Fire resistive A type of construction in which the structural components are noncombustible and protected from fire.

Fire-safety education Efforts through training programs and public announcements to improve citizens' awareness and behavior with regard to the hazards associated with fire.

Fire suppression Action taken to control and extinguish a fire.

First-level supervisors, or first-line supervisors The first managerial rank in an organization.

Flashover A dramatic event in a room fire that rapidly leads to full involvement of all combustible materials present.

Floor plan A bird's-eye view of the structure with the roof removed showing walls, doors, stairs, and so on.

Free-burning phase The second phase of fire growth. There is sufficient fuel and oxygen to allow for continued fire growth. Also called steady-burning phase.

Fuel load The expected maximum amount of combustible material in a given fire area, usually expressed as the weight of a comparable amount of wood, per square foot.

Goals A target or other object by which achievement can be measured. In the context of management, a goal helps define purpose and mission.

Grievance A formal dispute between an employee and employer over some condition of employment.

Grievance procedure A formal process for handling disputed issues between an employees and employer. Where a union contract is in place, the grievance procedure is part of the contract.

Gripe The least severe form of discontent.

Harass To disturb, torment or pester.

Health and safety officer A person assigned as the manager of the department's health and safety program. The health and safety officer may be assigned duties as an incident safety officer.

Heat of combustion The amount of heat given off by spontaneous ignition of a particular substance.

Heating equipment A leading cause of accidental fires.

As used in this context, heating equipment includes misused and overloaded heating equipment and equipment that is installed or used too close to combustibles.

Heavy timber construction A type of building construction in which the exterior walls are usually made of masonry, and therefore, noncombustible. The interior structural members are of large, unprotected, wood members. To meet this definition, columns must be at least 8 inches square and beams must be at least 6 by 10 inches.

High-achievement needs According to the needs theory of motivation, these individuals accept challenges and work diligently.

High-affiliation needs According to the needs theory of motivation, these individuals desire to be accepted by others.

High-power needs According to the needs theory of motivation, these individuals like to be in charge and have influence over others.

High-status needs According to the needs theory of motivation, these individuals like to be in charge and have influence over others.

Hygiene factors As used by Frederick Herzberg, hygiene factors keep people satisfied with their work environment.

Identification power A recognition of authority by virtue of the other individual's charter or trust.

Incendiary A term used to describe a deliberately set fire. Arson is the result of an incendiary fire.

Incident commander The person who has authority over the management of an incident.

Incident command system (IMS) A management tool to deal with emergencies of all kinds.

Incident management system An organized system of roles, responsibilities, and standard operating procedures used to manage an emergency operation.

Incident priorities Incident priorities include life safety, incident stabilization, and property conservation.

Incident safety officer As part of the IMS, that person responsible for monitoring and assessing safety hazards and developing measures to ensure personnel safety.

Incident stabilization   The second priority in emergency scene management; efforts to reduce further damage.

Incipient phase   The first stage of fire growth. The fire is limited to the material originally ignited.

Information officer   As part of the IMS, that person who acts as the contact between the IC and the news media, and is responsible for gathering and releasing information to the media.

Initial report   A vivid but brief description of the on-scene conditions relevant to the emergency.

In-service company inspection   Using suppression companies to conduct fire safety inspections in selected occupancies, usually within the company's first-due assignment area.

Internal customers   The members of the organization.

Labor relations   A continuous relationship between a group of employees, represented by a union or association, and an employer.

Latent heat of vaporization   The amount of heat required to convert a substance from a liquid to a vapor.

Leadership   The personal actions of managers and supervisors to get subordinates to carry out certain actions. Leadership also refers to the act of leading and to those who lead an organization.

Legitimate power   A recognition of authority derived from the government or other appointing agency.

Liaison officer   As a part of the incident management system, the liaison officer is a point of contact between the incident commander and agencies not represented in the incident command structure.

Licenses   A formal permission from an authority to participate in an activity.

Life risks   An assessment of the number of people in danger, the immediacy of their danger, and their ability to provide for their own safety.

Life safety   The first priority in emergency operations. Life safety should be the focus of fire prevention and fire suppression. Life safety addresses the safety of occupants and emergency responders.

Line and staff   The line authority is the avenue of command of the organization and the path of accountability for the work of the organization. Line managers are supported by staff.

Line authority   A characteristic of organizational structures, denoting the relationship between supervisors and subordinates.

Line functions   Functions that refer to those activities that provide emergency services.

Line-item budget   A budgeting tool that collects similar items into a single account and presents them on one line in a budget document. An example might be charges for all of the telephones within an organization.

Lock box   A small, accessible, locked storage box on the premises that can be unlocked by the fire department. A lock box may contain preplanning information, material data sheets, names and phone numbers of key personnel, and keys that allow emergency responders access to the property.

Logistics section   As part of the incident management system, the logistics section provides equipment, services, material, and other resources to support the response to an incident.

Management   The accomplishment of the organization's goals by utilizing the resources available.

Management by exception (MBE)   A management approach whereby attention is focused only on the exceptional situations when performance expectations are not being met.

Management by objectives (MBO)   A management process whereby attention is focused on a joint process of meeting goals. MBO involves collaboration, sharing, and an ongoing evaluation of the results.

Maslow's Hierarchy of Needs   Abraham Maslow's management theory incorporating a five-tiered representation of human needs.

Mediation   In labor relations, a term that describes the process by which contract disputes are resolved. Mediation differs from arbitration in that in mediation the third party facilitates the communications process by suggesting procedures, scheduling meetings, and so on.

Medium   In the communication process, the medium is the means or method the sender uses to transmit the message.

Message   In the communication process, the message is the information being sent to another.

Mill construction   A type of building construction in which the exterior walls are usually made of masonry, and therefore, noncombustible. The interior structural members are of large, unprotected, wood members.

Minimum standards   As used in codes and standards, the least or lowest accepted level of attainment.

Mission statement   A formal document indicating the focus and values for an organizations.

Models   A representation or example of something.

Motivation   Providing an incentive or inducement to encourage a desired outcome.

Motivators   Factors that are regarded as work incentives such as pay, benefits, and vacation time.

Motivation factors   As used by Frederick Herzberg, motivation factors are those conditions or qualities in the workplace that encourage employees to rise above the satisfied level.

Motive   The goal or object of one's actions. In the context of deliberately setting a fire, the motive is the reason one sets the fire.

Mutual aid   Assistance provided by another fire department or agency.

Natural causes   Fires that are the result of the sun's heat, spontaneous combustion, chemicals, lightning, or static discharge.

Noncombustible construction   A type of building fabrication in which the structural elements are nonflammable or limited flammable. Even if the structural elements are not combustible, they may offer no fire resistance.

Objectives   Something that one's efforts are intended to accomplish. In the context of goals, an objective is an element or component of the overall process.

Occupancy factors   Factors assessed as to the risks associated with a particular structure based on its contents and activities.

Occupant services section   As a part of the incident management system, the occupant services section focuses on the needs of the citizens who are directly or indirectly affected by the event.

Offensive mode   Firefighting operations that make a direct attack on a fire for purposes of control and extinguishment. For structural situations, this usually means interior fire fighting.

Open shop   Of all of the labor-relations arrangements, this is the weakest from the union's point of view: There is no requirement for the employee to join a union. Open-shop arrangements are protected by right-to-work laws in twenty-one states.

Operating budget   A financial plan to acquire the goods and services needed to run an organization for a specific period of time, usually one year.

Operations section   In the incident management system, the operations section oversees the functions that are directly involved in rescue, fire suppression, or other activities that are within the mission of the organization.

Oral reprimand   The first step in a formal disciplinary process.

Ordinary construction   A type of building construction in which the exterior walls are usually made of masonry, and therefore, noncombustible. The interior structural members may be either combustible or noncombustible.

Organization   A group of people working together to accomplish a task.

Organizing   As the second function of management, organizing involves bringing together and arranging the essential resources to get a job done.

Origin and cause   A term used by fire professionals indicating the location and reason for the fire's ignition.

Overhaul   A systematic process of searching the fire scene for possible hidden fires or sparks that may rekindle. Overhaul is also used to assist in determining the origin and cause of the fire.

Oxidation   A chemical reaction in which oxygen combines with other substances. Fires, explosions, and rusting are all examples of oxidation reactions.

Performance codes   A code that assigns an objective to be met and establishes a criteria for determining compliance.

Performance evaluation   A formal review of an employee's performance. Performance evaluations can affect an employee's retention, termination, promotion, demotion, transfer, salary increase, or other personnel actions.

Permits   A fire prevention tool, permits are required where there is potential for life loss, or where there are hazardous materials, or hazardous processes.

Personal barrier   A communications barrier that may be the result of prejudice or bias.

Personal protective equipment (PPE)   The garments and associated equipment worn by firefighters while performing the job of firefighting.

Personal alert safety system (PASS)   A component of PPE, the PASS device is intended to alert others to a firefighter who is in some way incapacitated.

Personnel accountability   The tracking of personnel as to location and activity during an emergency event.

Physical barriers   A communications barrier or obstacle that results from environmental conditions.

Physical factors   An assessment of the conditions relevant to population, area, topography, and valuation of a given area.

Piloted ignition temperature   The minimum temperature to which a substance must be heated to start combustion after an ignition source is introduced.

Planning   The first of the several management steps. Planning involves looking into the future and determining objectives.

Planning section   As part of an incident management system, the planning section focuses on information management to support the incident command structure. Information management includes the collection, evaluation, dissemination and use of information. The planning section may also be involved with tracking the availability and deployment of resources.

Plot plan   A bird's-eye view of a property showing existing structures for the purpose of preemergency planning, such as primary access points, barriers to access, utilities, water supply, and so on.

Policy   A policy is a formal statement that defines a course or method of action. Policies tend to be broad in nature, allowing some flexibility depending on the situation.

Power   The command or control over others.

Preemergency planning   A process of preparing for operations at the scene of a given hazard or occupancy.

Preemergency survey   The fact-finding part of the preemergency planning process in which the facility is visited to gather information regarding the building and its contents.

Probable cause   A reasonable cause for belief in the existence of facts. An apparent state of facts that would induce a reasonable person to believe that a cause of action existed or that the accused had committed a crime.

Procedure   A defined course of action.

Program budget   A program budget gathers expenses (and possible income) related to the delivery of a specific program within an organization.

Property conservation   The third element or priority at the scene of a fire; the efforts to reduce primary and secondary (i.e., as a result of firefighting operations) damage.

Property risks or property risk factors   An assessment of the value and hazards associated with property that is at risk. Property includes structures that are involved, exposures, and their contents.

Public fire department   A part of local government.

Punishment power   A recognition of authority by virtue of the supervisor's ability to administer punishment.

Quick Access Prefire Plan   A document that provides emergency responders with vital information pertaining to a particular occupancy.

Rapid intervention team   A team or company of emergency personnel kept immediately available for the potential rescue of other emergency responders.

Rehabilitation   As applies to firefighting personnel, an opportunity to take a short break from firefighting duties to rest, cool off, and replenish liquids.

Reprimand   A step in a formal disciplinary process. A reprimand can be oral or written. A reprimand documents unsatisfactory performance and specifies the corrective action expected. The written reprimand usually follows an oral reprimand.

Receiver   In the communications process, the receiver is the intended recipient of the message.

Residential   As an occupancy class, the residential

category includes homes, apartment buildings, hotels and motels, dormitories, and manufactured housing.

Resource factors   An assessment of the resources available to mitigate a given situation.

Responsibility   Being accountable for actions and activities; having a moral and, perhaps, legal obligation to carry out certain activities.

Reward power   A recognition of authority by virtue of the supervisor's ability to give recognition.

Risk analysis   Understanding the hazards and possible consequences associated with a particular act.

Rollover   This fire-related phenomenon is seen during the first and second phases of a structural fire. The hot gases rise and travel along the ceiling, encountering fresh air. As the fresh air provides a new supply of oxygen, the gases will reignite if they are hot enough.

Safety officer   As a part of the incident management system, the safety officer assesses safety hazards and takes steps to enhance firefighter safety during emergency operations.

Scaler principle   An organizational concept. The scaler principle refers to the interrupted series of steps or layers within an organization.

Search warrant   A legal process, a legal writ issued by a judge, magistrate, or other legal officer, that directs certain law enforcement officers to conduct a search of certain property for certain things or persons, and if found, bring them to court.

Sector, section, division, group   Terms that designate tactical-level management groups that command companies. Section and sector denote a supervisory component of the IMS; division indicates a *geographical* assignment, whereas the term group indicates a *functional* assignment. Not all terms are used in all areas of the country.

Self-contained breathing apparatus (SCBA)   Part of a firefighter's personal protective equipment. Given that the respiratory system is at risk during firefighting and overhaul activities and other emergency operations such as confined-space rescue and any incident involving hazardous materials, SCBA should be worn whenever one is engaged in such activities.

Semantic barriers   A communications barrier or obstacle that results from language differences or the lack of a common understanding of words.

Sender   A part of the communications process. The sender transmits a thought or message to the receiver.

Size-up   A continuous ongoing mental assessment of the situation; an ongoing process of gathering and analyzing information that is critical to the outcome of an event.

Smoking materials   A leading cause of fires. Carelessly discarded smoking materials, including the ash from cigarettes, can ignite other combustible materials, ranging from bed coverings to dry leaves.

Smoldering phase   The third stage of fire growth. In an enclosed environment, such as would be found in a structure, the fire will consume much of the oxygen in the air. Once the oxygen has been reduced, visible fire will diminish.

Span of control   An organizational principle that addresses the number of personnel a supervisor can effectively manage.

Specification codes   Specification codes spell in detail the type of construction or the materials to be used.

Specific heat   The heat absorbing capacity of a substance.

Staff assistants   Those designated to aid the IC in fulfilling his/her responsibilities, typically, an incident safety officer, an information officer, and a liaison officer.

Staffing   Hiring personnel to fill various positions.

Staff functions   Functions that refer to the activities of those who provide support to emergency personnel.

Standard   A rule for measuring or a model to be followed.

Standard operating procedures (SOP)   An organized directive that establishes a standard course of action.

Standard of performance   A defined level of accomplishment or achievement.

Standpipe systems   Plumbing installed in a building or other structure to facilitate firefighting operations.

Strategy   Strategy defines the scope of activities, selects the basis for excellence, sets goals, and selects objectives to reach the goals. Within the context of commanding incident response, strategy sets broad goals, and outlines the overall plan to control the incident.

Structural factors   An assessment of the age, condition, and structure type of a building, and the proximity of exposures.

Stuff   Items, material, or objects of an unspecified nature.

Supporting   As a leadership style, the supporting process involves open and continuous communications and a sharing in the decision-making process.

Survival factors   An assessment of the safety hazards for both civilians and firefighters in a particular occupancy.

Suspension   A disciplinary action in which the employee is relieved from duties, possibly with a partial or complete loss of pay, for a specified period.

Tactics   Various maneuvers that can be used to achieve a strategy while fighting a fire or dealing with a similar emergency.

Target hazards   Locations in which there are unusual hazards, or in which an incident would likely overload the department's resources, or in which there is a need for interagency cooperation to mitigate the hazard.

Tasks   The duties and activities performed by individuals, companies, or teams that lead to successful accomplishment of assigned tactics.

Termination   The final step in the disciplinary process or the final step in the incident command process.

Theoretical fire flow   The water flow requirements expressed in gallons per minute needed to control a fire in a given area.

Theory X   A management style, labeled by Douglas McGregor, cued by the supervisor's expectations about subordinates. In Theory X, the manager believes that people dislike work and cannot be trusted.

Theory Y   A management style, labeled by Douglas McGregor, cued by the supervisor's expectations about subordinates. In Theory Y, the manager believes that people like work and can be trusted.

Theory Z   A management style, labeled by William Ouchi, cued by the supervisor's expectations about subordinates. In Theory Z, the manager believes that people not only like work and can be trusted, but

that they want to be collectively involved in the management process and recognized when successful.

Thermal stratification   A fire in an interior space heats the gases present. Because of the laws of physics, the hotter gases rise.

Total quality management   A form of management in which the focus of the organization is on continuous improvement geared to customer satisfaction.

Transfer   A possible step in the disciplinary process. A transfer provides the employee a fresh start in another venue.

Transition mode   During firefighting operations, the critical process of shifting from the offensive to the defensive mode or from the defensive to the offensive mode.

Union shop   In labor relations, a term used to describe the situation in which an employee must agree to join the union after a specified period of time, usually 30 days after employment.

Unity of command   The organizational principle whereby one has but one supervisor.

Vented   Opened to the atmosphere so that heat and fire by-products are released, and fresh air enters.

Ventilation   A systematic process to enhance the removal of smoke and fire by-products and the entry of cooler air to facilitate rescue and fire fighting operations.

Vision   An imaginary concept, usually favorable, of the result of an effort. The ability to see the outcome of a situation.

Visualize   To recall the past or to anticipate the future through mental images.

Warrant   A legal process, in this case a legal writ issued by a magistrate or other legal officer commanding a person to search property, seize property, and/or arrest a person.

Water supply   The quantity of water available for fire fighting.

Woodframe construction   A type of construction in which the entire structure is made of wood or other combustible material.

Written reprimand   A step in a formal disciplinary process. A written reprimand documents unsatisfactory performance and specifies the corrective action expected. The written reprimand usually follows an oral reprimand.

Index